Advances in Delays and Dynamics

Volume 14

Delay systems are largely encountered in modeling propagation and transportation phenomena, population dynamics and representing interactions between interconnected dynamics through material, energy and communication flows. Thought as an open library on delays and dynamics, this series is devoted to publish basic and advanced textbooks, explorative research monographs as well as proceedings volumes focusing on delays from modeling to analysis, optimization, control with a particular emphasis on applications spanning biology, ecology, economy and engineering. Topics covering interactions between delays and modeling (from engineering to biology and economic sciences), control strategies (including also control structure and robustness issues), optimization and computation (including also numerical approaches and related algorithms) by creating links and bridges between fields and areas in a delay setting are particularly encouraged.

Indexed by zbMATH.

More information about this series at https://link.springer.com/bookseries/11914

Jean Auriol · Joachim Deutscher ·
Guilherme Mazanti · Giorgio Valmorbida
Editors

Advances in Distributed Parameter Systems

Springer

Editors
Jean Auriol
Laboratoire des Signaux et Systèmes (L2S)
Université Paris-Saclay, CNRS,
CentraleSupélec
Gif-sur-Yvette, France

Joachim Deutscher
Institute for Measurement, Control
and Microtechnology
Ulm University
Ulm, Germany

Guilherme Mazanti
Laboratoire des Signaux et Systèmes (L2S)
Université Paris-Saclay, CNRS,
CentraleSupélec, Inria
Gif-sur-Yvette, France

Giorgio Valmorbida
Laboratoire des Signaux et Systèmes (L2S)
Université Paris-Saclay, CNRS,
CentraleSupélec, Inria
Gif-sur-Yvette, France

ISSN 2197-117X ISSN 2197-1161 (electronic)
Advances in Delays and Dynamics
ISBN 978-3-030-94765-1 ISBN 978-3-030-94766-8 (eBook)
https://doi.org/10.1007/978-3-030-94766-8

This Springer imprint is published by the registered company Springer Nature Switzerland AG
The registered company address is: Gewerbestrasse 11, 6330 Cham, Switzerland

Preface

Distributed parameter systems provide a natural representation for infinite-dimensional dynamical systems with variables depending on time and space. Among these systems, we find energy systems (hydraulic channels, heat exchangers, drilling devices), transportation systems (traffic networks, cable-operated elevators), biology, and robotics. The transport and propagation phenomena appearing in these systems are the source of complex control engineering problems. Furthermore, their design has to cope with stringent environmental, safety, and economic feasibility constraints.

The transport of information, matter, or energy in these systems may cause instabilities and deteriorate performances. It thus appears essential to compensate for these effects in real time to ensure the efficiency of operations. However, compensating for these undesirable effects is challenging due to the spatial distribution, propagation phenomena, and the physical/economic infeasibility of placing sensors and actuators everywhere along the spatial domain. Similar difficulties arise when considering reaction–diffusion systems.

The stringent operating requirements and the high mathematical complexity of these systems explain why traditional control methods find limited applicability and poor technology readiness. Indeed, despite breakthroughs achieved in the last decades for the analysis and control of distributed systems, the state-of-the-art approaches suffer from several significant limitations: they usually neglect the effect of delays in the actuation or measurement chain, of actuators devices (dynamics and nonlinearities), of uncertainties and disturbances, and the impact of model reduction (approximating the control law is necessary for implementation). Moreover, since most of the current control systems operate in networks across several domains, the properties of the interconnections should be assessed in terms of the properties of individual systems. Specific tools when dealing with the complete interconnection tailored for distributed parameter systems are therefore required.

The purpose of this volume in the established series *Advances in Delays and Dynamics* (ADD@S) is to provide a collection of recent results on the analysis and control design for *distributed parameter systems*. The proposed contributions correspond to the most recent advances in the field and focus on methodological aspects (modeling, analysis, controller, and observer design), practical applications (traffic networks, robotic, mechanical systems), and numerical implementation. Cutting-edge experts in the field wrote this volume, making it a valuable reference for control practitioners, graduate students, and scientists researching practical and theoretical solutions to the challenging problems raised by distributed parameter systems.

The proposed book spans the full range from theoretical research to practical implementation. It follows three transverse axes. The first part concerns fundamental problems in distributed parameter systems, including *stability analysis and stabilization*, input saturation, stochastic input delays. The different contributions introduce various control methodologies for stabilizing and regulating such distributed parameter systems: energy shaping, backstepping methods, prediction-based algorithms, PI regulators (based on Lyapunov arguments), numerical approaches (with pseudospectral methods). The second part focuses on several classes of *interconnected systems* for which stabilization becomes more difficult due to the underlying under-actuation. The proposed methods combine different strategies, including flatness designs, Legendre polynomial functions, Linear Matrices Inequalities, backstepping, and time-delay inspired techniques. Finally, the third part sheds light on *practical applications* (traffic networks, cable-operated elevators) and numerical implementation. The questions of modeling, simulating, and avoiding useless solicitations of the actuator are in the center of this axis.

All in all, this volume particularly highlights how the control design can consider fundamentally different constraints. These constraints consist of structural constraints on the controller (e.g., imposing a PI controller structure), input and state constraints, or constraints related to the implementation such as the sampled-data nature inherent to digital communication, the robustness to disturbances, or sensor noise, or saturation in the actuation.

The contributions presented in the volume reflect the contents of talks presented in the specialized DECOD workshop—Delays and Constraints In Distributed Parameter Systems that took place on November 23–26, 2021, in the Laboratoire des Signaux et Systèmes, Gif-sur-Yvette, France. This workshop was the last one in a series of International Scientific Coordination Network meetings on Delay and Constraints in Distributed Parameter Systems (SPa-DisCo), funded by the French Center for Scientific Research (CNRS), the Laboratoire des Signaux et Systèmes and Université Paris-Saclay.

The contents of the book are organized into three parts:

Part I. Stabilization and Control Strategies

Part one of the volume exposes several of the most recent and innovative strategies to cope with the analysis and control of distributed parameter systems. Different situations are considered, including the case of saturated inputs or stochastic input delays.

The part starts with the chapter *Energy Shaping Control of 1D Distributed Parameter Systems* authored by Yann Le Gorrec, Hector Ramirez, Yongxin Wu, Ning Liu, and Alessandro Macchelli. In this chapter, the authors present an overview of the energy-shaping approach for controlling 1D distributed parameter systems using the port-Hamiltonian approach. The proposed approach covers the case of distributed sensors and actuators and the case of boundary control systems (the actuator is located at the boundary of the spatial domain). Asymptotic stability is guaranteed using damping injection and by partially shaping the closed-loop energy function.

The second chapter in this part *Prediction Control for Nonlinear Systems with Stochastic Input Delay*, authored by Sijia Kong and Delphine Bresch-Pietri, deals with the stabilization of nonlinear systems with a stochastic input delay. Stochastic delays can arise from communication lags and can be modeled by continuous-time Markov processes with finite values. Using a partial differential equations representation and backstepping transformations, the authors propose an extension of classical prediction-based control strategies to compensate for the stochastic delay robustly.

The third chapter *Output Feedback Stabilization of a Reaction-Diffusion PDE in the Presence of Saturations of the Input and Its Time Derivatives* provided by Hugo Lhachemi and Christophe Prieur, investigates the problem of output feedback stabilization for a reaction–diffusion equation in the presence of input saturation and saturations on the time derivatives of the input. The proposed strategy consists in designing a finite-dimensional observer and state-feedback to stabilize the unstable part of the infinite-dimensional system. The exponential stability is guaranteed under Linear Matrix Inequalities-based sufficient conditions. The proof uses a Lyapunov approach while invoking sector conditions to deal with the saturation.

The last chapter *Numerical Control of a Semilinear Wave Equation on an Interval* in this first part is authored by Marcelo Cavalcanti, Valeria Domingos Cavalcanti, Carole Rosier, and Lionel Rosier. It investigates the numerical exact controllability of a semilinear wave equation using a Picard iterative scheme. Here, the authors introduce collocation pseudospectral methods on Chebychev grids to discretize the second-order derivative. The efficiency of the algorithm is validated through several simulations.

Part II. Stabilization and Tracking of Interconnected Systems

Part two of the volume focuses on the design of stabilizing and tracking control laws for interconnected systems. As a matter of fact, distributed systems are usually coupled with ordinary differential equations that can model actuator dynamics or load charges. This part contains four chapters covering theoretical contributions for the control of such types of interconnections.

This part opens with the chapter *Insight into the Stability Analysis of the Reaction-Diffusion Equation Interconnected with a Finite-Dimensional System Taking Support on Legendre Orthogonal Basis* authored by Mathieu Bajodek, Alexandre Seuret, and

Frédéric Gouaisbaut. This chapter aims to design a linear matrix inequality criterion to guarantee the stability of the considered interconnected system. The authors propose an original approach that combines a Lyapunov functional and Fourier–Legendre series. They also highlight a link with Padé approximation of the delays.

The second chapter of this part *Tracking Control for 2×2 Linear Heterodirectional Hyperbolic PDEs that are Bidirectionally Coupled with Nonlinear ODEs* is authored by Abdurrahman Irscheid, Nicole Gehring, Joachim Deutscher, and Joachim Rudolph and deals with the design of tracking control laws for this class of interconnected systems, under flatness assumptions. By means of the backstepping methodology, the system is expressed as a nonlinear ODEs system with a delayed input. A prediction of the ODE state can be obtained by solving a general nonlinear Volterra integrodifferential to design the stabilizing controller.

The third chapter *Practical Output Regulation and Tracking for Linear ODE-hyperbolic PDE-ODE Systems* is authored by Jeanne Redaud, Federico Bribiesca-Argomedo, and Jean Auriol; it deals with the output regulation for a class of hyperbolic equation with load and actuation dynamics. The problem presented here arises in multiple engineering systems for which the main focus is on load tracking. The proposed methodology combines the backstepping approach with classical results on approximate output regulations.

This part closes with the chapter *A Systematic Backstepping Design of Tracking Controllers for ODE-PDE-ODE Systems with Nonlinear Actuator Dynamics* authored by Nicole Gehring that tackles the stabilization and tracking problems for such class of interconnection, the PDE being either of hyperbolic type or parabolic type. The analysis relies on successive backstepping transformations to iteratively stabilize the different subsystems by means of virtual feedback laws.

Part III. Applications and Numerical Aspects

Part three of the volume focuses on various applications that can be modeled by distributed systems and that present specific and challenging modeling and control constraints. The four different chapters in this part cover applications as well as theoretical contributions. The questions related to the computational efforts are at the center of this part.

This part opens with the chapter *Cable-Operated Elevators and Deep-Sea Construction: 4×4 Hyperbolic PDE-ODE Control with Moving Boundary* authored by Miroslav Krstic and Ji Wang, where the authors solve two challenging engineering control problems that can be modeled using an infinite-dimensional framework. Contrary to most applications, the spatial domain is here time-varying since the cable length may vary with time. The authors adjust the backstepping approach to overcome this additional difficulty.

The second chapter *Ramp Metering: Modeling, Simulations and Control Issues* authored by Cédric Join, Hassane Abouaïssa and Michel Fliess investigates the modeling of highway traffic that is regulated by ramp metering. The authors present original simulation techniques that do not require any heuristic fundamental laws. The final objective is to implement a model-free control that is easy to implement and presents excellent robustness properties with respect to model mismatch.

The third chapter in this part *Event-Triggered Output Feedback Control of Traffic Flow on Cascaded Roads* is authored by Nicolas Espitia, Jean Auriol, Huan Yu, and Miroslav Krstic. This chapter aims to simultaneously stabilize the traffic flow on connected roads while avoiding useless solicitations of the actuator. The control signal is only updated when an event triggering condition is satisfied, corresponding to a more realistic setting by ramp metering on a digital platform.

Part three closes with the chapter *Computational Modeling for One-Dimensional Distributed Parameter Systems* authored by Daniela Danciu, which considers parabolic and hyperbolic equations with non-standard boundary conditions. These boundary conditions arise from natural sciences and control engineering applications and make the system difficult to emulate. The author introduces a computational procedure based on a "convergent Method of Lines" that guarantees the convergence of the approximate numerical solution. The computational efficiency of the procedure is ensured using parallel structured devices. The presentation is illustrated over a control engineering application.

Gif-sur-Yvette, France — Jean Auriol
Ulm, Germany — Joachim Deutscher
Gif-sur-Yvette, France — Guilherme Mazanti
Gif-sur-Yvette, France — Giorgio Valmorbida
October 2021

Contents

Acronyms

ARZ	Aw–Rascle-Zhang
BC	Boundary Condition
BCS	Boundary Control System
BVP	Boundary Value Problem
CNN	Cellular Neural Network
DCV	Deep-sea Construction Vessel
DPCS	Distributed Parameter Control Systems
DSP	Distributed Parameter System
ePDE	elliptic Partial Differential Equation
hPDE	hyperbolic Partial Differential Equation
HUM	Hilbert Uniqueness Method
IBVP	Initial Boundary Problem
IC	Initial Condition
IDA	Interconnection and Damping Assignment
IDBC	In-domain/Boundary controlled
IP	Intelligent Proportional
IVP	Initial Value Problem
LMI	Linear Matrix Inequality
LTI	Linear Time Invariant
LWR	Lighthill–Whitham–Richards
MoL	Method of Lines
NN	Neural Network
nsBCs	non-standard boundary conditions
ODE	Ordinary Differential Equation
PBC	Passivity-Based Control
PDE	Partial Differential Equation
PHS	Port Hamiltonian System
PID	Proportional Integral Derivative
pPDE	parabolic Partial Differential Equation
RNN	Recurrent Neural Networks
SDE	Stochastic Delay Equations

SISO	Single-Input Single-Output
UAV	Unmanned Aerial Vehicles
VSL	Varying Speed Limits

Stabilization and Control Strategies

Energy Shaping Control of 1D Distributed Parameter Systems

Yann Le Gorrec, Hector Ramirez, Yongxin Wu, Ning Liu, and Alessandro Macchelli

Abstract In this chapter we give an overview on energy shaping control for Distributed Parameter Systems defined on a 1D spatial domain using the port Hamiltonian framework. We consider two different cases: when actuators and sensors are located within the spatial domain and when the actuator is situated at the boundary of the spatial domain, leading to a boundary control system (BCS). In the first case we show how dynamic extensions and structural invariants can be used to change the internal properties of the system when the system is fully actuated, and how it can be done in an approximate way when the system is actuated using piecewise continuous actuators stemming from the use of patches. Asymptotic stability is achieved using damping injection. In the boundary controlled case we show how the closed loop energy function can be partially shaped, modifying the minimum and a part of the shape of this function and how damping injection can be used to guarantee asymptotic convergence.

Y. Le Gorrec (✉) · Y. Wu · N. Liu
FEMTO-ST UMR CNRS 6174, Université Bourgogne Franche Comté, 26 chemin de l'épitaphe, 25030 Besançon, France
e-mail: legorrec@femto-st.fr

Y. Wu
e-mail: yongxin.wu@femto-st.fr

N. Liu
e-mail: ning.liu@femto-st.fr

H. Ramirez
Departamento de Electrónica, Universidad Técnica Federico Santa María, Valparaiso, Chile
e-mail: hector.ramireze@usm.cl

A. Macchelli
Department of Electrical, Electronic and Information Engineering "Guglielmo Marconi" (DEI), University of Bologna, viale del Risorgimento 2, 40136 Bologna, Italy
e-mail: alessandro.macchelli@unibo.it

J. Auriol et al. (eds.), *Advances in Distributed Parameter Systems*, Advances in Delays and Dynamics 14, https://doi.org/10.1007/978-3-030-94766-8_1

1 Introduction

Control of distributed parameter systems driven by Partial Differential Equations (PDEs) has raised a major attention in the last decades due to the increasing use of compliant structures, system's information networks, smart materials and structures, multiscale and multiphysics systems etc. in engineering applications. Port Hamiltonian formulations which are an extension of Hamiltonian formulations to open multiphysics systems have shown to be particularly suited for the modelling and control of non linear systems [1–3]. They have been extended to distributed parameter systems in [4] using differential geometry, and in [5, 6] in the case of one dimensional linear systems using functional analysis. This approach has shown to be very efficient to prove existence of solution, asymptotic or exponential stability in the case the system is controlled at the boundary of its spatial domain with a linear or non-linear static or dynamic feedback [7–10]. The extension of these results to 2D and 3D systems has been initiated in [11] and is still subject to many ongoing studies.

Even if the stabilization of non linear and infinite dimensional systems is quite well established in the literature, the control design consisting in assigning the closed loop performances of the system is much less developed even if it is of particular interest from an engineering point of view. In this respect, the natural control design technique that has been proposed for non linear finite dimensional port Hamiltonian systems is Energy Shaping [2, 3, 12–14]. This control design technique consists in modifying the closed loop energy function using control by interconnection with a dynamic controller and structural invariants or a state feedback. This approach has been improved in order to cope with systems with pervasive diffusion and the so called dissipation obstacle leading to the Interconnection and Damping Assignment Passivity-Based Control (IDA-PBC) methodology. These results have been generalized, at least partially, to one dimensional port Hamiltonian systems in [9].

In this chapter we give an overview on energy shaping control for Distributed Parameter Systems defined on a 1D spatial domain using the port Hamiltonian framework. We consider two different cases: when actuators and sensors are located within the spatial domain and when the actuator is situated at the boundary of the spatial domain, leading to a boundary control system (BCS). In the first case we show how dynamic extensions and structural invariants can be used to change the internal properties of the system when the system is fully actuated, and how it can be done in an approximate way when the system is actuated using piecewise continuous actuators stemming from the use of patches. Asymptotic stability is achieved using damping injection. In the boundary controlled case we show how the closed loop energy function can be partially shaped, modifying the minimum and a part of the shape of this function and how damping injection can be used to guarantee asymptotic convergence.

The chapter is organized as follows. In Sect. 2 we define the considered class of systems. The problem formulation is recalled in Sect. 3 and the control design is detailed in Sect. 4 when the system is actuated at the boundary of its spatial domain and in Sect. 5 when the system is actuated in the domain. The chapter ends with some conclusions and perspectives.

2 Considered Class of Systems

In this chapter we consider one dimensional distributed parameter systems ($\zeta \in [a, b]$) with distributed and/or boundary control and observation of the form

$$\frac{\partial}{\partial t}\begin{bmatrix} x_1(\zeta,t) \\ x_2(\zeta,t) \end{bmatrix} = \begin{bmatrix} 0 & \mathcal{G} \\ -\mathcal{G}^* & -R \end{bmatrix}\begin{bmatrix} \mathcal{L}_1(\zeta)x_1(\zeta,t) \\ \mathcal{L}_2(\zeta)x_2(\zeta,t) \end{bmatrix} + \begin{bmatrix} 0 \\ I \end{bmatrix} u_d(\zeta,t) \tag{1}$$

$$y_d(\zeta,t) = \begin{bmatrix} 0 & I \end{bmatrix}\begin{bmatrix} \mathcal{L}_1(\zeta)x_1(\zeta,t) \\ \mathcal{L}_2(\zeta)x_2(\zeta,t) \end{bmatrix} \tag{2}$$

$$u_\partial = \mathcal{B}\begin{bmatrix} \mathcal{L}_1(\zeta)x_1(\zeta,t) \\ \mathcal{L}_2(\zeta)x_2(\zeta,t) \end{bmatrix}, \; y_\partial = \mathcal{C}\begin{bmatrix} \mathcal{L}_1(\zeta)x_1(\zeta,t) \\ \mathcal{L}_2(\zeta)x_2(\zeta,t) \end{bmatrix} \tag{3}$$

where $x = [x_1^T, x_2^T]^T \in X := L^2([a,b], \mathbb{R}^n) \times L^2([a,b], \mathbb{R}^n)$, $\mathcal{L} = diag(\mathcal{L}_1, \mathcal{L}_2)$ is a bounded and Lipschitz continuous matrix-valued function such that $\mathcal{L}(\zeta) = \mathcal{L}^T(\zeta)$ and $\mathcal{L}(\zeta) \geq \eta$ with $\eta > 0$ for all $\zeta \in [a,b]$, $R \in \mathbb{R}^{(n,n)}$, $R = R^T > 0$, $\mathcal{B}(\cdot)$ and $\mathcal{C}(\cdot)$ are some boundary input and boundary output mapping operators that will be defined later. The state space X is endowed with the inner product $\langle x|\tilde{x}\rangle_{\mathcal{L}} = \langle x|\mathcal{L}\tilde{x}\rangle$ and norm $\|x\|_{\mathcal{L}}^2 = \langle x|x\rangle_{\mathcal{L}}$ where $\langle\cdot|\cdot\rangle$ denotes the natural L^2-inner product. $X \ni x$ is the space of energy variables and $\mathcal{L}x$ denotes the co-energy variable associated to the energy variable x. Furthermore

$$\mathcal{G} = \sum_{i=0}^{N} G_i \frac{\partial^i}{\partial \zeta^i}, \text{ and } \mathcal{G}^* = \sum_{i=0}^{N} (-1)^i G_i^T \frac{\partial^i}{\partial \zeta^i}$$

with $G_i \in \mathbb{R}^{(n,n)}$. For a sake of compactness we shall use the following notation

$$P_i = \begin{bmatrix} 0 & G_i \\ (-1)^{i+1}G_i^T & 0 \end{bmatrix}, \; R_0 = \begin{bmatrix} 0 & 0 \\ 0 & R \end{bmatrix} \tag{4}$$

and the formulation of (1)

$$\frac{\partial x}{\partial t}(\zeta,t) = \sum_{i=0}^{N} P_i \frac{\partial^i}{\partial \zeta^i}\left(\mathcal{L}(\zeta)x(\zeta,t)\right) - R_0\mathcal{L}(\zeta)x(\zeta,t) + \begin{bmatrix} 0 \\ I \end{bmatrix} u_d(\zeta,t) \tag{5}$$

$$y_d(\zeta,t) = \begin{bmatrix} 0 & I \end{bmatrix}\mathcal{L}(\zeta)x(\zeta,t) \tag{6}$$

$$u_\partial = \mathcal{B}\left(\mathcal{L}(\zeta)x(\zeta,t)\right), \; y_\partial = \mathcal{C}\left(\mathcal{L}(\zeta)x(\zeta,t)\right) \tag{7}$$

The total energy of the system $H(x)$ is defined by

$$H(x) = \int_a^b \left(x^T(\zeta,t)\mathcal{L}(\zeta)x(\zeta,t)\right) d\zeta$$

Definition 1 (*Mixed in-domain/boundary controlled port Hamiltonian systems (IDBC-PHS)*) A mixed in-domain/boundary controlled port Hamiltonian system is an infinite dimensional system of the form (5–7) where the input mapping $u_\partial = \mathcal{B}(\mathcal{L}(\zeta)x(\zeta,t))$ and the output mapping $y_\partial = C(\mathcal{L}(\zeta)x(\zeta,t))$ are defined by

$$u_\partial = W_B \begin{bmatrix} \mathcal{L}(b)x(b,t) \\ \vdots \\ \frac{\partial^{N-1}(\mathcal{L}x)}{\partial\zeta^{N-1}}(b,t) \\ \mathcal{L}(a)x(a,t) \\ \vdots \\ \frac{\partial^{N-1}(\mathcal{L}x)}{\partial\zeta^{N-1}}(a,t) \end{bmatrix}, \text{ and } y_\partial = W_C \begin{bmatrix} \mathcal{L}(b)x(b,t) \\ \vdots \\ \frac{\partial^{N-1}(\mathcal{L}x)}{\partial\zeta^{N-1}}(b,t) \\ \mathcal{L}(a)x(a,t) \\ \vdots \\ \frac{\partial^{N-1}(\mathcal{L}x)}{\partial\zeta^{N-1}}(a,t) \end{bmatrix} \tag{8}$$

with

$$W_B = \left[\tfrac{1}{\sqrt{2}}(\Xi_2 + \Xi_1 P_e) \; \tfrac{1}{\sqrt{2}}(\Xi_2 - \Xi_1 P_e)\right], \tag{9}$$

$$W_C = \left[\tfrac{1}{\sqrt{2}}(\Xi_1 + \Xi_2 P_e) \; \tfrac{1}{\sqrt{2}}(\Xi_1 - \Xi_2 P_e)\right], \tag{10}$$

where

$$P_e = \begin{bmatrix} P_1 & \cdots & (-1)^{N-1}P_N \\ \vdots & \ddots & 0 \\ (-1)^{N-1}P_N & 0 & 0 \end{bmatrix} \tag{11}$$

and Ξ_1 and Ξ_2 in dimension is $N \times N$ satisfy

$$\Xi_2^\top \Xi_1 + \Xi_1^\top \Xi_2 = 0, \text{ and } \Xi_2^\top \Xi_2 + \Xi_1^\top \Xi_1 = I \tag{12}$$

The energy balance associated to the system reads

$$\frac{dH}{dt} = \int_a^b y_d^T u_d d\zeta - \int_a^b \left(x_2^T(\zeta,t)\mathcal{L}_2^T(\zeta)R\mathcal{L}_2(\zeta)x_2(\zeta,t)\right) d\zeta + y_\partial^T u_\partial \tag{13}$$

$$\leq \int_a^b y_d^T u_d d\zeta + y_\partial^T u_\partial \tag{14}$$

The existence of solutions of (5–7) under the conditions (9–10) and (12) has been investigated in [5, 6, 15] as stated in Theorem 1.

Theorem 1 ([6]) *The operator*

$$\mathcal{J}x(\Xi,t)=\sum_{i=0}^{N}P_i\frac{\partial^i}{\partial\zeta^i}\left(\mathcal{L}(\zeta)x(\zeta,t)\right)-R_0\mathcal{L}(\zeta)x(\zeta,t)$$

with domain

$$D(\mathcal{J})=\left\{\mathcal{L}x\in H^N\left(a,b;\mathbb{R}^n\right)\mid\begin{bmatrix}\mathcal{L}(b)x(b,t)\\ \vdots\\ \frac{\partial^{N-1}(\mathcal{L}x)}{\partial\zeta^{N-1}}(b,t)\\ \mathcal{L}(a)x(a,t)\\ \vdots\\ \frac{\partial^{N-1}(\mathcal{L}x)}{\partial\zeta^{N-1}}(a,t)\end{bmatrix}\in Ker\,W_B\right\}$$

where W_B is defined by (9) and Ξ_1 and Ξ_2 satisfy (12), generates a contraction semigroup on X. Furthermore the system (5–7) with (9–10) and (12) defines a boundary control system.

The general formulation (1) allows to model a large class of systems. For example:

- The 1D wave equation where $n=1$, $N=1$, $G_0=0$, $G_1=1$.
- The Euler Bernouilli beam equation. In this case $n=1$, $N=2$, $G_0=0$, $G_1=0$, $G_2=1$.
- The Timoshenko beam equation. In this case $n=2$, $N=1$, and

$$G_0=\begin{bmatrix}0&-1\\0&0\end{bmatrix},\ G_1=\begin{bmatrix}1&0\\0&1\end{bmatrix}$$

Example 1 (*The Timohenko beam*) The Timoshenko beam model describes the behavior of a thick beam in a one dimensional spatial domain. We choose as state variables $x=(x_1,x_2,x_3,x_4)^\top$, where $x_1(\zeta,t)=w_\zeta(\zeta,t)-\phi(\zeta,t)$ is the shear strain, $x_2(\zeta,t)=\phi_\zeta(\zeta,t)$ the angular strain, $x_3(\zeta,t)=\rho(\zeta)w_t(\zeta,t)$ the transverse momentum distribution, and $x_4(\zeta,t)=I_\rho(\zeta)\phi_t(\zeta,t)$ the angular momentum distribution where $w(\zeta,t)$ and $\phi(\zeta,t)$ are respectively the transverse displacement of the beam and the rotation angle of the neutral fiber of the beam.[1] The balance equations on these state variables lead to the dynamic model

[1] Note that we have used the lower indexes ζ and t to refer to the partial derivative with respect to that index.

$$\frac{\partial}{\partial t}\begin{bmatrix} x_1(\zeta,t) \\ x_2(\zeta,t) \\ x_3(\zeta,t) \\ x_4(\zeta,t) \end{bmatrix} = \begin{bmatrix} 0 & 0 & \frac{\partial}{\partial \zeta} & -1 \\ 0 & 0 & 0 & \frac{\partial}{\partial \zeta} \\ \frac{\partial}{\partial \zeta} & 0 & 0 & 0 \\ 1 & \frac{\partial}{\partial \zeta} & 0 & 0 \end{bmatrix} \begin{bmatrix} T(\zeta)x_1(\zeta,t) \\ EI(\zeta)x_2(\zeta,t) \\ \rho(\zeta)^{-1}x_3(\zeta,t) \\ I_\rho(\zeta)^{-1}x_4(\zeta,t) \end{bmatrix} \tag{15}$$

where $T(\zeta)$ is the shear modulus, $\rho(\zeta)$ is the mass per unit length, $EI(\zeta)$ is the Youngs modulus of elasticity E multiplied by the moment of inertia of a cross section I, and $I_\rho(\zeta)$ is the rotational momentum of inertia of a cross section. Note that, $T(\zeta)x_1(\zeta,t)$ is the shear force, $EI(\zeta)x_2(\zeta,t)$ the torque, $\rho(\zeta)^{-1}x_3(\zeta,t)$ the transverse velocity, and $I_\rho(\zeta)^{-1}x_4(\zeta,t)$ the angular velocity. This formulation is equivalent to (1) with $N=1, n=2$

$$G_1 = \begin{bmatrix} 1 & 0 \\ 0 & 1 \end{bmatrix}, G_0 = \begin{bmatrix} 0 & -1 \\ 0 & 0 \end{bmatrix}, \mathcal{L}(\zeta) = \begin{bmatrix} T(\zeta) & 0 & 0 & 0 \\ 0 & EI(\zeta) & 0 & 0 \\ 0 & 0 & \rho(\zeta)^{-1} & 0 \\ 0 & 0 & 0 & I_\rho(\zeta)^{-1} \end{bmatrix}.$$

Using $\Xi_1 = \begin{pmatrix} -\frac{\sqrt{2}}{2} & 0 & 0 & 0 \\ 0 & -\frac{\sqrt{2}}{2} & 0 & 0 \\ 0 & 0 & \frac{\sqrt{2}}{2} & 0 \\ 0 & 0 & 0 & \frac{\sqrt{2}}{2} \end{pmatrix}$ and $\Xi_2 = \begin{pmatrix} 0 & 0 & \frac{\sqrt{2}}{2} & 0 \\ 0 & 0 & 0 & \frac{\sqrt{2}}{2} \\ \frac{\sqrt{2}}{2} & 0 & 0 & 0 \\ 0 & \frac{\sqrt{2}}{2} & 0 & 0 \end{pmatrix}$ satisfying the condition (12) we obtain the following inputs and outputs

$$u_\partial(t) = \begin{pmatrix} \rho(a)^{-1}x_2(a,t) \\ I_\rho(a)^{-1}x_4(a,t) \\ T(b)x_1(b,t) \\ EI(b)x_3(b,t) \end{pmatrix}, \quad y_\partial(t) = \begin{pmatrix} -T(a)x_1(a,t) \\ -EI(a)x_3(a,t) \\ \rho(b)^{-1}x_2(b,t) \\ I_\rho(b)^{-1}x_4(b,t) \end{pmatrix} \tag{16}$$

and $u_d = 0$. The energy balance is given by $\dot{H}(t) = u_\partial(t)^\top y_\partial(t)$. The reader is refered to [16] for more details on the model, to [6, 15] for the well-posedness of this class of systems and to [17] for stability analysis.

3 Control by Interconnection and Energy Shaping

In what follows we consider control by interconnection [14], meaning that the system is connected to a dynamic controller in a power preserving way, as shown in Fig. 1.

If the input and output of the controller are denoted u_c and y_c respectively, the interconnection reads $u = u' - y_c$ and $u_c = y$ where u' is a reference signal. This controller can be very simple and may reduce to a simple positive gain in order to dissipate energy. In this case the total energy function can be used as a Lyapunov function and the controller can only be used for stabilization purposes with limited

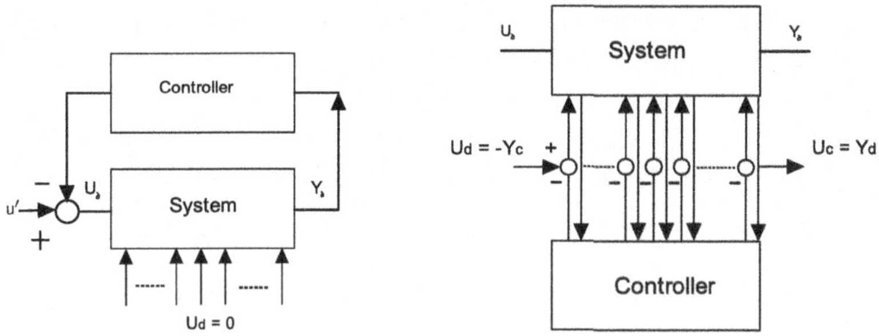

Fig. 1 Control by interconnection. Boundary control (left), in domain control (right)

performances. Damping injection only allows to stabilize the system with the main drawback of slowing it down. This is the reason why in general the controller is searched for as a dynamic system of state x_c and energy $H_c(x_c)$. The nature and dimensions of x_c depends on the nature of the control and will be made precise later on. From the power preserving interconnection the energy of the closed loop system is the sum of the energy of the open loop system and the one of the controller

$$H_{cl}(x, x_c) = H(x) + H_c(x_c)$$

Energy shaping control design consists in modifying the closed loop energy function by choosing appropriately the parameters of the controller. For that purpose, we first look for structural invariants $C(x, x_c)$, *i.e.* functions of the system and controller's state variables that do not vary along the system trajectories, $\dot{C} = 0$. They are looked for under the form

$$C(x, x_c) = x_c + F(x) = \kappa$$

where $F(x)$ is a smooth function. If these functions exist, it is possible to choose $\kappa = 0$ by an appropriate choice of the initial conditions. In this case the closed loop energy function reads

$$H_{cl}(x, x_c) = H_{cl}(x) = H(x) + H_c(-F(x))$$

and the shape of the energy function can be modified by choosing appropriately the function $H_c(\cdot)$. At the end asymptotic stability of the closed loop system in x^* is achieved using damping injection such that

$$\frac{dH_{cl}}{dt} < 0, \forall x \neq x^*.$$

In this chapter two different cases are considered, when the control is applied at the boundary of the spatial domain (boundary control) and when the control is applied within the spatial domain through a network of actuators and sensors. For a sake of simplicity we restrict the considered studies to the case $N = 1$, *i.e.*

$$\frac{\partial x}{\partial t}(\zeta,t) = P_1 \frac{\partial}{\partial \zeta}\left(\mathcal{L}(\zeta)x(\zeta,t)\right) + (P_0 - R_0)\,\mathcal{L}(\zeta)x(\zeta,t) + \begin{bmatrix} 0 \\ I \end{bmatrix} u_d(\zeta,t) \quad (17)$$

$$y_d(\zeta,t) = \begin{bmatrix} 0 & I \end{bmatrix} \mathcal{L}(\zeta)x(\zeta,t) \quad (18)$$

$$u_\partial = \mathcal{B}\left(\mathcal{L}(\zeta)x(\zeta,t)\right),\ y_\partial = C\left(\mathcal{L}(\zeta)x(\zeta,t)\right) \quad (19)$$

4 Boundary Control Design

We first consider boundary control design, *i.e.* $u_d = 0$. The system is connected to the controller at the boundaries of the spatial domain in the following way

$$\begin{pmatrix} u_\partial \\ y_\partial \end{pmatrix} = \begin{pmatrix} 0 & -I \\ I & 0 \end{pmatrix} \begin{pmatrix} u_C \\ y_C \end{pmatrix} + \begin{pmatrix} u' \\ 0 \end{pmatrix}. \quad (20)$$

where u' is the new reference input.

We consider the controller as a passive linear control system in port-Hamiltonian form [18]

$$\begin{cases} \dot{x}_C = (J_C - R_C)\,Q_C x_C + (G_C - P_C)\,u_C \\ y_C = (G_C + P_C)^T Q_C x_C + (M_C + S_C)\,u_C \end{cases} \quad (21)$$

where $x_C \in \mathbb{R}^{nc}$ and $u_C,\ y_C \in \mathbb{R}^n$, while $J_C = -J_C^T$, $M_C = -M_C^T$, $R_C = R_C^T$, $Q_C = Q_C^T > 0$ and $S_C = S_C^T$. The passivity is guaranteed as soon as

$$\begin{pmatrix} R_C & P_C \\ P_C^T & S_C \end{pmatrix} \geq 0\,. \quad (22)$$

For the sake of compactness, this system can be easily written in standard $(A_C,\ B_C,\ C_C,\ D_C)$ form, being

$$\begin{aligned} A_C &= (J_C - R_C)\,Q_C & B_C &= G_C - P_C \\ C_C &= (G_C + P_C)^T Q_C & D_C &= M_C + S_C\,. \end{aligned} \quad (23)$$

The closed loop energy function reads

$$H_{cl}(x(t), x_C(t)) = \frac{1}{2}\,\|x(t)\|_{\mathcal{L}}^2 + \underbrace{\frac{1}{2} x_C^T(t) Q_C x_C(t)}_{=:H_C(x_C(t))} \quad (24)$$

and the closed loop system can be written

$$\begin{cases} \dot{z} = \mathcal{J}_{cl} z \\ u' = \left(\mathcal{B} + D_C C\, C_C\right) z =: \mathcal{B}' z, \end{cases} \tag{25}$$

where

$$z = \begin{pmatrix} x \\ x_C \end{pmatrix} \in Z := X \times \mathbb{R}^{nc}$$

is the state variable of the closed loop system and $\mathcal{J}_{cl} : D(\mathcal{J}_{cl}) \subset Z \to Z$ is the following linear operator

$$\mathcal{J}_{cl} z := \begin{pmatrix} \mathcal{J} & 0 \\ B_C C & A_C \end{pmatrix} \begin{pmatrix} x \\ x_C \end{pmatrix} \tag{26}$$

with domain

$$D(\mathcal{J}_{cl}) = D(\mathcal{J}) \times \mathbb{R}^{nc}. \tag{27}$$

Z is endowed with the inner product defined as $\langle z_1, z_2 \rangle_Z = \langle x_1, x_2 \rangle_{\mathcal{L}} + x_{C,1}^T Q_C x_{C,2}$ which means that $H_{cl}(z) = \frac{1}{2} \|z\|_Z^2$.

Proposition 1.1 ([19]) *Consider the port-Hamiltonian system resulting from the power-conserving interconnection (20) of (17–19) and (21), which results in (25). Then, (25) with $\mathcal{J}_{cl}$ defined in (26) with domain (27) is a boundary control system. Moreover, the operator $\bar{\mathcal{J}}_{cl}$ given by*

$$\bar{\mathcal{J}}_{cl} z := \begin{pmatrix} \mathcal{J} & 0 \\ B_C C & A_C \end{pmatrix} \begin{pmatrix} x \\ x_C \end{pmatrix}$$

with domain

$$D(\bar{\mathcal{J}}_{cl}) = \left\{ \begin{pmatrix} x \\ x_C \end{pmatrix} \in Z \mid x \in D(\mathcal{J}),\ and\ \mathcal{B}' \begin{pmatrix} x \\ x_C \end{pmatrix} = 0 \right\}$$

with $\mathcal{B}'$ defined in (25) generates a contraction semigroup.

Definition 2 (*Casimir functions*) Consider the boundary control system defined in Proposition 1.1 with $u' = 0$ in (20). A function $C : X \times \mathbb{R}^{nc} \to \mathbb{R}$ is a Casimir function if $\dot{C} = 0$ along the (classical) solutions for every possible choice of $\mathcal{L}(\cdot)$ and Q_C [20–22].

As stated in Sect. 3 we look for Casimir functions in the form

$$C(x(t), x_C(t)) = \Gamma^T x_C(t) + \int_a^b \Psi^T(z) x(t, z)\, dz \tag{28}$$

with $\Gamma \in \mathbb{R}^{nc}$ and $\Psi \in L^2(a, b; \mathbb{R}^n)$. Note that they are not (yet) in the form assumed above.

Proposition 1.2 ([19]) *Consider the boundary control system introduced in Proposition 1.1 with $u' = 0$ in (20). Then, (28) is a Casimir function for this system if and only if $\Psi \in H^1(a, b; \mathbb{R}^n)$,*

$$P_1 \frac{d\Psi}{dz}(z) + (P_0 + R_0)\Psi(z) = 0 \tag{29}$$

$$(J_C + R_C)\Gamma + (G_C + P_C)\tilde{W}R \begin{pmatrix} \Psi(b) \\ \Psi(a) \end{pmatrix} = 0 \tag{30}$$

$$(G_C - P_C)^T\Gamma + \left[W + (M_C - S_C)\,\tilde{W}\right] R \begin{pmatrix} \Psi(b) \\ \Psi(a) \end{pmatrix} = 0 \tag{31}$$

We now suppose there exist n_c independent Casimir functions. In this case the state of the controller is related to the state of the system through the relation

$$x_C(t) = -\hat{\Gamma}^{-\mathrm{T}} \int_a^b \hat{\Psi}^T(z)x(t, z)\,dz + \kappa \tag{32}$$

It is then possible to verify that the closed loop dynamics is given by the boundary control system:

$$\begin{aligned} \frac{\partial x}{\partial t}(t, z) &= P_1 \frac{\partial}{\partial z}\frac{\delta H_{cl}}{\delta x}(x(t))(z) + (P_0 - G_0)\frac{\delta H_{cl}}{\delta x}(x(t))(z) \\ u'(t) &= W'R \begin{pmatrix} \left(\frac{\delta H_{cl}}{\delta x}(x(t))\right)(b) \\ \left(\frac{\delta H_{cl}}{\delta x}(x(t))\right)(a) \end{pmatrix} \end{aligned} \tag{33}$$

where δ denotes the Fréchet derivative and the closed loop energy function is given by:

$$\begin{aligned} H_{cl}(x(t)) = \frac{1}{2}\,\|x(t)\|^2_{\mathcal{L}} + \frac{1}{2}\left(\int_a^b \hat{\Psi}^T(z)x(t, z)\,dz\right)^T \times \\ \times\, \hat{\Gamma}^{-1} Q_C \hat{\Gamma}^{-\mathrm{T}} \int_a^b \hat{\Psi}^T(z)x(t, z)\,dz \end{aligned} \tag{34}$$

and W' is a $n \times 2n$ full rank, real matrix such that $W'\Sigma W'^T \geq 0$.

The main drawback of such approach is that existence of Casimir function is subject to the fact that there is no dissipation in the considered coordinate. This is known as the dissipation obstacle. It is then not possible to use the dynamic extension and its reduction to shape the closed loop energy function. Yet, as stated in Proposition 1.3 it is possible to directly shape the closed loop energy by using a state feedback having a similar form than the one obtained using the immersion/reduction approach.

Proposition 1.3 (Energy-shaping [9]) *Consider the system (17) with boundary control given by (19). Denote by* $H(x) = \frac{1}{2}\|x\|^2_{\mathcal{L}}$ *its Hamiltonian function. Then, the feedback law* $u = \beta(x) + u'$, *with* u' *an auxiliary boundary input, maps (17), (19) into the target dynamical system*

$$\begin{aligned}\frac{\partial x}{\partial t}(t,z) &= P_1 \frac{\partial}{\partial z}\frac{\delta H_d}{\delta x}(x(t))(z) + (P_0 - G_0)\frac{\delta H_d}{\delta x}(x(t))(z)\\ u'(t) &= W_B \begin{pmatrix} \left(\frac{\delta H_d}{\delta x}(x(t))\right)(b) \\ \left(\frac{\delta H_d}{\delta x}(x(t))\right)(a) \end{pmatrix}\end{aligned} \tag{35}$$

with $H_d(x) = H(x) + H_a(x)$, *provided that*

$$P_1 \frac{\partial}{\partial z}\frac{\delta H_a}{\delta x}(x) + (P_0 - G_0)\frac{\delta H_a}{\delta x}(x) = 0 \tag{36}$$

$$\beta(x) + W_B \begin{pmatrix} \left(\frac{\delta H_a}{\delta x}(x)\right)(b) \\ \left(\frac{\delta H_a}{\delta x}(x)\right)(a) \end{pmatrix} = 0. \tag{37}$$

Once H_d is defined a natural choice for the output is

$$y'(t) = W_C \begin{pmatrix} \left(\frac{\delta H_d}{\delta x}(x(t))\right)(b) \\ \left(\frac{\delta H_d}{\delta x}(x(t))\right)(a) \end{pmatrix} \tag{38}$$

which implies that $\frac{\mathrm{d}}{\mathrm{d}t} H_d(x(t)) \leq y'^T(t)u'(t)$. In order to get asymptotic stability one has to consider damping injection:

$$u'(t) = -\Xi y'(t), \qquad \Xi = \Xi^T \geq 0. \tag{39}$$

The additional Hamiltonian H_a is constructed in such a way that $\mathcal{L}^{-1}\frac{\delta H_a}{\delta x}(x)$ are equilibrium states of (17). Furthermore, since the system has to reach a non-zero state, H_d is chosen with a global minimum in this non-zero state. In the following lemma, a construction for H_a which achieves this is illustrated. Since, in this paper, the linear case is treated, the focus is on quadratic Hamiltonian functions.

Lemma 1 ([9]) *Let* $\Phi_i \in H^1(a,b;\mathbb{R}^n)$, $i = 1,\ldots,n$ *be independent solutions of*

$$P_1 \frac{\mathrm{d}\Phi_i}{\mathrm{d}\zeta}(\zeta) + (P_0 - G_0)\Phi_i(\zeta) = 0, \tag{40}$$

and define $\hat{\Phi}(\zeta) = \big(\Phi_1(\zeta), \ \ldots, \ \Phi_n(\zeta)\big)$. *Furthermore, let* $x_\star$ *be an equilibrium state of (17), i.e.* $\mathcal{L}x_\star \in H^1(a,b;\mathbb{R}^n)$ *and*

$$P_1 \frac{\partial(\mathcal{L}x_\star)}{\partial\zeta}(\zeta) + (P_0 - G_0)(\mathcal{L}x_\star)(\zeta) = 0. \tag{41}$$

Then

$$H_a(x) = \frac{1}{2}\left[\int_a^b \hat{\Phi}^T (x - x_\star)\, d\zeta\right]^T \times Q_a \left[\int_a^b \hat{\Phi}^T (x - x_\star)\, d\zeta\right] - \\ - \int_a^b x_\star^T \mathcal{L} x \, d\zeta + \kappa, \quad (42)$$

with $Q_a = Q_a^T > 0$ *and* $\kappa \in \mathbb{R}$ *some constant, satisfies (36) and* $H_d = H + H_a$ *has a global minimum in* $x_\star$.

Define the bounded linear operator $K_\Phi : X \to \mathbb{R}^n$ as

$$K_\Phi x = \int_a^b \hat{\Phi}^T(\zeta) x(\zeta)\, d\zeta, \quad (43)$$

and $\mathcal{L}_d$ as

$$\mathcal{L}_d = \mathcal{L} + K_\Phi^* Q_a K_\Phi, \quad (44)$$

where $K_\Phi^* : \mathbb{R}^n \to X$ is the adjoint operator of K_Φ. Clearly, $K_\Phi^* = \hat{\Phi}$, and $\mathcal{L}_d$ is a bounded, coercive operator on $L^2(a, b; \mathbb{R}^n)$. Furthermore, H_a being given by (42) we find

$$H_d(x) = \frac{1}{2} \langle (x - x_\star) \mid \mathcal{L}_d (x - x_\star) \rangle_{L^2} + H_d(x_\star). \quad (45)$$

In order to prove the asymptotic stability (Proposition 2) one has first to prove existence and uniqueness of solutions as well as precompactness of trajectories (Propositions 1.4 and 1.5) in order to use LaSalle's invariance principle

Proposition 1.4 ([9]) *The closed loop system (35), (38) with (39) in which* H_a *is defined by (42) admits a unique solution. Furthermore, the mapping from the initial error state at time* $t = 0$, $x_0 - x_\star$ *to the error state at time* t, $x(t) - x_\star$ *defines a contraction semigroup in the norm* $\frac{1}{2} \langle (x - x_\star) \mid \mathcal{L}_d (x - x_\star) \rangle_{L^2}$.

Proposition 1.5 *The operator* $\mathcal{J}_d$ *defined as*

$$\mathcal{J}_d x := P_1 \frac{\partial(\mathcal{L}_d x)}{\partial \zeta} + (P_0 - G_0)(\mathcal{L}_d x) \quad (46)$$

with domain

$$D(\mathcal{J}_d) = \left\{ x \in L^2(a, b; \mathbb{C}^n) \mid \mathcal{L}_d x \in H^1(a, b; \mathbb{C}^n) \right. \\ \left. \text{and } 0 = [W_B + \Xi W_C] \begin{pmatrix} (\mathcal{L}_d x)(b) \\ (\mathcal{L}_d x)(a) \end{pmatrix} \right\} \quad (47)$$

is the infinitesimal generator of a contraction semigroup and has a compact resolvent.

Theorem 2 (Asymptotic stability [9]) *Consider the linear, infinite dimensional, port-Hamiltonian system (17) and the equilibrium state* $x_\star$ *satisfying (41). Then, the control action* $u = \beta(x) + u'$ *with* β *defined in (37),* H_a *chosen as in (42), and with* u' *defined in (39) with* $\Xi > 0$*, makes* $x_\star$ *asymptotically stable.*

It has been shown for the vibrating string example [9] that the energy shaping control plus damping injection using closed loop invariants or integral feedback forms is equivalent to a proportional derivative control acting at the boundary. The overall stability is achieved while the performances at the end point of the string are assignable. The main advantage of using this control technique lies in its physical interpretation and its potential extension to the case of a PDE connected to a non linear ODE.

5 In Domain Distributed Control

In this section the distributed parameter system is controlled within its spatial domain meaning that $u_\partial = 0$

$$\begin{pmatrix} u_d(\zeta,t) \\ y_d(\zeta,t) \end{pmatrix} = \begin{pmatrix} 0 & -I \\ I & 0 \end{pmatrix} \begin{pmatrix} u_C(\zeta,t) \\ y_C(\zeta,t) \end{pmatrix} + \begin{pmatrix} u'(\zeta,t) \\ 0 \end{pmatrix}, \tag{48}$$

as depicted in Fig. 2.

We consider two cases. The ideal case where the control is infinite dimensional and uniformly distributed along the spatial domain. In this case we show how the dynamic controller parameters and structural invariants can be used to perfectly shape the closed loop energy function. This control strategy being not realistic from an engineering point of view, the existence of solution and the precompactness of the closed loop trajectories are not further investigated to the benefit to the second case when the control is achieved through a continuous piecewise network of actuators. In this case an early lumping approach is proposed and the shaping of the closed

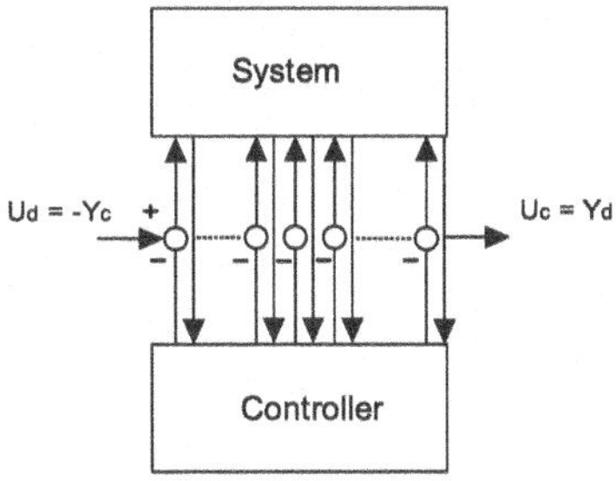

Fig. 2 Distributed control

loop energy function is achieved in an approximate and optimal way. The finite dimensional controller is shown to achieve closed loop asymptotic stability.

5.1 Ideal Case

We first consider the ideal case where the distributed parameter system is actuated at any point of its spatial domain. In this case the state of the controller can be distributed in space and the controller has the general formulation

$$\begin{cases} \frac{\partial x_C}{\partial t}(\zeta,t) = \mathcal{J}_c \mathcal{Q}_c x_C(\zeta,t) + \mathcal{B}_c u_C(\zeta,t) \\ y_C(\zeta,t) = \mathcal{B}_c{}^* \mathcal{Q}_c x_C(\zeta,t) + \mathcal{S}_c u_C(\zeta,t) \end{cases} \tag{49}$$

where $\mathcal{Q}_c$ is a bounded and Lipschitz continuous matrix-valued function such that $\mathcal{Q}_c(\zeta) = \mathcal{Q}_c^T(\zeta)$ and $\mathcal{Q}_c(\zeta) \geq \eta_c$ with $\eta_c > 0$ for all $\zeta \in [a,b]$, $\mathcal{S}_c$ is a bounded and Lipschitz continuous function such that $\mathcal{S}_c(\zeta) = \mathcal{S}_c^T(\zeta)$ and $\mathcal{S}_c(\zeta) \geq \eta_s$ with $\eta_s > 0$ for all $\zeta \in [a,b]$. $\mathcal{B}_c$ and $\mathcal{J}_c$ are differential operators of the form:

$$\mathcal{B}_c = B_{c0} + B_{c1}\frac{\partial}{\partial \zeta} \tag{50}$$

$$\mathcal{J}_c = J_{c0} + J_{c1}\frac{\partial}{\partial \zeta} \tag{51}$$

with $B_{c0}, B_{c1} \in \mathbb{R}^{(n_c,1)}$, $J_{c0} = -J_{c0}^T$, $J_{c1} = J_{c1}^T \in \mathbb{R}^{(n_c,n_c)}$. The closed loop system corresponding to (17) interconnected to Fig. 2 through (48) is given by:

$$\frac{\partial x_e}{\partial t} := \begin{pmatrix} \frac{\partial x_1}{\partial t} \\ \frac{\partial x_2}{\partial t} \\ \frac{\partial x_c}{\partial t} \end{pmatrix} = \begin{pmatrix} 0 & \mathcal{G} & 0 \\ -\mathcal{G}^* & -(\mathcal{S}_c + R) & -\mathcal{B}_C^* \\ 0 & \mathcal{B}_C & \mathcal{J}_c \end{pmatrix} \begin{pmatrix} \mathcal{L}_1 x_1 \\ \mathcal{L}_2 x_2 \\ \mathcal{Q}_c x_c \end{pmatrix} \tag{52}$$

Proposition 1.6 *The closed loop system (52) admits structural invariants of the form*

$$\kappa_0 = C(x_e) = \int_a^b \Psi^T x_e d\zeta \tag{53}$$

with $\Psi = (\psi_1, \psi_2, \psi_3)$ *if and only if*

$$-\mathcal{G}\psi_2(\zeta) = 0 = -\mathcal{B}_c\psi_2(\zeta) + \mathcal{J}_c^*\psi_3(\zeta) \tag{54}$$

$$(\mathcal{S}+R)\,\psi_2(\zeta) = 0 \tag{55}$$

$$\mathcal{G}\psi_1(\zeta) + \mathcal{B}_c^*\psi_3(\zeta) = 0 \tag{56}$$

$$\left.\begin{pmatrix} 0 & G_1 & 0 \\ -G_1^T & 0 & -B_{c1} \\ 0 & B_{c1}^T & J_{c1} \end{pmatrix}\begin{pmatrix} \psi_1(\zeta) \\ \psi_2(\zeta) \\ \psi_3(\zeta) \end{pmatrix}\right|_{a,b} = 0 \tag{57}$$

Proposition 1.7 ([19]) *Choosing $\mathcal{B}_c = \mathcal{G}$ the closed loop system (52) admits as structural invariants the function $C(x_e)$ defined by (53) and*

$$\Psi = (\Psi_1, 0, \Psi_1)$$

In this case the hyperbolic system (1) connected to the dynamic controller (49) of the form

$$\begin{cases} \dfrac{\partial x_C}{\partial t}(\zeta,t) = \mathcal{G}u_C(\zeta,t) \\ y_C(\zeta,t) = -\mathcal{G}^*\mathcal{Q}_c x_C(\zeta,t) + \mathcal{S}_c u_C(\zeta,t) \end{cases} \tag{58}$$

is equivalent to the system (choosing $u' = 0$)

$$\frac{\partial}{\partial t}\begin{bmatrix} x_1(\zeta,t) \\ x_2(\zeta,t) \end{bmatrix} = \begin{bmatrix} 0 & \mathcal{G} \\ -\mathcal{G}^* & -(R+\mathcal{S}) \end{bmatrix}\begin{bmatrix} (\mathcal{L}_1(\zeta) + \mathcal{Q}_c(\zeta))\,x_1(\zeta,t) \\ \mathcal{L}_2(\zeta)x_2(\zeta,t) \end{bmatrix} \tag{59}$$

$$u_\partial = \mathcal{B}\begin{bmatrix} \mathcal{L}_1(\zeta)x_1(\zeta,t) \\ \mathcal{L}_2(\zeta)x_2(\zeta,t) \end{bmatrix}, \quad y_\partial = \mathcal{C}\begin{bmatrix} \mathcal{L}_1(\zeta)x_1(\zeta,t) \\ \mathcal{L}_2(\zeta)x_2(\zeta,t) \end{bmatrix} \tag{60}$$

One can see that using an ideal distributed feedback the equivalent stiffness and the overall damping coefficient of the closed loop system can be modified using $\mathcal{Q}_c$ and $\mathcal{S}$ respectively. The former allows to modify the closed loop performances of the system while the latter guarantees its stability. In this respect existence and precompactness of solutions as well as closed loop asymptotic stability can be proven using passivity arguments. The proof is omitted here and will be proposed in the more realistic case when the control is implemented using a network of actuators.

5.2 *Under Actuated Case*

We consider now that the distributed parameter system is actuated through piecewise constant elements (cf. Fig. 3).

The controller is in this case finite dimensional and the control design is implemented using an early lumping approach. The first step is to discretize the system (1) using a structure preserving discretization method. We apply the mixed finite element method [23] such that the approximation of (1) is again a PHS with p elements:

$$\begin{pmatrix} \dot{x}_{1d} \\ \dot{x}_{2d} \end{pmatrix} = (J_n - R_n) \begin{pmatrix} Q_1 x_{1d} \\ Q_2 x_{2d} \end{pmatrix} + B_b u_b + \begin{pmatrix} 0 \\ B_{0d} \end{pmatrix} \mathbf{u}_d, \tag{61a}$$

$$y_b = B_b^T \begin{pmatrix} Q_1 x_{1d} \\ Q_2 x_{2d} \end{pmatrix} + D_b u_b, \tag{61b}$$

$$\mathbf{y}_d = \begin{pmatrix} 0 & B_{0d}^T \end{pmatrix} \begin{pmatrix} Q_1 x_{1d} \\ Q_2 x_{2d} \end{pmatrix}, \tag{61c}$$

where $x_{id} = \left(x_i^1 \cdots x_i^p\right)^T$ for $i \in \{1, \cdots, 2n\}$,

$$J_n = \begin{pmatrix} 0 & J_i \\ -J_i^T & 0 \end{pmatrix} \quad \text{and} \quad R_n = \begin{pmatrix} 0 & 0 \\ 0 & R_d \end{pmatrix},$$

are the discretized matrices of the operators $\mathcal{J}$ and $\mathcal{R}$ with J_i and R_d the discretized matrices of the operators $\mathcal{G}$ and R. $B_{0d} \in \mathbb{R}^{np \times p}$, $Q_1 \in \mathbb{R}^{np \times np}$ and $Q_2 \in \mathbb{R}^{np \times np}$ are the discretized matrices of B_0, $\mathcal{L}_1$ and $\mathcal{L}_2$, respectively. D_b is feedthrough term for boundary control. The input u_b denotes the boundary input which corresponds to the boundary actuation or/and conditions. Since the distributed actuation of the system is considered, we assume that there is no energy changes (actuation) at the boundary of the spatial domain, *i.e.* $u_b = 0$ and the discretized system (61) can therefore be simplified. The Hamiltonian of the discretized model (61) writes:

$$H_d(x_{1d}, x_{2d}) = \frac{1}{2} \left(x_{1d}^T Q_1 x_{1d} + x_{2d}^T Q_2 x_{2d}\right). \tag{62}$$

Remark 1 In what follows the patches are supposed to cover the entire spatial domain of the system in an homogeneous way. Yet the network of patched may present discontinuities. In this case the high frequency modes are not controlled by the actuators but have a guaranteed stability due to the internal dissipation of the open loop system.

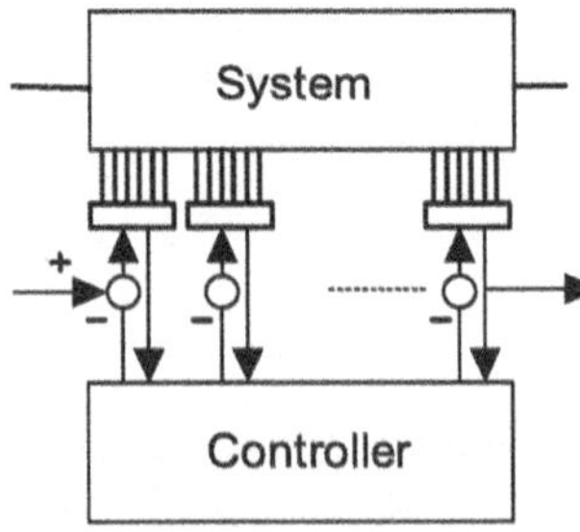

Fig. 3 In domain control using piecewise constant elements

5.2.1 Control Design

The controller in Fig. 3 is designed to be a finite dimensional PHS, which is expressed as follows:

$$\begin{aligned} \dot{x}_c &= (J_c - R_c)\, Q_c x_c + B_c u_c, \\ y_c &= B_c^T Q_c x_c + D_c u_c, \end{aligned} \tag{63}$$

where $x_c \in \mathbb{R}^{m\times 1}$, $J_c = -J_c^T \in \mathbb{R}^{m\times m}$, $R_c = R_c^T \geq 0$ and $Q_c = Q_c^T \geq 0$, $B_c \in \mathbb{R}^{m\times m}$, $\mathbb{R}^{m\times m} \ni D_c > 0$, $u_c \in \mathbb{R}^{m\times 1}$ and $y_c \in \mathbb{R}^{m\times 1}$. Matrices Q_c and D_c work as the energy shaping and the damping injection or the diffusion operator, respectively. Without considering external signals, the interconnection between the discretized plant system (61) and the controller (63) is:

$$\begin{pmatrix} \mathbf{u}_d \\ u_c \end{pmatrix} = \begin{pmatrix} 0 & -M \\ M^T & 0 \end{pmatrix} \begin{pmatrix} \mathbf{y}_d \\ y_c \end{pmatrix}, \quad \text{where } M = I_m \otimes \mathbf{1}_{k\times 1} \in \mathbb{R}^{p\times m}, \tag{64}$$

as illustrated in Fig. 3, k being the number of elements sharing the same input.[2] The passive interconnection (64) keeps the passivity of the closed loop. It results in a new PHS in closed loop:

$$\dot{x}_{cl} = (J_{cl} - R_{cl})\, Q_{cl} x_{cl}, \tag{65}$$

where $x_{cl} = \left(x_{1d}^T, x_{2d}^T, x_c^T\right)^T$, $Q_{cl} = \text{diag}\left(Q_1, Q_2, Q_c\right)$,

$$J_{cl} = \begin{pmatrix} O & J_i & 0 \\ -J_i^T & 0 & -B_{0d} M B_c^T \\ 0 & B_c M^T B_{0d}^T & J_c \end{pmatrix}, \quad R_{cl} = \begin{pmatrix} 0 & 0 & 0 \\ 0 & R_d + B_{0d} M D_c M^T B_{0d}^T & 0 \\ 0 & 0 & R_c \end{pmatrix}.$$

The Hamiltonian of the controller (63) is:

$$H_c(x_c) = \frac{1}{2} x_c^T Q_c x_c. \tag{66}$$

Therefore, the closed loop Hamiltonian function reads:

$$H_{cld}(x_{1d}, x_{2d}, x_c) = H_d(x_{1d}, x_{2d}) + H_c(x_c). \tag{67}$$

The next step is to design controller matrices J_c, R_c, B_c, Q_c, and D_c in order to shape the closed loop Hamiltonian (67). We first give the following Proposition 1.8 that links the state variables of the discretized plant (61) and that of the controller (63) through a Casimir function.

Proposition 1.8 ([24]) *Choosing $J_c = 0$, and $R_c = 0$, the closed loop system (65) admits the Casimir function $C(x_{1d}, x_c)$ defined by:*

---

[2] $\otimes$ is the Kronecker product and $\mathbf{1}_{k\times 1}$ the vector of dimension k containing only ones.

$$C(x_{1d}, x_c) = B_c M^T B_{0d}^T J_i^{-1} x_{1d} - x_c \tag{68}$$

as structural invariant, i.e. $\dot{C}(x_{1d}, x_c) = 0$ *along the closed loop trajectories. If the initial conditions of* $x_{1d}(0)$ *and* $x_c(0)$ *satisfy* $C(x_{1d}(0), x_c(0)) = 0$*, the controller is a proportional-integral control, and the control law (64) is equivalent to the state feedback:*

$$\mathbf{u}_d = -B_c^T Q_c B_c M^T B_{0d}^T J_i^{-1} x_{1d} - D_c M^T B_{0d}^T Q_2 x_{2d}. \tag{69}$$

Therefore, the closed loop system yields:

$$\begin{pmatrix} \dot{x}_{1d} \\ \dot{x}_{2d} \end{pmatrix} = \begin{pmatrix} 0 & J_i \\ -J_i^T & -(R_d + B_{0d} M D_c M^T B_{0d}^T) \end{pmatrix} \begin{pmatrix} \tilde{Q}_1 x_{1d} \\ Q_2 x_{2d} \end{pmatrix}, \tag{70}$$

where

$$\tilde{Q}_1 = Q_1 + Q_c, \text{ with } Q_c = J_i^{-T} B_{0d} M B_c^T Q_c B_c M^T B_{0d}^T J_i^{-1} \tag{71}$$

is the new closed loop energy matrix associated to x_{1d}*.*

From a physical point of view, (68) implies that with the dynamic controller (63) equivalent to the state feedback (69), it is possible to change, at least partially (depending on p and the range of B_{0d}), the energy matrix related to x_{1d}. Actually, one can only shape the energy matrix related to the first p elements of x_{1d}, i.e. $\left(\tilde{Q}_1\right)_{p\times p}$. For a given number of distributed inputs m, the objectives of the energy shaping is to find matrices B_c and Q_c such that the *distance* (considered here in the Frobenius norm, see Definition 6.4 of [25]) between the desired energy matrix $\tilde{Q}_{1d}$ and the closed loop one $\tilde{Q}_1$ is minimal:

$$\min_{B_c^T Q_c B_c} \left\| J_i^{-T} B_{0d} M B_c^T Q_c B_c M^T B_{0d}^T J_i^{-1} + Q_1 - \tilde{Q}_{1d} \right\|_F. \tag{72}$$

If we consider p and eliminate B_{0d}, (72) is equivalent to:

$$\min_{B_c^T Q_c B_c} \left\| (J_i)_{p\times p}^{-T} M B_c^T Q_c B_c M^T (J_i)_{p\times p}^{-1} - Q_m \right\|_F, \tag{73}$$

where $(J_i)_{p\times p}$ is the first p lines p columns of J_i and $Q_m = \left(\tilde{Q}_{1d} - Q_1\right)_{p\times p} \geq 0$. Furthermore, (73) can be formalized by the optimization Problem 1.

Problem 1 The energy related to first p elements of x_{1d} in closed loop is shaped in an optimal way if and only if $X = B_c^T Q_c B_c \in SR_0^{m\times m}$ minimizes the criterion

$$f(X) = \left\| A X A^T - Q_m \right\|_F, \tag{74}$$

where $A = (J_i)_{p\times p}^{-T} M \in \mathbb{R}^{p\times m}$ and $SR_0^{m\times m}$ represents the set of symmetric and positive semi-definite matrices.

The solution of Problem 1 depends on the number of actuators. The system can be fully actuated ($m = p$) or under-actuated ($m < p$). In what follows we consider the realistic under-actuated case. In this case A in (74) is not invertible and one has to consider the numerical optimization of $f(X)$ in Problem 1. The solution of the optimization Problem 1 is given in Proposition 1.9.

Proposition 1.9 *$f(X)$ is convex and the minimization of $f(X)$ is equivalent to the minimization of $f^2(X)$, which has a unique minimum given by*

$$\hat{X} = V\Sigma_0^{-1}U_1^T Q_m U_1 \Sigma_0^{-1} V^T \tag{75}$$

with V, Σ_0 and U_1 the matrices of the singular value decomposition (SVD) of the matrix A i.e.

$$A = U\Sigma V^T = \begin{pmatrix} U_1 & U_2 \end{pmatrix} \begin{pmatrix} \Sigma_0 \\ 0 \end{pmatrix} V^T, \tag{76}$$

where $U \in \mathbb{R}^{p\times p}$ and $V \in \mathbb{R}^{m\times m}$ are unitary matrices, $U_1 \in \mathbb{R}^{p\times m}$, $U_2 \in \mathbb{R}^{p\times q}$, $q = p - m$, and $\Sigma_0 = \Sigma_0^T \geq 0$ is the diagonal matrix of singular values of A.

The choice of the controller matrices B_c and Q_c is not unique and only has to satisfy the condition (75). It is done in order to modify the shape of the closed loop energy function of the system in the x_1 coordinate. The choice of the controller matrix D_c follows a similar procedure, with the optimization of the difference between the approximate dissipation and the desired one.

5.2.2 Stability Analysis

We now consider the closed loop stability analysis when the finite-dimensional controller (63) elaborated on the finite dimensional approximation of the system is connected to the infinite-dimensional system (17). In this case the closed loop system reads:

$$\dot{\mathcal{X}} = \underbrace{\begin{pmatrix} (\mathcal{J} - \mathcal{R} - \mathcal{B}D_c\mathcal{B}^*) & -\mathcal{B}B_c^T \\ B_c\mathcal{B}^* & 0 \end{pmatrix} \begin{pmatrix} \mathcal{L} & 0 \\ 0 & Q_c \end{pmatrix}}_{\mathcal{A}_{cl}} \mathcal{X}, \tag{77}$$

where $\mathcal{X} = \begin{pmatrix} x^T & x_c^T \end{pmatrix}^T \in X_s$ is the state defined on the state space $X_s = L_2\left([0, L], \mathbb{R}^{2n}\right) \times \mathbb{R}^m$.

Theorem 3 *The operator $\mathcal{A}_{cl}$ defined in (77) generates a contraction semigroup on $X_s = L_2\left([0, L], \mathbb{R}^{2n}\right) \times \mathbb{R}^m$.*

Theorem 4 *The operator $\mathcal{A}_{cl}$ has a compact resolvent.*

Theorem 5 *For any $\mathcal{X}(0) \in L_2\left([0, L], \mathbb{R}^{2n}\right) \times \mathbb{R}^m$, the unique solution of (77) tends to zero asymptotically, and the closed loop system (77) is globally asymptotically stable.*

Example 2 The discretization of the Timoshenko beam with dissipation in Example 1 in p elements with clamped-free boundary conditions leads to the finite dimensional system:

$$\frac{d}{dt}\underbrace{\begin{pmatrix} x_{1d} \\ x_{2d} \\ x_{3d} \\ x_{4d} \end{pmatrix}}_{x_d} = (J_n - R_n)\begin{pmatrix} Q_1 x_{1d} \\ Q_2 x_{2d} \\ Q_3 x_{3d} \\ Q_4 x_{4d} \end{pmatrix} + \begin{pmatrix} 0 \\ 0 \\ 0 \\ I \end{pmatrix}\mathbf{u}_d, \tag{78a}$$

$$\mathbf{y}_d = Q_4 x_{4d}, \tag{78b}$$

where $x_{id} = \left(x_i^1 \cdots x_i^p\right)^T$ for $i \in \{1, 2, 3, 4\}$,

$$J_n = \begin{pmatrix} 0 & J_i & 0 & -S^T \\ -J_i^T & 0 & 0 & 0 \\ 0 & 0 & 0 & J_i \\ S & 0 & -J_i^T & 0 \end{pmatrix}, \; S = \operatorname{diag}\left(L_{ab}\right)_{p\times p},$$

$$J_i = \begin{pmatrix} \frac{1}{\beta} & & & \\ -\frac{1}{\beta^2} & \frac{1}{\beta} & & \\ \vdots & \ddots & \ddots & \\ (-1)^{n-1}\frac{(\beta')^{n-2}}{\beta^n} & \cdots & -\frac{1}{\beta^2} & \frac{1}{\beta} \end{pmatrix}_{p\times p},$$

$$R_n = \begin{pmatrix} 0 & 0 & 0 & 0 \\ 0 & R_{td} & 0 & 0 \\ 0 & 0 & 0 & 0 \\ 0 & 0 & 0 & R_{rd} \end{pmatrix}, \quad L_{ab} = \frac{L}{p},$$

$Q_1 = \operatorname{diag}\left(\frac{T}{L_{ab}}\right) \in \mathbb{R}^{p\times p}$, $Q_2 = \operatorname{diag}\left(\frac{1}{\rho L_{ab}}\right) \in \mathbb{R}^{p\times p}$, $Q_3 = \operatorname{diag}\left(\frac{EI}{L_{ab}}\right) \in \mathbb{R}^{p\times p}$, $Q_4 = \operatorname{diag}\left(\frac{1}{I\rho L_{ab}}\right) \in \mathbb{R}^{p\times p}$, $R_{td} = \operatorname{diag}\left(R_t L_{ab}\right) \in \mathbb{R}^{p\times p}$, and $R_{rd} = \operatorname{diag}\left(R_r L_{ab}\right) \in \mathbb{R}^{p\times p}$. β denotes the co-energy variables mapping parameter and $\beta' = 1 - \beta$ [23]. They are chosen in this case equal to $\frac{1}{2}$ in order to get a centered scheme. The Hamiltonian of the discretized model (78) is given as follows:

$$H_d(x_d) = \frac{1}{2}\left(x_{1d}^T Q_1 x_{1d} + x_{2d}^T Q_2 x_{2d} + x_{3d}^T Q_3 x_{3d} + x_{4d}^T Q_4 x_{4d}\right). \tag{79}$$

It is important to notice that the interconnection matrix M in (64) depends on the considered case, whether the system is fully, *i.e.* $m = p$, or under-actuated, *i.e.*

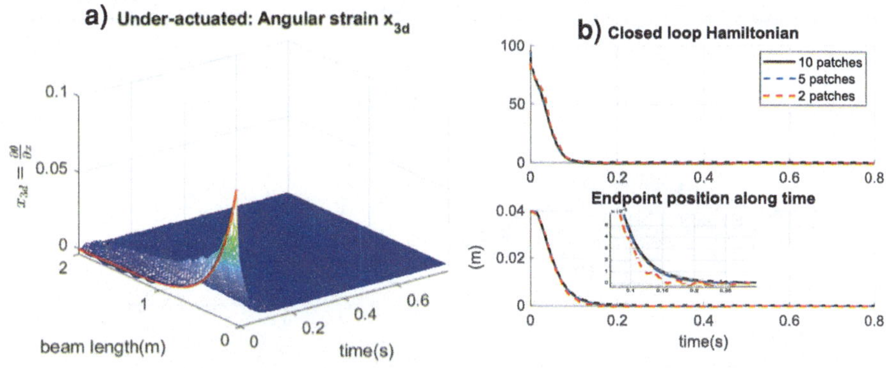

Fig. 4 Closed loop evolution of the angular strain for $m = 10$ (**a**), Hamiltonian function and endpoint position (**b**), in the under-actuated case for $m = 10$, $m = 5$ and $m = 2$

$m < p$. We consider that the control is achieved using m patches as depicted in Fig. 3. The aim of the control design is to modify as far as possible the closed loop Young's modulus $\tilde{E}$ of the beam. $J_m \in \mathbb{R}^{m\times m}$ stems from the discretization of $\partial/\partial\zeta$, B_c is chosen to be $B_c = J_m$. According to (71), $Q_c = J_m^{-T} V \Sigma_0^{-1} U_1^T Q_m U_1 \Sigma_0^{-1} V^T J_m^{-1}$. D_c is chosen such that the time derivative of the Hamiltonian behaves similarly as in the fully-actuated case, *i.e.* in order to satisfy $\min\limits_{D_c\in\mathbb{R}^{m\times m}} \left\| M D_c M^T - \text{diag}\,(\alpha L_{ab})\right\|_F$, where M is given in (64). This optimization problem is similar to Problem 1, and the optimal D_c is given by $\hat{D}_c = \text{diag}\left(\frac{\alpha L_{ab}}{k}\right)$.

We first consider the case with 10 patches, *i.e.* $m = 10$, $p = 50$ and $k = 5$. The initial conditions are set to a spatial distribution $x_3(\zeta, 0) = 0.1e^{-4\zeta}$ for the angular strain and to zero for other state variables. In this case the angular strain evolution is quite similar to that obtained in the fully-actuated case as depicted in Fig. 4(a). This indicates that if the controller matrices B_c, Q_c and D_c are adequately selected, the achievable performances in the under-actuated case can be optimized in order to be close to the ones obtained in the fully-actuated case. When the number of patches is reduced to 5 and to 2, *i.e.* $p = 50$, $k = 10$ and $p = 50$, $k = 25$ respectively, these performances are slightly deteriorated in the high frequencies as shown in Fig. 4(b).

In order to illustrate the effect of the neglected dynamics on the achievable performances we implement the controller designed considering 10 patches on the discretized system where $p = 50$ in a more precise model of the beam derived using $p = 200$. In Fig. 5 we can see that, due to the damping injection and the associated closed loop bandwidth, the neglected dynamics do not impact significantly the closed loop response of the system.

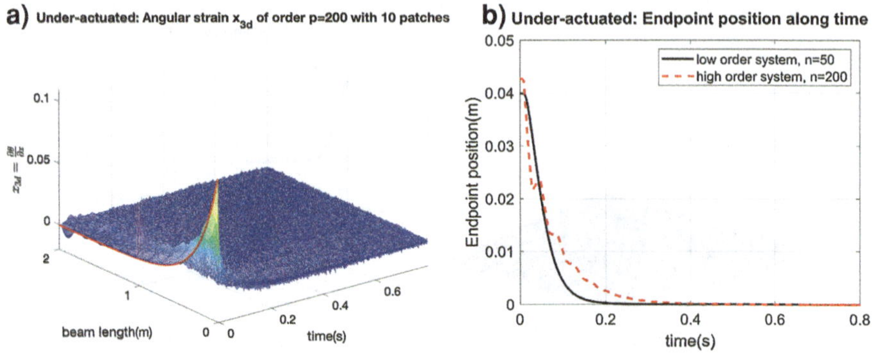

Fig. 5 Closed loop evolution of the angular strain of the high order system (**a**), and comparison of the endpoint position of the low order and high order systems using the same controller (**b**)

6 Conclusions, Future Works

In this chapter we give an overview on energy shaping control design techniques for linear infinite dimensional port Hamiltonian systems defined on a one dimensional spatial domain. This allows to cope with a large class of physical systems with spatially varying parameters such as the wave equation, the Timoshenko beam or the elastic string. Regarding the controller implementation, two cases are considered, when the system is actuated/sensed at the boundaries of its spatial domain and when the system is actuated/sensed through a network of actuators/sensors situated along its spatial domain. In the two cases we show how a dynamic extension and structural invariants can be used to shape, at least partially in the case of boundary control, the closed loop energy function of the system. This means that through the use of a dynamic feedback with proper initialization or of a state feedback one can modify the dynamic properties of the closed loop system, *i.e.* its stiffness from the shaping of its potential energy for example. In the distributed control case an early lumping approach is used and the closed loop properties are modified in an optimal way. The passivity of the controller is used to achieve the asymptotic stability on the infinite dimensional system. This work has been illustrated on the boundary control and in domain control of a clamped-free Timoshenko beam. Future works will consist in implementing the dynamic controller using an observer in order to get rid of the constraint of perfect initialization. Other extensions are the generalization of the control strategy and stability results to the 2D case and to a class of non linear systems.

References

1. Maschke, B., van der Schaft, A., Breedveld, P.: An intrinsic Hamiltonian formulation of network dynamics: non-standard Poisson structures and gyrators. J. Franklin Inst. **329**(5), 923–966 (1992)
2. Ortega, R., van der Schaft, A., Maschke, B., Escobar, G.: Interconnection and damping assignment passivity based control of port-controlled Hamiltonian systems. Automatica **38**, 585–596 (2002)
3. Duindam, V., Macchelli, A., Stramigioli, S., Bruyninckx, H. (eds.): Modeling and control of complex physical systems—the Port-Hamiltonian approach. Springer, Berlin, Germany (2009)
4. van der Schaft, A., Maschke, B.: Hamiltonian formulation of distributed parameter systems with boundary energy flow. J. Geom. Phys. **42**, 166–194 (2002)
5. Le Gorrec, Y., Zwart, H., Maschke, B.: A semigroup approach to port Hamiltonian systems associated with linear skew symmetric operator. In: 16th International Symposium on Mathematical Theory of Networks and Systems (MTNS 2004) (2004)
6. Le Gorrec, Y., Zwart, H., Maschke, B.: Dirac structures and boundary control systems associated with skew-symmetric differential operators. SIAM J. Control Optim. **44**(5), 1864–1892 (2005)
7. Villegas, J., Zwart, H., Le Gorrec, Y., Maschke, B.: Exponential stability of a class of boundary control systems. IEEE Trans. Autom. Control **54**, 142–147 (2009)
8. Ramirez, H., Le Gorrec, Y., Macchelli, A., Zwart, H.: Exponential stabilization of boundary controlled port-Hamiltonian systems with dynamic feedback. IEEE Trans. Autom. Control **99** 1 (2014)
9. Macchelli, A., Le Gorrec, Y., Ramirez, H., Zwart, H.: On the synthesis of boundary control laws for distributed port-Hamiltonian. IEEE Trans. Autom. Control **62**(5) (2017)
10. Ramirez, H., Zwart, H., Le Gorrec, Y.: Stabilization of infinite dimensional port-Hamiltonian systems by non-linear dynamic boundary control. Automatica **85**, 61–69 (2017)
11. Kurula, M., Zwart, H.: Linear wave systems on n-d spatial domains. Int. J. Control **88** (2015)
12. Ortega, R., van der Schaft, A., Maschke, B., Escobar, G.: Energy-shaping of port-controlled hamiltonian systems by interconnection. In: 38th IEEE Conference on Decision and Control, vol. 2, pp. 1646–1651 (1999)
13. Ortega, R., van der Schaft, A., Mareels, I., Maschke, B.: Energy shaping control revisited. In: Baños, A., Lamnabhi-Lagarrigue, F., Montoya, F.J. (eds.). Advances in the Control of Nonlinear Systems, pp. 277–307. Springer, Berlin/Heidelberg (2001). https://doi.org/10.1007/BFb0110388. ISBN 1-85233-378-2
14. Ortega, R., van der Schaft, A., Castanos, F., Astolfi, A.: Control by interconnection and standard passivity-based control of port-Hamiltonian systems. IEEE Trans. Autom. Control **53**(11), 2527–2542 (2008)
15. Jacob, B., Zwart, H.: Linear Port-Hamiltonian Systems on Infinite-dimensional Spaces. Operator Theory: Advances and Applications, vol. 223. Birkhäuser, Basel, Switzerland (2012)
16. Macchelli, A., Melchiorri, C.: Modeling and control of the Timoshenko beam. The distributed port Hamiltonian approach. SIAM J. Control Optim. **43**(2), 743–767 (2004)
17. Villegas, J.A.: A port-Hamiltonian approach to distributed parameter systems. Ph.D. dissertation, Universiteit Twente (2007)
18. van der Schaft, A.: Port-Hamiltonian systems. In: Modeling and Control of Complex Physical Systems: The Port-Hamiltonian Approach, pp. 53–130. Springer Berlin Heidelberg (2009)
19. Trenchant, V., Ngoc Minh, T.V., Ramirez, H., Lefevre, L., Le Gorrec, Y.: On the use of structural invariants for the discributed control of infinite dimensional port-hamiltonian systems. In: Conference on Decision and Control, CDC'17, Melbourne—Australia (2017)
20. Dalsmo, M., van der Schaft, A.: On representation and integrability of mathematical structures in energy-conserving physical systems. SIAM J. Control Optim. **37**, 54–91 (1999)
21. van der Schaft, A.: L_2-Gain and Passivity Techniques in Nonlinear Control. Communication and Control Engineering. Springer (2000)

22. Macchelli, A., Melchiorri, C.: Modeling and control of the Timoshenko beam. The distributed port Hamiltonian approach. SIAM J. Control Optim. **43**(2), 743–767 (2005)
23. Golo, G., Talasila, V., van der Schaft, A., Maschke, B.: Hamiltonian discretization of boundary control systems. Automatica **40**(5), 757–771 (2004)
24. Liu, N., Wu, Y., Le Gorrec, Y., Lefèvre, L., Ramirez, H.: In domain energy shaping control of distributed parameter port-Hamiltonian systems. In: Proceedings of the SIAM Conference on Control Applications, pp. 70–77. SIAM, Washington, USA (2021)
25. Shores, T.: Applied linear algebra and matrix analysis, vol. 2541. Springer (2007)

Prediction Control for Nonlinear Systems with Stochastic Input Delay

Sijia Kong and Delphine Bresch-Pietri

Abstract This chapter focuses on nonlinear systems with a stochastic input delay, which can arise from communication lags for instance. For input-delay systems, the use of prediction-based control methods is quite standard. This method was first used for constant delays before being extended to time-varying delays. Yet, the design of a prediction-based controller to handle stochastic delays is still an open problem. In this chapter, we propose to describe the stochastic input delay as a continuous-time Markov process with a finite number of values. This random delay is then robustly compensated through a constant-horizon state prediction. Applying the method of backstepping, we obtain that exponential stabilization of the closed-loop system is achieved, under a sufficient condition. This condition bears on the range of the possible values of the time lag, which should be sufficiently small and close enough to the prediction horizon.

1 Introduction

Time delays have always represented an important research topic, due to their frequent appearance and the performance limitation they often induce for control systems [29]. In the case where the delay affects the input of a dynamical system, active control techniques generically referred to as Smith Predictor [28, 32] have been developed to counteract these limitations. Grounding on a prediction of the system state, this so-called Finite Spectrum Assignment [21, 24] aims at compensating the input delay in the closed-loop dynamics and recovering the nominal delay-free performances. This technique, originally designed for constant delays, has gradually been extended to time-varying ones [3, 27] or more intricate cases such as state-

S. Kong (✉) · D. Bresch-Pietri
MINES ParisTech, PSL Research University CAS-Centre Automatique et Systèmes, 60 Boulevard Saint Michel, 75006 Paris, France
e-mail: sijia.kong@mines-paristech.fr

D. Bresch-Pietri
e-mail: delphine.bresch-pietri@mines-paristech.fr

J. Auriol et al. (eds.), *Advances in Distributed Parameter Systems*, Advances in Delays and Dynamics 14, https://doi.org/10.1007/978-3-030-94766-8_2

dependent delays [1, 2], multiple delays [5] or additive disturbances [22]. In these cases, the prediction time horizon has to be carefully computed to account for future delay variations or implicit dependencies.

Recently, the rise of communication technologies has intensified the need to cope with transmission lags [11, 30, 31]. They result for instance from congestion in the transmission path(s) or changes in the routing algorithm and are usually modeled as stochastic variables to account for their sudden and random variations [6, 10, 26, 34].

However, while in recent years many monographs have studied Stochastic Delay Equations (SDEs), such as [12, 15, 25], the case where the delay is itself a random variable has been seldomly studied. Up to the authors' knowledge, one of the few examples to do so are [16, 17, 23, 33]. While [33] studies a piecewise constant process and [23] analyzes a deterministic delay term multiplied by a random variable, [16, 17] consider stochastic state delays modeled as a Markov process with a finite number of states. The authors then consider each delay value separately, following the so-called technique of probabilistic delay averaging. This constant delay reasoning inspired the core of the analysis methodology proposed in this chapter.

In this paper, we investigate the problem of prediction-based control design for nonlinear systems subject to stochastic input delay and extend the scope of our previous work [18] focusing on linear systems. We model the random input delay as a Markov process with a finite number of values. In lieu of computing a state prediction exactly compensating for the input delay (which cannot be achieved as future delay values are random and thus unknown), we propose to use a constant prediction horizon. Similarly to [18], we prove that such a controller achieves exponential stabilization of the origin, provided that the range of possible delay values is sufficiently narrow and that the prediction horizon is chosen close enough to these values.

This chapter is organized as follows. In Sect. 2, the problem statement and the main stabilization result are given. We then propose a PDE representation of the stochastic delay and a backstepping reformulation of the closed-loop system in Sect. 3, in view of Lyapunov stability analysis in Sect. 4.

Notations. In the following sections, for a signal $v : (x, t) \in [0, 1] \times \mathbb{R} \mapsto v(x, t) \in \mathbb{R}$, we denote $\|v(t)\|$ its spatial $\mathcal{L}_2$-norm

$$\|v(t)\| = \sqrt{\int_0^1 v(x,t)^2 dx} \tag{1}$$

and $\|v\|_\infty$ its $\mathcal{L}_\infty$-norm for a measurable essentially bounded signal as

$$\|v(t)\|_\infty = \inf\{C \geq 0 \mid |v(x,t)| \leq C \text{ f.a.a. } x \in [0,1]\} = \lim_{p\to\infty} \|v(t)\|_p \tag{2}$$

in which $\|v(t)\|_p$ denotes the $\mathcal{L}_p$-norm of signal v. Additionally, for a real matrix A, $|A|$ denotes its Euclidean norm

$$|A| = \sqrt{max(\lambda(A^T A))} \tag{3}$$

in which λ represents the eigenvalues of the matrix, and A^T denotes the transpose of A.

$\mathbb{E}(x)$ denotes the expectation of a random variable x. For a random signal $x(t)$ ($t \in \mathcal{T} \subset \mathbb{R}$), the conditional expectation of $x(t)$ at the instant t knowing that $x(s) = x_0$ at the instant $s < t$ is denoted $\mathbb{E}_{[s,x_0]}(x(t))$.

Finally, $e_i \in \mathbb{R}^r$ ($r \in \mathbb{N}_+$ and $i \in \{1, \ldots, r\}$) denotes the unit vector, that is, $e_1 = \begin{pmatrix}1 & 0 & \cdots & 0\end{pmatrix}^T$, $e_2 = \begin{pmatrix}0 & 1 & \cdots & 0\end{pmatrix}^T$, ..., $e_r = \begin{pmatrix}0 & 0 & \cdots & 1\end{pmatrix}$. Besides, for $(x_1, x_2) \in \mathbb{R}^r \times \mathbb{R}^r$, we denote the Hadamard product of x_1 and x_2 as $x_1 \bullet x_2 = (x_{1i}x_{2i})_{1 \le i \le r} \in \mathbb{R}^r$.

2 Problem Statement and Control Design

We consider the following controllable nonlinear dynamics

$$\dot{X}(t) = f(X(t), U(t - D)) \tag{4}$$

in which $X \in \mathbb{R}^n$ and $U \in \mathbb{R}$ are the state and control input, respectively. f is a nonlinear function of class C^1 such that $f(0, 0) = 0$. The stochastic delay D is a Markov process with the following properties:

(1) $D(t) \in \{D_i, i \in \{1, \ldots, r\}\}$, $r \in \mathbb{N}$ with $0 < \underline{D} \le D_1 < D_2 < \cdots < D_r \le \overline{D}$.
(2) The transition probabilities $P_{ij}(t_1, t_2)$, which quantify the probability to switch from D_i at time t_1 to D_j at time t_2 ($(i, j) \in \{1, \ldots, r\}^2$, $t_2 \ge t_1 \ge 0$), are differentiable functions $P_{ij} : \mathbb{R}^2 \to [0, 1]$ satisfying

$$\sum_{j=1}^{r} P_{ij}(t_1, t_2) = 1, \quad (0 \le t_1 \le t_2) \tag{5}$$

(3) The realizations of D are continuous from the right.

The control objective is to stabilize plant (4) following a prediction-based approach, despite the random nature of the input delay. This will be achieved by a robust delay compensation approach. Before detailing our control design, we first further characterize the delay-free characteristics of the plant under consideration.

Assumption 1 There exists a feedback law κ of class C^1 such that, the plant

$$\dot{X} = f(X, \kappa(X)) \tag{6}$$

is exponentially stable with basin of attraction $\mathcal{A} \subset \mathbb{R}^n$, i.e. there exists a positive definite function $W : \mathbb{R}^n \to \mathbb{R}_+$ such that, for all compact $C \subset \mathcal{A}$, there exist positive constants λ, C_1, C_2 and C_3 such that, for all $x \in C$,

$$\frac{dW}{dX}(X) f(X, \kappa(X)) \le -\lambda |X|^2 \tag{7}$$

$$C_1|X|^2 \leq W(X) \leq C_2|X|^2 \tag{8}$$

$$\left|\frac{dW}{dX}(X)\right| \leq C_3|X| \tag{9}$$

Assuming exponential stability of the delay-free closed-loop dynamics in lieu of asymptotic stability is not restrictive, in the sense that some dynamical systems for which the origin is only asymptotically stable can have a zero-delay robustness property. This prohibits us from using a robust compensation controller as the one we aim at designing.

We now detail two additional technical assumptions.

Assumption 2 The function $f(X, U)$ is globally Lipschitz, i.e. there exists $C_L > 0$ such that for $\forall(u_1, u_2) \in \mathbb{R}^2$ and $\forall(X_1, X_2) \in (\mathbb{R}^n)^2$

$$|f(X_1, u_1) - f(X_2, u_2)| \leq C_L|(X_1, u_1) - (X_2, u_2)| \tag{10}$$

Assumption 3 The feedback law κ defined in Assumption 1 is such that $\kappa(0) = 0$ and there exists a positive constant C_4 such that

$$\left|\frac{d\kappa}{dX}\right| \leq C_4 \tag{11}$$

Contrary to Assumption 1, these last two assumptions are more restrictive and more technical. They aim at simplifying the stability analysis detailed in the following, but should be relaxed in future works.

Note that Assumption 2 jointly with the fact that $f(0, 0) = 0$ guarantee that the plant does not escape in finite time, which is a necessary condition for stabilization to be achieved for any delay values. Therefore, we do not need to assume strong forward completeness here, as usually done for prediction-based control for nonlinear dynamics [3], but, in all likelihood, future works will have to replace Assumption 2 by this property.

The control law is chosen here as

$$U(t) = \kappa(\hat{P}(t)) \tag{12}$$

in which the predictor $\hat{P}$ is defined for $\theta \in [t - D_0, t]$ as

$$\hat{P}(\theta, t) = X(t) + \int_{t-D_0}^{\theta} f(\hat{P}(s, t), U(s))ds, \quad t - D_0 \leq \theta \leq t \tag{13}$$

To exactly compensate for the input delay, one would need the function $\phi : t \rightarrow t - D(t)$ to be invertible in order to define the predictor. However, the stochastic nature of the delay makes the computation of the inverse function $\phi^{-1}(t)$ (if it exists) impossible as it depends on future delay variations and is therefore not causal.

For this reason, we hence propose to use a constant time horizon prediction in the controller (12)–(13). Note that, if the time lag D were constant and equal to D_0, the control law would then involve the exact prediction of the state X over a time window of D_0 units of time. Consistently, we now formulate the main result of the paper which states that robust compensation is achieved if the time delay remains sufficiently close to D_0.

Theorem 1 *Consider the closed-loop system consisting of the system (4) satisfying Assumptions 1–3 and the control law (12). Let C be a compact set of $\mathcal{A}$ containing the origin. Then, there exist positive constants $\epsilon^\star$ and $\rho^\star$ such that, if*

$$|D_0 - D_j| \leq \epsilon^\star, \quad j \in \{1, \ldots, r\} \tag{14}$$

and if

$$\Upsilon(0) \leq \rho^\star \tag{15}$$

there exist positive constants R and γ (independent of the initial condition) such that

$$\mathbb{E}_{[0,\Upsilon(0)]}(\Upsilon(t)) \leq R\Upsilon(0)e^{-\gamma t} \tag{16}$$

with

$$\Upsilon(t) = |X(t)|^2 + \int_{t-\overline{D}-D_0}^{t} U(s)^2 ds \tag{17}$$

Condition (14) guarantees that the prediction performed in (13) remains sufficiently accurate in the case of a stochastic delay. In details, (14) requires the sequence of the random delay $\mathbf{D} = \begin{pmatrix} D_1 & \cdots & D_i & \cdots & D_r \end{pmatrix}^T$ to be limited in a vicinity $\epsilon^\star$ of the constant D_0.

Note that this result is consistent with the delay-robustness results obtained in the deterministic delay case. Indeed, [4, 19] provide a similar robust compensation result for a time-differentiable delay function, under the assumptions that both the range of variation of the delay and its variation rate are sufficiently limited. A similar result was obtained in [13] but through a small-gain approach enabling to avoid restricting the delay rate. Hence, Theorem 1 falls within this framework and extends it to the stochastic context.

We now provide the proof of this theorem in the following sections.

3 PDE Representation of the Delay and Backstepping Transformation

First, to represent the control input which is subject to a stochastic delay, we define a distributed actuator vector as, for $x \in [0, 1]$, $\mathbf{v}(x,t) = \left(v_1(x,t) \cdots v_k(x,t) \cdots v_r(x,t)\right)^T$ with $v_k(x,t) = U(t + D_k(x-1))$. This enables to rewrite (4) as

$$\begin{cases} \dot{X}(t) = f(X(t), \delta(t)^T \mathbf{v}(0,t)) \\ \Lambda_D \mathbf{v}_t(x,t) = \mathbf{v}_x(x,t) \\ \mathbf{v}(1,t) = U(t)\mathbf{1} \end{cases} \tag{18}$$

in which $\Lambda_D = diag(D_1, \ldots, D_r)$, $\mathbf{1}$ is a r-by-1 all-ones vector and $\delta \in \mathbb{R}^r$ is such that, if $D(t) = D_j$, then $\delta(t) = e_j$, that is,

$$\delta_i(t) = \begin{cases} 1 & \text{if } i = j \\ 0 & \text{otherwise} \end{cases} \tag{19}$$

Hence, δ is a random process with the same transition probabilities as the stochastic process D, but with the finite number of states (e_i) instead of (D_i). In the sequel, δ and D will thus be equivalently used.

Now, we introduce $\hat{v}(x,t)$ to represent the control input $U(t)$ within the interval $[t - D_0, t]$, and the corresponding input estimation error $\tilde{\mathbf{v}}(x,t)$ defined as

$$\begin{cases} \hat{v}(x,t) = U(t + D_0(x-1)) \\ \tilde{\mathbf{v}}(x,t) = \mathbf{v}(x,t) - \hat{v}(x,t)\mathbf{1} \end{cases} \tag{20}$$

Then, we define $\mu(x,t) = U(t - D_0 + \overline{D}(x-1))$ to represent the controller within the interval $[t - D_0 - \overline{D}, t - D_0]$. The extended state $(X(t), \hat{v}(x,t), \tilde{\mathbf{v}}(x,t), \mu(x,t))$ satisfies

$$\begin{cases} \dot{X}(t) = f(X(t), \hat{v}(0,t) + \delta(t)^T \tilde{\mathbf{v}}(0,t)) \\ D_0 \hat{v}_t(x,t) = \hat{v}_x(x,t) \\ \hat{v}(1,t) = U(t) \\ \Lambda_D \tilde{\mathbf{v}}_t(x,t) = \tilde{\mathbf{v}}_x - \Sigma_D \hat{v}_x \\ \tilde{\mathbf{v}}(1,t) = \mathbf{0} \\ \overline{D}\mu_t(x,t) = \mu_x(x,t) \\ \mu(1,t) = \hat{v}(0,t) \end{cases} \tag{21}$$

in which $\Sigma_D = \left(\frac{D_1 - D_0}{D_0} \cdots \frac{D_r - D_0}{D_0}\right)^T$ and $\mathbf{0}$ is a r-by-1 all-zeros vector.

In view of stability analysis, we introduce the backstepping transformation (see [20])

$$w(x,t) = \hat{v}(x,t) - \kappa(\hat{p}(x,t)), \ 0 \leq x \leq 1 \tag{22}$$

in which the distributed predictor $\hat{p}$ is defined as

$$\hat{p}(x,t) = X(t) + D_0 \int_0^x f(\hat{p}(y,t), \hat{v}(y,t))dy, \ 0 \leq x \leq 1 \tag{23}$$

Notice that this backstepping transformation is similar to the one used in the deterministic context (see [20]) but applied here to our estimated distributed input $\hat{v}$.

Lemma 1 *The backstepping transformation (22), jointly with the control law (12)–(13), transform the plant (21) into the target system* $(X, w, \tilde{\mathbf{v}}, \mu)$

$$\begin{cases} \dot{X}(t) = f(X(t), \kappa(X(t)) + w(0,t) + \delta(t)^T \tilde{\mathbf{v}}(0,t)) \\ D_0 w_t(x,t) = w_x(x,t) + r_w(x,t) \\ w(1,t) = 0 \\ \Lambda_D \tilde{\mathbf{v}}_t(x,t) = \tilde{\mathbf{v}}_x - \Sigma_D h(t + D_0(x-1)) \\ \tilde{\mathbf{v}}(1,t) = \mathbf{0} \\ \overline{D}\mu_t(x,t) = \mu_x(x,t) \\ \mu(1,t) = \kappa(X(t)) + w(0,t) \end{cases} \tag{24}$$

in which r_w *is defined for* $x \in [0, 1]$ *as*

$$r_w(x,t) = -D_0 \frac{d\kappa}{dX}(\hat{p}(x,t))\Phi(x,0,t)\tilde{f}(t) \tag{25}$$

with

$$\tilde{f}(t) = f(X(t), \kappa(X(t)) + w(0,t) + \delta(t)^T \tilde{\mathbf{v}}(0,t)) - f(X(t), \kappa(X(t)) + w(0,t)) \tag{26}$$

Φ *is the state-transition matrix associated with the equation*

$$r_x(x,t) = D_0 \frac{\partial f}{\partial X}\big(\hat{p}(x,t), w(x,t) + \kappa(\hat{p}(x,t))\big) r(x,t) \tag{27}$$

and h *is defined for* $t \geq 0$ *as*

$$h(t) = D_0 \frac{d\kappa}{dX}(\hat{p}(1,t)) f(\hat{p}(1,t), w(0,t) + \kappa(X(t)) + \delta(t)^T \tilde{\mathbf{v}}(0,t)) \tag{28}$$

with $\hat{p}(x,t)$ *defined in* (23) *a function of* $X(t)$ *and* $w(y,t)$ *for* $y \in [0, x]$ *and* $x \in [0, 1]$.

Proof Define $q(x,t) = D_0\hat{p}_t(x,t) - \hat{p}_x(x,t)$. Taking time- and space-derivatives of $\hat{p}$ defined in (23), one obtains

$$\begin{aligned} & q(x,t) \\ =& D_0\Big[f(X(t), \hat{v}(0,t) + \delta(t)^T \tilde{\mathbf{v}}(0,t)) + \int_0^x \frac{\partial f}{\partial X}(\hat{p}(y,t), \hat{v}(y,t)) D_0 \hat{p}_t dy \\ & + \int_0^x \frac{\partial f}{\partial U}(\hat{p}(y,t), \hat{v}(y,t))\hat{v}_x dy - f(X(t), \hat{v}(0,t)) \\ & - \int_0^x \frac{\partial f}{\partial X}(\hat{p}(y,t), \hat{v}(y,t))\hat{p}_x dy - \int_0^x \frac{\partial f}{\partial U}(\hat{p}(y,t), \hat{v}(y,t))\hat{v}_x dy\Big] \\ =& D_0 \int_0^x \frac{\partial f}{\partial X}(\hat{p}(y,t), \hat{v}(y,t)) q(y,t) dy + D_0 \tilde{f}(t) \end{aligned} \tag{29}$$

with $\tilde{f}(t)$ defined in (26). This integral equation can be rewritten under the differential form

$$\begin{cases} q_x(x,t) = D_0 \dfrac{\partial f}{\partial X}(\hat{p}(x,t), \hat{v}(x,t)) q(x,t) \\ q(0,t) = D_0 \tilde{f}(t) \end{cases} \tag{30}$$

Thus, introducing the state-transition matrix Φ associated with (30), it follows that

$$q(x,t) = D_0 \Phi(x,0,t)\tilde{f}(t) \tag{31}$$

As the time-derivative of the backstepping transformation (22) is given as

$$w_t(x,t) = \hat{v}_t(x,t) - \frac{d\kappa}{dX}(\hat{p}(x,t))\hat{p}_t(x,t) \tag{32}$$

and its space-derivative as

$$w_x(x,t) = \hat{v}_x(x,t) - \frac{d\kappa}{dX}(\hat{p}(x,t))\hat{p}_x(x,t) \tag{33}$$

the dynamics of the backstepping variable w then follows.

In addition, the expression of $h(t)$ for $t \geq 0$ is obtained by taking a time derivative of (12)

$$h(t + D_0(x-1)) = \hat{v}_x(x,t) = D_0 \dot{U}(t + D_0(x-1)) = D_0 \frac{d\kappa}{dX}(\hat{p}(1,t))\dot{\hat{P}}(t) \tag{34}$$

We are now ready to analyze the closed-loop system stablization.

4 Stability Analysis of the Closed-Loop System

Let us define the state of the target system (24) as $\Psi = (X, w, \tilde{\mathbf{v}}, \mu) \in \mathbb{R}^n \times \mathcal{L}_\infty([0,1],\mathbb{R}) \times \mathcal{L}_\infty([0,1],\mathbb{R}^r \times \mathcal{L}_\infty([0,1],\mathbb{R}) \triangleq \mathcal{D}_\Psi$. Note that (24) was reformulated as a dynamical system involving a random parameter, as studied in [14] or [7]. We now formulate a well-posedness result for this system.

Following [14], by a weak solution to the closed-loop system (4) and (12)–(13), we refer to a $\mathbb{R}^n \times \mathcal{L}_\infty([-\overline{D}, 0], \mathbb{R}) \times \mathbb{R}$-valued random variable $(X(X_0, t), U_t(U_0, \cdot), D(t))$, the realizations of which satisfy an integral form of (4) and (12)–(13).

Similarly, by a weak solution to (24), we refer to a $\mathcal{D}_\Psi \times \mathbb{R}$-valued random variable $(X(X_0, t), w(w_0, \cdot, t), \tilde{\mathbf{v}}(\tilde{\mathbf{v}}_0, \cdot, t), \mu(\mu_0, \cdot, t), D(t))$, the realizations of which are a weak solution of (24), that is, in the standard weak solution sense of [8] for the transport PDEs and under an integral form for the ODE.

Lemma 2 *For every initial condition $(X_0, U_0) \in \mathbb{R}^n \times \mathcal{L}_\infty([-\overline{D}, 0], \mathbb{R})$, the closed-loop system consisting of the plant (4) satisfying Assumptions 1–2 and the control law defined in (12)–(13) has a unique weak solution.*

Consequently, for each initial condition in $\mathcal{D}_\Psi$, the target system (24) also has a unique weak solution.

Proof The proof is omitted due to space limitation.

From Lemma 2, (Ψ, δ) thus defines a continuous-time Markov process and we can therefore introduce the following elements for stability analysis.

In the sequel, we consider the following Lyapunov functional candidate

$$V(\Psi) = W(X) + bD_0 \int_0^1 (1+x) w(x)^2 dx + c \sum_{l=1}^{r} \int_0^1 (1+x)(e_l \bullet \mathbf{D})^T \tilde{\mathbf{v}}(x)^2 dx + d\overline{D} \int_0^1 (1+x)\mu(x)^2 dx \tag{35}$$

with $b, c, d > 0$, and $\mathbf{D} = \begin{pmatrix} D_1 & \cdots & D_i & \cdots & D_r \end{pmatrix}^T$ and where $\bullet$ denotes the Hadamard multiplication and the square in $\tilde{\mathbf{v}}(x)^2$ should be understood component-wise.

As the functional V is not differentiable with respect to time t when evaluated at $\Psi(t)$, we introduce the infinitesimal generator L (see [15, 17]) as

$$LV(\Psi) = \limsup_{\Delta t \to 0^+} \frac{1}{\Delta t} \left(\mathbb{E}_{(t,\Psi)}[V(\Psi(t+\Delta t))] - V(\Psi) \right) \tag{36}$$

We also define L_j, the infinitesimal generator of the Markov process Ψ obtained by fixing the value $\delta(t) = e_j$, as

$$L_j V(\Psi) = \frac{dV}{d\Psi}(\Psi) F_j(\Psi) \tag{37}$$

in which F_j denotes the operator corresponding to the dynamics of the target system (24) with the fixed value $\delta(t) = e_j$, that is, for $\Psi = (X, w, \tilde{\mathbf{v}}, \mu)$,

$$F_j(\Psi)(x) = \begin{pmatrix} f(X(t), \kappa(X(t)) + e_j^T \tilde{\mathbf{v}}(0) + w(0)) \\ \frac{1}{D_0}\left[w_x(x) + r_w(x)\right] \\ \Lambda_D^{-1}\left[\tilde{\mathbf{v}}_x(x) - \Sigma_D h(\cdot + D_0(x-1))\right] \\ \frac{1}{D}\mu_x(x) \end{pmatrix} \tag{38}$$

For the sake of conciseness, in the sequel, we denote $V(t)$, $LV(t)$ and $L_jV(t)$, for short, instead of $V(\Psi(t), \delta(t))$, $LV(\Psi(t), \delta(t))$ and $L_jV(\Psi(t), t)$ respectively.

It is worth noticing that, due to the fact that V does not depend explicitly on δ, the infinitesimal generators are related as follows

$$\begin{aligned} LV(t) =& \sum_{j=1}^{r} P_{ij}(0,t)\frac{dV}{d\Psi}(\Psi(t))F_j(\Psi(t)) + \sum_{j=1}^{r} \frac{\partial P_{ij}}{\partial t}(0,t)V(\Psi(t)) \\ =& \sum_{j=1}^{r} P_{ij}(0,t)L_jV(t) \end{aligned} \tag{39}$$

Therefore, in view of stability analysis, as a first step, one can focus on the derivative of the Lyapunov functional evaluated for a dynamic with a fixed delay, that is, L_jV. This is the approach we follow in the sequel.

4.1 Lyapunov Analysis

Before detailing the Lyapunov analysis, we provide a technical lemma, useful in the following computations.

Lemma 3 *Consider the function h defined in (28). There exists $M > 0$ such that*

$$\|h(t + D_0(\cdot - 1))\|^2 \leq MV(t), \quad t \geq D_0 \tag{40}$$

Proof From the definition of the control law (12)–(13), (34) and Assumptions 2–3, then

$$\begin{aligned} |h(t)|^2 =& D_0^2\left(\frac{d\kappa}{dX}(\hat{p}(1,t))\hat{p}_t(1,t)\right)^2 \\ \leq& C_4^2|f(\hat{p}(1,t), \hat{v}(1,t)) + \Phi(1,0,t)\tilde{f}(t)|^2 \\ \leq& C_4^2\Big(|f(\hat{p}(1,t), \hat{v}(1,t))| + e^{D_0C_L}|\tilde{f}(t)|\Big)^2 \end{aligned} \tag{41}$$

Hence, using (26) and Assumption 2, one obtains

$$
\begin{aligned}
&\|h(t+D_0(\cdot-1))\|^2 = \int_0^1 h(t+D_0(x-1))^2dx \\
\leq& 2C_4^2\Big(\int_0^1 \left|f\left(\hat{p}(1,t+D_0(x-1)),\hat{v}(1,t+D_0(x-1))\right)\right|^2 dx \\
&+ e^{2D_0C_L}\int_0^1 |\tilde{f}(t+D_0(x-1))|^2dx\Big) \\
\leq& 2C_4^2\Big(C_L^2\int_0^1 \left(|\hat{P}(t+D_0(x-1))|+|U(t+D_0(x-1))|\right)^2dx \\
&+ 2C_L^2e^{2D_0C_L}\int_0^1 U(t+D_0(x-2))^2dx \\
&+ 2C_L^2e^{2D_0C_L}\int_0^1 U(t+D_0(x-1)-D(t+D_0(x-1)))^2dx\Big)
\end{aligned} \tag{42}
$$

Besides, from (13) and Assumption 2, for $\theta \in [t-D_0, t]$,

$$
\begin{aligned}
|\hat{P}(\theta)| \leq& |X(t)| + C_L\int_{t-D_0}^{\theta}(|\hat{P}(s)|+|U(s)|)ds \\
\leq& \Big(|X(t)| + C_L\int_{t-D_0}^{\theta}|U(s)|ds\Big) + C_L\int_{t-D_0}^{\theta}|\hat{P}(s)|ds
\end{aligned} \tag{43}
$$

Using Grönwall's inequality, it follows that

$$
\begin{aligned}
|\hat{P}(\theta)| \leq& |X(t)| + C_L\int_{t-D_0}^{\theta}|U(s)|ds \\
&+ C_L\int_{t-D_0}^{\theta}\Big(|X(t)| + \int_{t-D_0}^{s}|U(\xi)|d\xi\Big)e^{C_L(\theta-s)}ds \\
=& e^{C_L(\theta-t+D_0)}|X(t)| + C_L\int_{t-D_0}^{\theta}e^{C_L(\theta-s)}|U(s)|ds
\end{aligned} \tag{44}
$$

Replacing (44) into (42), one has

$$
\begin{aligned}
\|h(t+D_0(\cdot-1))\|^2 \leq& 2C_L^2D_0^2C_4^2\int_0^1\Big(e^{C_LD_0x}|X(t)| + |U(t+D_0(x-1))| \\
&+ C_L\int_{t-D_0}^{t+D_0(x-1)}e^{C_L(t+D_0(x-1)-s)}|U(s)|ds\Big)^2dx \\
&+ 4C_L^2D_0^2C_4^2e^{2D_0C_L}\Big(\int_0^1 U(t+D_0(x-2))^2dx \\
&+ \int_0^1 U(t+D_0(x-1)-D(t+D_0(x-1)))^2dx\Big)
\end{aligned} \tag{45}
$$

Now, define the inverse of the backstepping transformation of (22), which is given as

$$\hat{v}(x,t) = w(x,t) + \kappa(\hat{\rho}(x,t)) \tag{46}$$

where $\hat{\rho}$ satisfies

$$\hat{\rho}(x,t) = X(t) + D_0 \int_0^x f(\hat{\rho}(y,t), w(y,t) + \kappa(\hat{\rho}(y,t)))dy \tag{47}$$

Similarly, using Assumption 2 and Grönwall's inequality, one can also bound $\hat{\rho}(x,t)$ as

$$\hat{\rho}(x,t) \leq e^{D_0 C_L(1+C_4)x}|X(t)| + D_0 C_L \int_0^x e^{D_0 C_L(1+C_4)(x-y)}|w(y,t)|dy \tag{48}$$

Then, from (46), one obtains

$$\|\hat{v}(t)\|^2 \leq 3C_4^2 e^{2D_0 C_L(1+C_4)}|X(t)|^2 + 3\big(C_4^2 D_0^2 C_L^2 e^{2D_0 C_L(1+C_4)} + 1\big)\|w(t)\|^2 \tag{49}$$

Hence, observing that

$$\begin{aligned}\int_0^1 U(t + D_0(x-1) - D(t + D_0(x-1)))^2 dx &\leq \sum_{j=1}^r \int_0^1 U(t + D_0(x-1) - D_j)^2 dx \\ &\leq 2r\left(\|\hat{v}(t)\|^2 + \|\mu(t)\|^2\right)\end{aligned} \tag{50}$$

and using (49) in (45), it follows that

$$\|h(t + D_0(\cdot - 1))\|^2 \leq M_X|X(t)|^2 + M_w\|w(t)\|^2 + M_\mu\|\mu(t)\|^2 \tag{51}$$

in which (M_X, M_w, M_μ) are positive constants defined as follows

$$\begin{cases} M_X = 6C_L^2\overline{D}^2 C_4^2\big(e^{2C_L\overline{D}} + C_4^2 e^{2C_L\overline{D}(1+C_4)}(3 + 3C_L^2 e^{2C_L\overline{D}} + 2(1+2r)e^{2C_L\overline{D}})\big) \\ M_w = 6C_L^2\overline{D}^2 C_4^2\big(C_4^2\overline{D}^2 C_L^2 e^{2\overline{D}C_L(1+C_4)} + 1\big)(3 + 3C_L^2 e^{2C_L\overline{D}} + 2(1+2r)e^{2C_L\overline{D}}) \\ M_\mu = 4(1+2r)C_L^2\overline{D}^2 C_4^2 e^{2C_L\overline{D}} \end{cases} \tag{52}$$

Consequently, from the definition of the Lyapunov functional V in (35), one finally gets the desired result with $M = \overline{D}^2 \max\left\{\frac{M_X}{\min(\lambda(P))}, \frac{M_w}{b\underline{D}}, \frac{M_\mu}{d\overline{D}}\right\}$.

With this last ingredient, we are now ready to develop the Lyapunov analysis.

Lemma 4 *Consider the closed-loop system consisting of (4) and the control law (12)–(13) satisfying Assumptions 1–3. Let C be a compact set of $\mathcal{A}$ containing the origin. Assume there exists a positive constant ϵ such that*

$$|D_0 - D_j| \leq \epsilon, \quad j \in \{1, \ldots, r\} \tag{53}$$

and that $X(t) \in C$ *for a certain* $t \geq \overline{D}$, *then, there exist* $b, c, d \in \mathbb{R}_+^{*\,3}$ *which are independent of* ϵ *such that the Lyapunov functional* V *defined in (35) satisfies*

$$LV(t) \leq -(\eta - g(\epsilon))V(t) \tag{54}$$

with $\eta > 0$ *independent of* ϵ *and the function* $g : \mathbb{R}_+ \to \mathbb{R}_+$ *satisfying* $\lim_{\epsilon \to 0} g(\epsilon) = 0$.

Proof Applying integrations by parts and Young's inequality, one obtains

$$\begin{aligned}\frac{dV}{d\Psi}(\Psi)F_j(\Psi) \leq& \frac{dW}{dX} f(X(t), \kappa(X(t)) + w(0,t) + \tilde{v}_j(0,t)) \\ &- b(1 - 2\gamma_1)\|w(t)\|^2 - bw(0,t)^2 + 2b\frac{1}{\gamma_1}\|r_w(x,t)\|^2 \\ &- c\sum_{l=1}^{r}\left(1 - 2\left|1 - \frac{D_l}{D_0}\right|\gamma_2\right)\|\tilde{v}_l(t)\|^2 - c\sum_{l=1}^{r}\tilde{v}_l(0,t)^2 \\ &+ 2c\sum_{l=1}^{r}\left|1 - \frac{D_j}{D_0}\right|\frac{1}{\gamma_2}\|h(t + D_0(\cdot - 1))\|^2 \\ &- d\mu(0,t)^2 - d\|\mu(t)\|^2 + 2d\mu(1,t)^2\end{aligned} \tag{55}$$

From the definition of $\tilde{f}$ in (26) with $\delta(t) = e_j$ and Assumptions 2–3, it holds

$$\|r_w(x,t)\|^2 = D_0^2 C_4^2 e^{2D_0 C_L} C_L^2 \tilde{v}_j(0,t)^2 \tag{56}$$

Then, with Assumption 2, we obtain the following inequality

$$\begin{aligned}\frac{dV}{d\Psi}(\Psi)F_j(\Psi) \leq& \frac{dW}{dX}\Big(f(X(t), \kappa(X(t)) + C_L|w(0,t) + \tilde{v}_j(0,t)|\Big) \\ &- b(1 - 2\gamma_1)\|w(t)\|^2 - bw(0,t)^2 \\ &+ 2b\frac{1}{\gamma_1}D_0^2 C_L^2 C_4^2 e^{2D_0 C_L}\tilde{v}_j(0,t)^2 \\ &- c\sum_{l=1}^{r}\left(1 - 2\left|1 - \frac{D_l}{D_0}\right|\gamma_2\right)\|\tilde{v}_l(t)\|^2 - c\sum_{l=1}^{r}\tilde{v}_l(0,t)^2 \\ &+ 2c\sum_{l=1}^{r}\left|1 - \frac{D_l}{D_0}\right|\frac{1}{\gamma_2}\|h(t + D_0(\cdot - 1))\|^2 \\ &- d\mu(0,t)^2 - d\|\mu(t)\|^2 + 2d\mu(1,t)^2\end{aligned} \tag{57}$$

Therefore, applying Assumptions 2–3, Lemma 3 and Young's inequality, $LV(t)$ satisfies pagination

$$\begin{aligned} LV(t) &= \sum_{j=1}^{r} P_{ij}(0,t)\frac{dV}{d\Psi}(\Psi)F_j(\Psi) \\ &\leq -\left(\frac{\lambda}{2} - 4dC_4^2\right)|X(t)|^2 - b(1-2\gamma_1)\|w(t)\|^2 \\ &\quad - c\sum_{l=1}^{r}\left(1 - 2\left|1 - \frac{D_l}{D_0}\right|\gamma_2\right)\|\tilde{v}_l(t)\|^2 - d\|\mu(t)\|^2 \\ &\quad - \left(b - 4d - \frac{C_3^2C_L^2}{\lambda}\right)w(0,t)^2 \\ &\quad - \left(c - 2b\frac{1}{\gamma_1}D_0^2e^{2D_0C_L}C_4^2C_L^2 - \frac{C_3^2C_L^2}{\lambda}\right)\sum_{j=1}^{r}P_{ij}(0,t)\tilde{v}_j(0,t)^2 \\ &\quad - c\sum_{j=1}^{r}P_{ij}(0,t)\sum_{l\neq j}\tilde{v}_l(0,t)^2 - d\mu(0,t)^2 + \frac{2cr\epsilon}{D_0}\frac{1}{\gamma_2}MV(t) \end{aligned} \tag{58}$$

in which M is a positive constant which does not depend on ϵ and is defined in Lemma 3.

Observing that $D_0 \in [\underline{D}, \overline{D}]$, let us choose $(b, c, d, \gamma_1, \gamma_2) \in (\mathbb{R}_+^*)^5$ as follows

(a) $d < \frac{\lambda}{8C_4^2}$
(b) $b \geq 4d + \frac{C_3^2C_L^2}{\lambda}$,
(c) $\gamma_1 \leq \frac{1}{4}$,
(d) $\gamma_2 \leq \frac{1}{4}\min\left\{(1 - \frac{D_1}{\overline{D}})^{-1}, (\frac{D_r}{\underline{D}} - 1)^{-1}\right\}$,
(e) $c \geq 2b\frac{1}{\gamma_1}\overline{D}^2e^{2\overline{D}C_L}C_4^2C_L^2 + \frac{C_3^2C_L^2}{\lambda}$.

From (58), one then obtains (54) with the well-defined positive constant $\eta = \min\left\{\frac{\lambda - 8dC_4^2}{2C_2}, \frac{1}{4\overline{D}}\right\}$, and the function

$$g(\epsilon) = \frac{2cr\epsilon}{D_0}\frac{1}{\gamma_2}M \tag{59}$$

which satisfies $\lim_{\epsilon\to 0} g(\epsilon) = 0$.

The proof of Theorem 1 now follows.

4.2 *Proof of Theorem 1*

Firstly, as $\lim_{\epsilon\to 0} g(\epsilon) = 0$, there exists $\epsilon^\star > 0$ such that $\eta - g(\epsilon) = \gamma > 0$ for $\epsilon < \epsilon^\star$. Therefore, according to Dynkin's formula [9, Theorem 5.1, p. 133], assuming temporally that $X(t) \in C$ for all $t \geq \overline{D}$, from (54), one obtains for $\epsilon < \epsilon^\star$

$$\mathbb{E}_{[0,\Phi(0)]}(V(t)) - \mathbb{E}_{[0,\Phi(0)]}(V(\overline{D})) \leq \mathbb{E}_{[0,\Phi(0)]}\left[\int_{\overline{D}}^{t} -\gamma V(s)ds\right] \tag{60}$$

Consequently, applying Gronwall's inequality to (60) (see [16])

$$\mathbb{E}_{[0,\Phi(0)]}(V(t)) \leq \mathbb{E}_{[0,\Phi(0)]}(V(\overline{D}))e^{-\gamma(t-\overline{D})} \tag{61}$$

Similarly to [19, Lemma 4], by applying Young's and Gronwall's inequalities to the backstepping transformation (22) and its inverse (46), there exist positive constants q_1 and q_2 such that for $\forall t \geq 0$, $q_1 V(t) \leq \Upsilon(t) \leq q_2 V(t)$. It thus follows that $\mathbb{E}_{[0,\Upsilon(0)]}(\Upsilon(t)) \leq \frac{q_2}{q_1}\Upsilon(\overline{D})e^{-\gamma(t-\overline{D})}$. To relate $\Upsilon(\overline{D})$ to $\Upsilon(0)$, we formulate the following lemma.

Lemma 5 *Consider the dynamics (4) satisfying Assumptions 1–3. There exists a constant R_0 such that the function Υ defined in (17) satisfies*

$$\Upsilon(t) \leq R_0\Upsilon(0), \quad t \in [0, \overline{D}] \tag{62}$$

Proof Firstly, with Assumption 2, notice that

$$\begin{aligned} |X(t)| \leq& e^{C_L t}|X(0)| + C_L \int_0^t e^{C_L(t-s)}|U(s - D(s))|ds \\ \leq& e^{C_L t}|X(0)| + C_L \sum_{j=1}^{r} \int_0^t e^{C_L(t-s)}|U(s - D_j)|ds \end{aligned} \tag{63}$$

For $t \in [0, \overline{D}]$, with $U(t) = U_0(t)$ for $t \leq 0$, (63) then gives

$$|X(t)|^2 \leq N_1|X(0)|^2 + N_2 \int_{-\overline{D}}^{\min\{t-\underline{D},0\}} U_0(s)^2 ds + N_2 \int_{\min\{t-\underline{D},0\}}^{t-\underline{D}} U(s)^2 ds \tag{64}$$

with $N_1 = 2e^{2C_L\overline{D}}$ and $N_2 = rC_L^2 N_1$. Besides, using Assumption 3 and (44), the prediction-based control law satisfies the inequality

$$|U(t)| = |\kappa(\hat{P}(t))| \leq C_4\left(e^{C_L D_0}|X(t)| + C_L \int_{t-D_0}^{t} e^{C_L(t-s)}|U(s)|ds\right) \tag{65}$$

from which one obtains

$$U(t)^2 \leq N_3|X(t)|^2 + N_4 \int_{t-D_0}^{t} U(s)^2 ds \tag{66}$$

with $N_3 = 2C_4^2 e^{2C_L D_0}$ and $N_4 = C_L^2 N_3$.

Replacing (66) into (64) and using again (64), one obtains

$$|X(t)|^2 \leq N_1|X(0)|^2 + N_2 \int_{-\overline{D}}^{\min\{t-\underline{D},0\}} U_0(s)^2 ds + N_2 N_3 \int_{\min\{t-\underline{D},0\}}^{t-\underline{D}} X(s)^2$$

$$+ N_2 N_4 D_0 \int_{\min\{t-\underline{D},0\}-D_0}^{t-\underline{D}} U(s)^2 ds \tag{67}$$

$$\leq N_5|X(0)|^2 + N_6 \int_{-\overline{D}}^{t-\underline{D}} U(s)^2 ds \tag{68}$$

with

$$N_5 = N_1 \left(1 + N_2 N_3(\overline{D} - \underline{D})\right) \tag{69}$$

$$N_6 = N_2 \left(1 + N_2 N_3(\overline{D} - \underline{D}) + N_4 D_0\right) \tag{70}$$

Hence, with (66), it follows that

$$U(t)^2 \leq N_3 N_5 |X(0)|^2 + (N_3 N_6 + N_4) \int_{-\overline{D}}^{t} U(s)^2 ds \tag{71}$$

and, applying Gronwall's lemma, that

$$U(t)^2 \leq N_7 |X(0)|^2 + N_8 \int_{-\overline{D}}^{t} U_0(s)^2 ds \tag{72}$$

with $N_7 = N_3 N_5 e^{(N_3 N_6 + N_4)\overline{D}}$ and $N_8 = (N_3 N_6 + N_4) e^{(N_3 N_6 + N_4)\overline{D}}$. Consequently, it also holds

$$X(t)^2 \leq N_9 |X(0)|^2 + N_{10} \int_{-\overline{D}}^{t} U_0(s)^2 ds \tag{73}$$

with $N_9 = N_5 + N_6(\overline{D} - \underline{D}) N_7$ and $N_{10} = N_6(1 + (\overline{D} - \underline{D}) N_8)$. Therefore, as from (17), it holds

$$\Upsilon(t) = |X(t)|^2 + \int_{t-\overline{D}-D_0}^{0} U_0(s)^2 ds + \int_0^t U(s)^2 ds \tag{74}$$

the conclusion follows from (72) and (67).

With Lemma 5, we can now conclude the proof of Theorem 1. Indeed, the function Υ thus satisfies $\mathbb{E}_{[0,\Upsilon(0)]}(\Upsilon(t)) \leq \frac{q_2}{q_1} R_0 \Upsilon(0) e^{-\gamma(t-\overline{D})}$, if $X(t) \in C$ for all $t \geq \overline{D}$.

Finally, let us denote $\sqrt{\rho_0^\star}$ such that $B(0, \sqrt{\rho_0^\star}) \subset C$ and observe that $|X(t)|^2 \leq \Upsilon(t)$ for all $t \geq 0$. Consequently, a sufficient condition for $X(t)$ belonging to C is $\Upsilon(t) \leq \rho_0^\star$ and the result follows with $\rho^\star = \frac{q_1}{q_2 R_0} \rho_0^\star$ and $R = \frac{q_2}{q_1} R_0 e^{\gamma \overline{D}}$. Theorem 1 is then proved.

5 Conclusion

In this chapter, we extended our previous work [18] to the case of nonlinear systems, using a controller based on a constant time horizon prediction to robustly compensate for stochastic input delays. A first direction of improvement of this work will be to relax the technical assumptions made in this chapter on the dynamics, in particular on the global lipschitzness of the vector field.

Besides, the sufficient condition we obtain for closed-loop stabilization bears on the range of the possible delay values, which should be small enough, and on the prediction horizon which should be close enough to this range. Hence, this condition does not distinguish between the delay probability distributions. It is thus likely to be somehow conservative. Relaxing this condition to involve the delay distribution into the stability analysis is an important direction of future work.

Similarly, to increase the closed-loop performances, other prediction horizons could be considered in the future, such as the current expected value of the delay.

Finally, our analysis methodology grounds on the probabilistic delay averaging approach [16], modeling the input delay as a Markov process with a finite number of states. Another important direction of future works is to extend this analysis methodology to the case where the delay takes values in a continuum.

References

1. Bekiaris-Liberis, N., Krstic, M.: Compensation of state-dependent input delay for nonlinear systems. IEEE Trans. Autom. Control **58**(2), 275–289 (2012)
2. Bekiaris-Liberis, N., Krstic, M.: Nonlinear control under delays that depend on delayed states. Eur. J. Control **19**(5), 389–398 (2013)
3. Bekiaris-Liberis, N., Krstic, M.: (Eds.). Nonlinear Control Under Nonconstant Delays. Society for Industrial and Applied Mathematics (2013)
4. Bekiaris-Liberis, N., Krstic, M.: Robustness of nonlinear predictor feedback laws to time-and state-dependent delay perturbations. Automatica **49**(6), 1576–1590 (2013)
5. Bekiaris-Liberis, N., Krstic, M.: Predictor-feedback stabilization of multi-input nonlinear systems. IEEE Trans. Autom. Control **62**(2), 516–531 (2016)
6. Bresch-Pietri, D., Del Vecchio, D.: Estimation for decentralized safety control under communication delay and measurement uncertainty. Automatica **62**, 292–303 (2015)
7. Cortés, J.C., Jódar, L., Villafuerte, L.: Random linear-quadratic mathematical models: computing explicit solutions and applications. Math. Comput. Simul. **79**(7), 2076–2090 (2009)
8. Curtain, R.F., Zwart, H.: An Introduction to Infinite-Dimensional Linear Systems Theory, vol. 21. Springer Science & Business Media (2012)
9. Dynkin, E.B.: Theory of Markov Processes. Courier Corporation (2012)
10. Gupta, V.: On the effect of stochastic delay on estimation. IEEE Trans. Autom. Control **56**(9), 2145–2150 (2011)
11. Gupta, V., Dana, A.F., Hespanha, J.P., Murray, R.M., Hassibi, B.: Data transmission over networks for estimation and control. IEEE Trans. Autom. Control **54**(8), 1807–1819 (2009)
12. Ito, K., Nisio, M.A.K.I.K.O.: On stationary solutions of a stochastic differential equation. J. Math. Kyoto Univ. **4**(1), 1–75 (1964)
13. Karafyllis, I., Krstic, M.: Delay-robustness of linear predictor feedback without restriction on delay rate. Automatica **49**(6), 1761–1767 (2013)

14. Kats, I.I., Krasovskii, N.N.: On the stability of systems with random parameters. J. Appl. Math. Mech. **24**(5), 1225–1246 (1960)
15. Kolmanovskii, V., Myshkis, A.: Introduction to the Theory and Applications of Functional Differential Equations, vol. 463. Springer Science & Business Media (2013)
16. Kolmanovsky, I., Maizenberg, T.L.: Mean-square stability of nonlinear systems with time-varying, random delay. Stoch. Anal. Appl. **19**(2), 279–293 (2001)
17. Kolmanovsky, I.V., Maizenberg, T.L.: Optimal control of continuous-time linear systems with a time-varying, random delay. Syst. Control Lett. **44**(2), 119–126 (2001)
18. Kong, S., Bresch-Pietri, D.: Prediction-based controller for linear systems with stochastic input delay. Working paper or preprint (2021)
19. Kong, S., Bresch-Pietri, D.: Constant time horizon prediction-based control for linear systems with time-varying input delay. IFAC-PapersOnLine **53**(2), 7665–7670 (2020)
20. Krstic, M., Smyshlyaev, A.: Boundary Control of PDEs: A Course on Backstepping Designs. Society for Industrial and Applied Mathematics (2008)
21. Kwon, W., Pearson, A.: Feedback stabilization of linear systems with delayed control. IEEE Trans. Autom. Control **25**(2), 266–269 (1980)
22. Léchappé, V., Moulay, E., Plestan, F., Glumineau, A., Chriette, A.: New predictive scheme for the control of LTI systems with input delay and unknown disturbances. Automatica **52**, 179–184 (2015)
23. Li, K., Mu, X.: Predictor-based H∞ leader-following consensus of stochastic multi-agent systems with random input delay. Optim. Control Appl. Methods **42**(5), 1349–1364 (2021)
24. Manitius, A.W.O.A.Z., Olbrot, A.: Finite spectrum assignment problem for systems with delays. IEEE Trans. Autom. Control **24**(4), 541–552 (1979)
25. Mohammed, S.E.A., Salah-El Din, A.M.: Stochastic Functional Differential Equations, vol. 99. Pitman Advanced Publishing Program (1984)
26. Molnár, T.G., Qin, W.B., Insperger, T., Orosz, G.: Application of predictor feedback to compensate time delays in connected cruise control. IEEE Trans. Intell. Transp. Syst. **19**(2), 545–559 (2017)
27. Nihtila, M.T.: Finite pole assignment for systems with time-varying input delays. In: Proceedings of the 30th IEEE Conference on Decision and Control, pp. 927–928. IEEE (1991)
28. Palmor, Z.J.: Time-delay compensation-Smith predictor and its modifications. In: The Control Handbook, pp. 224–237 (1996)
29. Richard, J.P.: Time-delay systems: an overview of some recent advances and open problems. Automatica **39**(10), 1667–1694 (2003)
30. Selivanov, A., Fridman, E.: Predictor-based networked control under uncertain transmission delays. Automatica **70**, 101–108 (2016)
31. Simon, D., Seuret, A., Sename, O.: Real-time control systems: feedback, scheduling and robustness. Int. J. Syst. Sci. **48**(11), 2368–2378 (2017)
32. Smith, O.J.: A controller to overcome dead time. ISA J. **6**, 28–33 (1959)
33. Sykora, H.T., Sadeghpour, M., Ge, J.I., Bachrathy, D., Orosz, G.: On the moment dynamics of stochastically delayed linear control systems. Int. J. Robust Nonlinear Control **30**(18), 8074–8097 (2020)
34. Zhang, L., Orosz, G.: Beyond-line-of-sight identification by using vehicle-to-vehicle communication. IEEE Trans. Intell. Transp. Syst. **19**(6), 1962–1972 (2017)

Output Feedback Stabilization of a Reaction-Diffusion PDE in the Presence of Saturations of the Input and Its Time Derivatives

Hugo Lhachemi and Christophe Prieur

Abstract This chapter tackles the output feedback stabilization of a reaction-diffusion PDE in the presence of saturations applying to the command input as well as a finite number of its time derivatives. The control strategy consists of a finite dimensional observer and a finite-dimensional state-feedback. We derive LMI-based sufficient conditions that ensure the local exponential stability of the closed-loop system while providing an estimation of the domain of attraction. These LMI conditions are shown to be feasible provided the order of the observer is selected large enough. The stability analysis is performed by using Lyapunov's direct method while invoking sector conditions commonly used for the analysis of saturated finite-dimensional systems.

1 Introduction

Due to the inherent physical limitations of the actuators, saturation mechanisms are commonly encountered in the practical implementation of control laws. Such saturation mechanisms introduce stringent constraints on the control strategies [1]. It is worth noting that even for finite-dimensional linear time invariant (LTI) systems, such saturation mechanisms are very well-known for introducing strong nonlinear phenomena characterized by multiple equilibrium points and bounded domains of attraction [2]. Due to its practical importance, the topic of feedback stabilization of finite-dimensional LTI systems in the presence of input saturations has been inten-

The work of Hugo Lhachemi was supported by ANR PIA funding: ANR-20-IDEES-0002. The work of Christophe Prieur has been partially supported by MIAI@Grenoble Alpes (ANR-19-P3IA-0003).

H. Lhachemi
Université Paris-Saclay, CNRS, CentraleSupélec, Laboratoire des signaux et systèmes, 91190 Gif-sur-Yvette, France
e-mail: hugo.lhachemi@centralesupelec.fr

C. Prieur (✉)
Université Grenoble Alpes, CNRS, Grenoble-INP, GIPSA-lab, 38000 Grenoble, France
e-mail: christophe.prieur@gipsa-lab.fr

J. Auriol et al. (eds.), *Advances in Distributed Parameter Systems*, Advances in Delays and Dynamics 14, https://doi.org/10.1007/978-3-030-94766-8_3

sively studied [3, 4]. One of the most fruitful approaches in this field relies on Lyapunov's direct method combined with the use of a generalized sector condition [3, Lem. 1.6] in order to derive sufficient linear matrix inequality (LMI) conditions ensuring the local stability of the closed-loop plant. A distinguished feature of this approach is that it is also employed to obtain an estimate of the domain of attraction of the origin.

We embrace in this chapter the topic of stabilization of infinite-dimensional systems, and particularly of partial differential equations (PDEs), in the presence of input saturation mechanisms. Among the first contributions in this field of research, such a problem was studied in [5, 6] with saturation mechanisms defined for control input functions evaluated in the norm of an abstract functional space (typically the space of square integrable functions $L^2(0, 1)$). Since then, the stabilization of PDEs in the presence of saturation mechanisms has attracted much attention with a particular focus on pointwise saturation mechanisms which are, in general, the most relevant for practical applications. In this context, stabilization of wave and Korteweg-de Vries PDEs under cone-bounded feedback were reported in [7–9] using Lyapunov's direct method. The case of reaction-diffusion PDEs in the presence of control input constraints was studied in [10] using a model predictive control approach while singular perturbation techniques were reported in [11]. Also in the case of reaction-diffusion PDEs but using spectral reduction methods [12–17], it was demonstrated in [18] how a state-feedback can be designed to achieve the local stabilization of the plant while deriving an explicit estimation of the domain of attraction using LMIs. In the same context, the possibility to achieve the local output feedback stabilization of a reaction-diffusion PDE using a either distributed or Dirichlet/Neumann boundary measurement, with estimation of the domain of attraction, was demonstrated in [19]. It is worth noting that for the state feedback setting studied in [18], the region of attraction constrains only a finite number of modes of the initial condition. This is essentially because the state-feedback setting allows to achieve the local exponential stabilization of the first modes of the system while preserving the stability of the residual infinite-dimensional dynamics. See [18, Proposition 1] for a precise statement of this result. In contrast, the output feedback setting does not allow such a strong separation between the to-be-stabilized modes and the residual ones. This is because the measurement is built based on the contribution of all the modes of the system. In this case, the region of attraction imposes constraints on all the modes of the initial condition [19].

We extend in this chapter the results obtained in [19] for the local stabilization of reaction-diffusion PDEs in the presence of an input saturation to the case of saturations applying on the input as well as on a finite number of its time derivatives. Typically, a saturation on the time derivative of the control input can be used to model the limitations in the rate of change of the actuator output. Similarly, a saturation on the second time derivative of the input is employed to model a limit in the acceleration of change of the actuator output, etc. In this context, the output feedback control design strategy adopted in this chapter consists of a finite-dimensional observer-based controller [20–25] that leverages a control architecture reported first in [21] along with a LMI-based approach introduced in [26] (see also [27]). We particularly adopt the enhanced and general procedures from [28, 29] which allow to design, in a generic and systematic manner, finite-dimensional observer-based control strategies for general 1-D reaction-diffusion PDEs, with Dirichlet/Neumann/Robin boundary control and Dirichlet/Neumann boundary measurement, in a great variety of settings. These include regulation control [30], nonlinear boundary control [29], and stabilization in the presence of an arbitrarily long input [31], output [32] or state [33] delay. We leverage here similar procedures but in the context of the saturation of the control input u and its n_s first time derivatives $\dot{u}, \ldots, u^{(n_s)}$. Compared to [19] where only a saturation in position was considered, hence the control design was performed directly on the input u, the stabilization problem in the presence of saturations on the n_s first time derivatives of u requires to consider $u^{(n_s)}$ as an auxiliary input for control design. Therefore we perform the control design on the system composed of the original PDE augmented with n_s integral components.

The reminder of this chapter is organized as follows. A general background on the stabilization of finite-dimensional LTI systems in the presence of an input saturation is placed in Sect. 2. The control design problem addressed in this work is introduced in Sect. 3. The control design procedure and the resulting stability theorems are reported in Sect. 4. Some numerical examples and simulations illustrating the main theoretical results are presented in Sect. 5. Finally, some concluding remarks are formulated in Sect. 6.

Notation. The real vector spaces $\mathbb{R}^n$ are endowed with the usual Euclidean norm $\|x\| = \sqrt{x^\top x}$. The associated induced norms of matrices are also denoted by $\|\cdot\|$. For any $x, y \in \mathbb{R}^n$, $x \leq y$ means that each component of x is less than or equal to the corresponding component of y. For any $x \in \mathbb{R}^n$ we denote by $|x|$ the vector of $\mathbb{R}^n$ obtained by replacing each component of x by its absolute value. Considering any two vectors X and Y of arbitrary dimensions, we denoted by $\mathrm{col}(X, Y)$ the vector $[X^\top, Y^\top]^\top$. Given a vector ℓ, $\mathrm{diag}(\ell)$ denotes the diagonal matrix with the entries of ℓ on the diagonal. $L^2(0, 1)$ stands for the space of square integrable functions on $(0, 1)$ and is endowed with the usual inner product $\langle f, g\rangle = \int_0^1 f(x)g(x)\,\mathrm{d}x$. The corresponding norm is denoted by $\|\cdot\|_{L^2}$. For any given integer $m \geq 1$, $H^m(0, 1)$ stands for the m-order Sobolev space and is equipped with its usual norm $\|f\|_{H^m} = \left(\sum_{k=0}^m \|f^{(k)}\|_{L^2}^2\right)^{1/2}$. For any given symmetric matrix $P \in \mathbb{R}^{n\times n}$, $P \succeq 0$ (resp. $P \succ 0$) means that P is positive semi-definite (resp. positive definite).

Let $(\phi_n)_{n\geq 1}$ be a Hilbert basis of $L^2(0, 1)$. For any two integers $1 \leq N < M$, we define the operators of projection:

$$\begin{aligned} \pi_N : L^2(0, 1) &\longrightarrow \mathbb{R}^N \\ f &\longmapsto \left[\langle f, \phi_1\rangle \ldots \langle f, \phi_N\rangle\right]^\top \end{aligned}$$

and

$$\begin{aligned} \pi_{N,M} : L^2(0, 1) &\longrightarrow \mathbb{R}^{M-N} \\ f &\longmapsto \left[\langle f, \phi_{N+1}\rangle \ldots \langle f, \phi_M\rangle\right]^\top . \end{aligned}$$

We finally define

$$\begin{aligned}\mathcal{R}_N : L^2(0,1) &\longrightarrow L^2(0,1)\\ f &\longmapsto f - \sum_{n=1}^{N} \langle f, \phi_n \rangle \phi_n = \sum_{n \geq N+1} \langle f, \phi_n \rangle \phi_n .\end{aligned}$$

2 Background on the Stabilization of Finite-Dimensional LTI Systems in the Presence of an Input Saturation

We start by reviewing in this section some classical results on the stabilization of finite-dimensional LTI systems in the presence of an input saturation. We refer the readers to [3, 4] for a deeper insight into this topic. Let us first introduce the scalar saturation map

$$\begin{aligned}\mathrm{sat}_l : \mathbb{R} &\longrightarrow \mathbb{R}\\ u &\longmapsto \mathrm{sign}(u) \min(|u|, l)\end{aligned}$$

where $l > 0$. Then, for any $\ell = (l_1, \ldots, l_m) \in (\mathbb{R}_{>0})^m$, we define the vector saturation map

$$\begin{aligned}\mathrm{sat}_\ell : \quad \mathbb{R}^m &\longrightarrow \mathbb{R}^m\\ (u_1, \ldots, u_m) &\longmapsto (\mathrm{sat}_{l_1}(u_1), \ldots, \mathrm{sat}_{l_m}(u_m)).\end{aligned}$$

With these notations, consider the system described by

$$\dot{z}(t) = Az(t) + B\mathrm{sat}_\ell(u(t)) \tag{1}$$

with $A \in \mathbb{R}^{n \times n}$ and $B \in \mathbb{R}^{n \times m}$. Here $z(t) \in \mathbb{R}^n$ represents the state at time $t \geq 0$ while $u(t) \in \mathbb{R}^m$ stands for the control input. We assume that the pair (A, B) is stabilizable, i.e., there exists a feedback gain $K \in \mathbb{R}^{m \times n}$ so that $A + BK$ is Hurwitz. Hence, in the absence of saturation, the system described by $\dot{z}(t) = Az(t) + Bu(t)$ can be exponentially stabilized by setting

$$u(t) = Kz(t) \tag{2}$$

because the closed-loop dynamics now reads $\dot{z}(t) = (A + BK)z(t)$. In the case of an input saturation, the situation is more complex since the application of the feedback law (2) to (1) gives the closed-loop dynamics:

$$\dot{z}(t) = Az(t) + B\mathrm{sat}_\ell(Kz(t)). \tag{3}$$

In order to study the stability of (3), a relevant idea is to consider the saturated term $\text{sat}_\ell(Kz(t))$ as a perturbation of the nominal term $Kz(t)$. The objective of this approach is to make explicitly appear the Hurwitz matrix $A + BK$ in the dynamics of the closed-loop system:

$$\begin{aligned}\dot{z}(t) &= (A + BK)z(t) + B\left\{\text{sat}_\ell(Kz(t)) - Kz(t)\right\} \\ &= (A + BK)z(t) + B\phi_\ell(Kz(t))\end{aligned} \tag{4}$$

where we introduce the deadzone nonlinearity defined by

$$\begin{aligned}\phi_\ell : \quad \mathbb{R}^m &\longrightarrow \mathbb{R}^m \\ v &\longmapsto \text{sat}_\ell(v) - v.\end{aligned}$$

When $z(t)$, hence $Kz(t)$, is sufficiently small, the perturbation introduced by the saturation map is such that $\phi_\ell(Kz(t)) = 0$. In that case, the dynamics of the closed-loop system (4) reduced to $\dot{z}(t) = (A + BK)z(t)$, ensuring the local exponential stability of the closed-loop system in a neighborhood of the origin. However, in order to estimate the domain of attraction of the origin of (4), one needs to capture more precisely the impact of the perturbation term $B\phi_\ell(Kz(t))$ on the dynamics of (4). A very efficient approach to achieve this objective is to embed the deadzone nonlinearity ϕ_ℓ into a generalized sector condition. Such a sector condition must be suitable for Lyapunov's direct method in order to derive LMI-based sufficient stability conditions. In this chapter, we leverage the following generalized sector condition that is borrowed from [3, Lemma 1.6].

Lemma 1 *Let $\ell \in (\mathbb{R}_{>0})^m$ be given. For any $v, \omega \in \mathbb{R}^m$ such that $|v_i - \omega_i| \leq \ell_i$ for all $i \in \{1, \dots, m\}$ and for any diagonal positive definite matrix $T \in \mathbb{R}^{m\times m}$ we have $\phi_\ell(v)^\top T(\phi_\ell(v) + \omega) \leq 0$.*

Let us now explain how Lemma 1 can be used to derive an estimation of the domain of attraction of (4). To do so, consider the classical quadratic Lyapunov function candidate

$$V(z) = z^\top Pz, \quad \forall z \in \mathbb{R}^n \tag{5}$$

for some matrix $P \succ 0$. The computation of the time-derivative of V along the system trajectories (4) reads

$$\begin{aligned}\dot{V} &= z^\top\left\{(A + BK)^\top P + P(A + BK)\right\}z + 2z^\top PB\phi_\ell(Kz) \\ &= \begin{bmatrix} z \\ \phi_\ell(Kz)\end{bmatrix}^\top \begin{bmatrix}(A + BK)^\top P + P(A + BK) & PB \\ * & 0\end{bmatrix}\begin{bmatrix} z \\ \phi_\ell(Kz)\end{bmatrix}.\end{aligned}$$

Here the symbol $*$ is used to denote a symmetric part. For any matrix $C \in \mathbb{R}^{m\times n}$ and any diagonal positive definite matrix $T \in \mathbb{R}^{m\times m}$, we obtain from Lemma 1 that

$$\forall z \in \mathbb{R}^n, \quad |(K - C)z| \leq \ell \Rightarrow \phi_\ell(Kz)^\top T(\phi_\ell(Kz) + Cz) \leq 0.$$

Combining the two latter results, we infer that, along the system trajectories (4), $|(K - C)z| \leq \ell$ gives

$$\dot{V} \leq \begin{bmatrix} z \\ \phi_\ell(Kz) \end{bmatrix}^\top \begin{bmatrix} (A+BK)^\top P + P(A+BK) & PB - C^\top T \\ * & -2T \end{bmatrix} \begin{bmatrix} z \\ \phi_\ell(Kz) \end{bmatrix}.$$

Assuming that

$$\Theta_1 = \begin{bmatrix} (A+BK)^\top P + P(A+BK) & PB - C^\top T \\ * & -2T \end{bmatrix} \prec 0,$$

we deduce the existence of $\alpha > 0$ so that $\Theta_1 \preceq -\alpha I$. Hence we have

$$\forall z \in \mathbb{R}^n, \quad |(K - C)z| \leq \ell \Rightarrow \dot{V} \leq -\alpha \|z\|^2 \leq -2\kappa V.$$

with $\kappa = \frac{\alpha}{2\lambda_{\max}(P)} > 0$. We now need to derive a set of initial conditions $\mathcal{E}$ containing the origin so that $\mathcal{E}$ is (1) positively invariant; and (2) included into the set $\{z \in \mathbb{R}^n : |(K - C)z| \leq \ell\}$. To do so we define

$$\mathcal{E} = \{z \in \mathbb{R}^n \, : \, V(z) = z^\top P z \leq 1\}$$

and we assume that

$$\Theta_2 = \begin{bmatrix} P & (K-C)^\top \\ * & \mathrm{diag}(\ell)^2 \end{bmatrix} \succeq 0.$$

Since $\mathrm{diag}(\ell) \succ 0$, the Schur complement shows that $P \succeq (K - C)^\top \mathrm{diag}(\ell)^{-2}(K - C)$. Hence $z \in \mathcal{E}$ implies that $|(K - C)z| \leq \ell$. Considering now a non zero initial condition $z_0 \in \mathcal{E} \setminus \{0\}$, we infer that $\dot{V}(z_0) \leq -\alpha \|z_0\| < 0$. From now, a simple contradiction argument shows that $z(t) \in \mathcal{E}$ for all $t \geq 0$ hence $\dot{V}(z(t)) \leq -2\kappa V(z(t))$. Overall, this reasoning shows that if there exist $P \succ 0$, $C \in \mathbb{R}^{m \times n}$, and a diagonal positive definite matrix $T \in \mathbb{R}^{m \times m}$ such that $\Theta_1 \prec 0$ and $\Theta_2 \succeq 0$, then for any initial condition $z_0 \in \mathcal{E}$, the trajectories of (3) exponentially converge to the origin.

Remark 1 It is worth noting that the constraints $\Theta_1 \prec 0$ and $\Theta_2 \succeq 0$ are always feasible for a suitable choice of the decision variables P, C, T. Indeed, setting $C = 0$, $P = \beta P_0$, and $T = \gamma I$ where $\beta, \gamma > 0$ and $P_0 \succ 0$ is the unique solution to the Lyapunov equation $(A + BK)^\top P_0 + P_0(A + BK) = -I$, one can see using Schur complement that $\Theta_2 \succeq 0$ is obtained provided $\beta > 0$ is selected large enough. Using the same argument, $\Theta_1 \prec 0$ can now be obtained provided $\gamma > 0$ is selected large enough.

Remark 2 In their original version, the constraints $\Theta_1 \prec 0$ and $\Theta_2 \succeq 0$ are nonlinear w.r.t. the decision variables. However, they can be equivalently reformulated as LMI constraints for which efficient solvers exist. See, e.g., [18, Sect. III.C].

3 Problem Description and Properties of Sturm-Liouville Operators

We introduce in this section the control design problem for 1-D reaction-diffusion PDEs. These PDEs are strongly related to the class of Sturm-Liouville operators. The main properties of these operators that will be useful to perform the control design are presented in the second part of this section.

3.1 Problem Description

We investigate the problem of feedback stabilization of the reaction-diffusion equation described by

$$z_t(t,x) = (p(x)z_x(t,x))_x - \tilde{q}(x)z(t,x) + b(x)\mathrm{sat}_{l_0}(u(t)) \tag{6a}$$
$$\cos(\theta_1)z(t,0) - \sin(\theta_1)z_x(t,0) = 0 \tag{6b}$$
$$\cos(\theta_2)z(t,1) + \sin(\theta_2)z_x(t,1) = 0 \tag{6c}$$
$$z(0,x) = z_0(x). \tag{6d}$$

Here $\theta_1, \theta_2 \in [0, \pi/2]$, $p \in C^1([0,1])$ with $p > 0$, $\tilde{q} \in C^0([0,1])$, and $l_0 > 0$. The shape function $b \in L^2(0,1)$ represents the way the saturated scalar control input $u_{\mathrm{sat}}(t) = \mathrm{sat}_{l_0}(u(t)) \in \mathbb{R}$ acts on the system, $z_0 \in L^2(0,1)$ is the initial condition, and $z(t,\cdot) \in L^2(0,1)$ is the state of the reaction-diffusion PDE. The system output takes the form of the measurement $y(t) \in \mathbb{R}$ described by

$$y(t) = \int_0^1 c(x)z(t,x)\,\mathrm{d}x \tag{7}$$

where $c \in L^2(0,1)$.

Remark 3 The approach presented in this chapter can be extended in a straightforward manner to the case of multiple control inputs for which the PDE (6a) is replaced by

$$z_t(t,x) = (p(x)z_x(t,x))_x - \tilde{q}(x)z(t,x) + \sum_{k=1}^{m} b_k(x)\mathrm{sat}_{l_{k,0}}(u(t))$$

where the shape function $b_k \in L^2(0,1)$, for $k \in \{1,\dots,m\}$, represents the way the saturated scalar control input $\mathrm{sat}_{l_{k,0}}(u(t)) \in \mathbb{R}$ acts on the system. However, we focus the presentation on the single input case ($m = 1$) in order to ease the notations.

Let $n_s \in \mathbb{N}_{>0}$ and $l_1, \dots, l_{n_s} > 0$ be arbitrarily given. The control design objective is to achieve the local exponential stabilization of (6), with explicit estimation of the domain of attraction, while ensuring that

$$|u^{(i)}(t)| \leq l_i, \quad \forall t \geq 0, \quad \forall i \in \{1, \dots, n_s\}. \tag{8}$$

3.2 Properties of Sturm-Liouville Operators

Reaction-diffusion PDEs such as described by (6) are strongly related to the so-called Sturm-Liouville operators. We gather in this subsection the key properties of the Sturm-Liouville operators which will be intensively used in the sequel for both control design and stability analysis. Let $\theta_1, \theta_2 \in [0, \pi/2]$, $p \in C^1([0,1])$, and $q \in C^0([0,1])$ with $p > 0$ and $q \geq 0$. We define the Sturm-Liouville operator by

$$\begin{array}{rcl} \mathcal{A} : D(\mathcal{A}) & \longrightarrow & L^2(0,1) \\ f & \longmapsto & -(pf')' + qf \end{array} \tag{9}$$

with the domain of the operator defined by

$$\begin{aligned} D(\mathcal{A}) = \{f \in H^2(0,1) \, : \, & \cos(\theta_1) f(0) - \sin(\theta_1) f'(0) = 0 \\ & \cos(\theta_2) f(1) + \sin(\theta_2) f'(1) = 0\}. \end{aligned}$$

Then it holds that the eigenvalues λ_n, $n \geq 1$, of $\mathcal{A}$ are simple, non negative, and form an increasing sequence with $\lambda_n \to +\infty$ as $n \to +\infty$. The corresponding unit eigenvectors $\phi_n \in L^2(0,1)$ form a Hilbert basis. The domain of the operator $\mathcal{A}$ is equivalently characterized by

$$D(\mathcal{A}) = \left\{ f \in L^2(0,1) \, : \, \sum_{n \geq 1} |\lambda_n|^2 |\langle f, \phi_n \rangle|^2 < +\infty \right\}.$$

Moreover we have $\mathcal{A}f = \sum_{n\geq 1} \lambda_n \langle f, \phi_n \rangle \phi_n$ for all $f \in D(\mathcal{A})$.

Let $p_*, p^*, q^* \in \mathbb{R}$ be such that $0 < p_* \leq p(x) \leq p^*$ and $0 \leq q(x) \leq q^*$ for all $x \in [0,1]$, then it holds

$$0 \leq \pi^2 (n-1)^2 p_* \leq \lambda_n \leq \pi^2 n^2 p^* + q^*$$

for all $n \geq 1$ (see, e.g., [34]). Moreover if $p \in C^2([0,1])$, we have (see, e.g., [34]) that $\phi_n(\xi) = O(1)$ and $\phi_n'(\xi) = O(\sqrt{\lambda_n})$ as $n \to +\infty$ for any given $\xi \in [0,1]$.

If we further assume that $q > 0$, an integration by parts and the continuous embedding $H^1(0,1) \subset L^\infty(0,1)$ imply the existence of constants $C_1, C_2 > 0$ such that

$$C_1 \|f\|_{H^1}^2 \leq \sum_{n \geq 1} \lambda_n \langle f, \phi_n \rangle^2 = \langle \mathcal{A}f, f \rangle \leq C_2 \|f\|_{H^1}^2, \quad \forall f \in D(\mathcal{A}). \tag{10}$$

Inequalities (10) and the Riesz-spectral nature [35] of $\mathcal{A}$ imply that the series expansion $f = \sum_{n\geq 1}\langle f, \phi_n\rangle\phi_n$ holds in $H^2(0,1)$ norm for any $f \in D(\mathcal{A})$. Invoking the continuous embedding $H^1(0,1) \subset L^\infty(0,1)$, we deduce in particular that $f(0) = \sum_{n\geq 1}\langle f, \phi_n\rangle\phi_n(0)$ and $f'(0) = \sum_{n\geq 1}\langle f, \phi_n\rangle\phi_n'(0)$.

4 Control Design and Main Results

4.1 Auxiliary Command Input and Spectral Reduction

Recalling that the objective is to achieve the local exponential stabilization of (6) while satisfying the constraints (8) on the n_s first time derivatives of the command input u, we introduce the auxiliary control input $v(t) \in \mathbb{R}$ along with $u_1(t), \ldots, u_{n_s-1}(t) \in \mathbb{R}$ as follows:

$$\begin{aligned}\dot{u}(t) &= \mathrm{sat}_{l_1}(u_1(t)),\\ \dot{u}_i(t) &= \mathrm{sat}_{l_{i+1}}(u_{i+1}(t)), \quad 1 \leq i \leq n_s - 2\\ \dot{u}_{n_s-1}(t) &= \mathrm{sat}_{l_{n_s}}(v(t)).\end{aligned}$$

We define the short notation $u_{i,\mathrm{sat}}(t) = \mathrm{sat}_{l_i}(u_i(t))$. Hence introducing

$$\mathcal{V} = \begin{bmatrix} u \\ u_1 \\ \vdots \\ u_{n_s-1}\end{bmatrix} \in \mathbb{R}^{n_s}, \qquad J_{n_s} = \begin{bmatrix} 0 & 1 & 0 & \ldots & 0 \\ 0 & 0 & 1 & \ldots & 0 \\ \vdots & \vdots & \vdots & \ddots & \vdots \\ 0 & 0 & 0 & \ldots & 1 \\ 0 & 0 & 0 & \ldots & 0 \end{bmatrix} \in \mathbb{R}^{n_s \times n_s}, \qquad F_{n_s} = \begin{bmatrix} 0 \\ \vdots \\ 0 \\ 1 \end{bmatrix} \in \mathbb{R}^{n_s}$$

we infer that

$$\begin{aligned}\dot{\mathcal{V}}(t) &= J_{n_s}\mathrm{sat}_{(l_0,\ldots,l_{n_s-1})}(\mathcal{V}(t)) + F_{n_s}\mathrm{sat}_{l_{n_s}}(v(t))\\ &= J_{n_s}\mathcal{V}_{\mathrm{sat}}(t) + F_{n_s}v_{\mathrm{sat}}(t)\end{aligned} \tag{11}$$

where $\mathcal{V}_{\mathrm{sat}}(t) = \mathrm{sat}_{(l_0,\ldots,l_{n_s-1})}(\mathcal{V}(t))$ and $v_{\mathrm{sat}}(t) = \mathrm{sat}_{l_{n_s}}(v(t))$.

We now introduce a function $q \in C^0([0,1])$ and a constant $q_c \in \mathbb{R}$ so that

$$\tilde{q} = q - q_c, \quad q \geq 0. \tag{12}$$

Therefore, the reaction-diffusion PDE described by (6) along with (11) can be equivalently represented under the abstract form:

$$z_t(t,\cdot) = \{-\mathcal{A} + q_c \mathrm{Id}_{L^2}\} z(t,\cdot) + b(x) u_{\mathrm{sat}}(t) \tag{13a}$$

$$\dot{\mathcal{V}}(t) = J_{n_s} \mathcal{V}_{\mathrm{sat}}(t) + F_{n_s} v_{\mathrm{sat}}(t) \tag{13b}$$

$$z(0,\cdot) = z_0 \tag{13c}$$

$$\mathcal{V}(0) = \mathcal{V}_0 = \begin{bmatrix} u_0 \; u_{1,0} \; \ldots \; u_{n_s-1,0} \end{bmatrix}^\top \tag{13d}$$

with the Sturm-Liouville operator $\mathcal{A}$ defined by (9) and where v is the auxiliary input to perform the control design. Introducing the coefficients of projection

$$z_n(t) = \langle z(t,\cdot), \phi_n \rangle, \quad b_n(t) = \langle b, \phi_n \rangle, \quad c_n(t) = \langle c, \phi_n \rangle,$$

the projection of (13), along with the measurement equation (7), into the Hilbert basis $(\phi_n)_n$ reads

$$\dot{z}_n(t) = (-\lambda_n + q_c) z_n(t) + b_n u_{\mathrm{sat}}(t), \quad n \geq 1 \tag{14a}$$

$$\dot{\mathcal{V}}(t) = J_{n_s} \mathcal{V}_{\mathrm{sat}}(t) + F_{n_s} v_{\mathrm{sat}}(t) \tag{14b}$$

$$y(t) = \sum_{n \geq 1} c_n z_n(t) \tag{14c}$$

4.2 Control Architecture

As in [19] dealing with the case of a saturation applying only on the control input but not on its time derivatives (i.e. $n_s = 0$), the control strategy takes the form of a finite-dimensional state-feedback coupled with a finite-dimensional observer. More precisely, let $\delta > 0$ be the desired exponential decay rate for the closed-loop system trajectories and let $N_0 \geq 1$ be such that $-\lambda_n + q_c < -\delta$ for all $n \geq N_0 + 1$. For any given integer $N \geq N_0 + 1$, which stands for the dimension of the observer and that will be specified later, the control strategy is described by:

$$\dot{\hat{z}}_n(t) = (-\lambda_n + q_c)\hat{z}_n(t) + b_n u_{\mathrm{sat}}(t) - L_n \left\{ \sum_{k=1}^{N} c_k \hat{z}_k(t) - y(t) \right\}, \quad 1 \leq n \leq N_0 \tag{15a}$$

$$\dot{\hat{z}}_n(t) = (-\lambda_n + q_c)\hat{z}_n(t) + b_n u_{\mathrm{sat}}(t), \quad N_0 + 1 \leq n \leq N \tag{15b}$$

$$\dot{u}(t) = u_{1,\mathrm{sat}}(t), \tag{15c}$$

$$\dot{u}_i(t) = u_{i+1,\mathrm{sat}}(t), \quad 1 \leq i \leq n_s - 2 \tag{15d}$$

$$\dot{u}_{n_s-1}(t) = v_{\mathrm{sat}}(t) \tag{15e}$$

$$v(t) = \sum_{k=1}^{N_0} k_{z,k} \hat{z}_k(t) + k_u u(t) + \sum_{l=1}^{n_s-1} k_{u,l} u_l(t) \tag{15f}$$

where $L_n \in \mathbb{R}$ and $k_{z,k}, k_u, k_{u,l} \in \mathbb{R}$ are the observer and feedback gains, respectively.

Remark 4 The idea of splitting the observer architecture, composed of one part with active correction of the error of estimation (15a) and one part without correction (15b), is borrowed from [21].

Introducing the error of observation $e_n = z_n - \hat{z}_n$, the residue of measurement $\zeta = \sum_{n \geq N+1} c_n z_n$, and the vectors

$$\hat{Z}^{N_0} = \begin{bmatrix} \hat{z}_1 \\ \vdots \\ \hat{z}_{N_0} \end{bmatrix}, \quad \hat{Z}^{N-N_0} = \begin{bmatrix} \hat{z}_{N_0+1} \\ \vdots \\ \hat{z}_N \end{bmatrix}, \quad E^{N_0} = \begin{bmatrix} e_1 \\ \vdots \\ e_{N_0} \end{bmatrix}, \quad E^{N-N_0} = \begin{bmatrix} e_{N_0+1} \\ \vdots \\ e_N \end{bmatrix},$$

we infer that

$$\dot{\hat{Z}}^{N_0} = A_0 \hat{Z}^{N_0} + B_0 E_1 \mathcal{V}_{\text{sat}} + L C_0 E^{N_0} + L C_1 E^{N-N_0} + L\zeta \tag{16a}$$

$$\dot{\mathcal{V}} = J_{n_s} \mathcal{V}_{\text{sat}} + F_{n_s} v_{\text{sat}} \tag{16b}$$

$$\dot{E}^{N_0} = (A_0 - LC_0) E^{N_0} - L C_1 E^{N-N_0} - L\zeta \tag{16c}$$

$$\dot{\hat{Z}}^{N-N_0} = A_1 \hat{Z}^{N-N_0} + B_1 E_1 \mathcal{V}_{\text{sat}} \tag{16d}$$

$$\dot{E}^{N-N_0} = A_1 E^{N-N_0} \tag{16e}$$

$$v = K_z \hat{Z}^{N_0} + K_u \mathcal{V} \tag{16f}$$

where $A_0 = \operatorname{diag}(-\lambda_1 + q_c, \ldots, -\lambda_{N_0} + q_c)$, $A_1 = \operatorname{diag}(-\lambda_{N_0+1} + q_c, \ldots, -\lambda_N + q_c)$, $B_0 = \begin{bmatrix} b_1 & \ldots & b_{N_0} \end{bmatrix}^\top$, $B_1 = \begin{bmatrix} b_{N_0+1} & \ldots & b_N \end{bmatrix}^\top$, $C_0 = \begin{bmatrix} c_1 & \ldots & c_{N_0} \end{bmatrix}$, $C_1 = \begin{bmatrix} c_{N_0+1} & \ldots & c_N \end{bmatrix}$, $E_1 = \begin{bmatrix} 1 & 0 & \ldots & 0 \end{bmatrix}$, $L = \begin{bmatrix} L_1 & \ldots & L_{N_0} \end{bmatrix}^\top$, $K_z = \begin{bmatrix} k_{z,1} & \ldots & k_{z,N_0} \end{bmatrix}$, and $K_u = \begin{bmatrix} k_u & k_{u,1} & \ldots & k_{u,n_s-1} \end{bmatrix}$. Introducing the augmented vector

$$\hat{Z}_a^{N_0} = \operatorname{col}\left(\hat{Z}^{N_0}, \mathcal{V}\right) \tag{17}$$

we obtain that

$$\dot{\hat{Z}}_a^{N_0} = \tilde{A}_0 \hat{Z}_a^{N_0} + \tilde{B}_0 \mathcal{V}_{\text{sat}} + \tilde{B}_1 v_{\text{sat}} + \tilde{L} C_0 E^{N_0} + \tilde{L} C_1 E^{N-N_0} + \tilde{L}\zeta \tag{18a}$$

$$v = K \hat{Z}_a^{N_0} \tag{18b}$$

where

$$\tilde{A}_0 = \begin{bmatrix} A_0 & 0 \\ 0 & 0 \end{bmatrix}, \quad \tilde{B}_0 = \begin{bmatrix} B_0 E_1 \\ J_{n_s} \end{bmatrix}, \quad \tilde{B}_1 = \begin{bmatrix} 0 \\ F_{n_s} \end{bmatrix}, \quad \tilde{L} = \begin{bmatrix} L \\ 0 \end{bmatrix}, \quad K = \begin{bmatrix} K_z & K_u \end{bmatrix}.$$

Now, since $\mathcal{V}_{\text{sat}} = \mathcal{V} + \phi_{(l_0,\ldots,l_{n_s-1})}(\mathcal{V})$ and $v_{\text{sat}} = v + \phi_{l_{n_s}}(v)$, we have

$$
\begin{aligned}
\dot{\hat{Z}}_a^{N_0} &= \tilde{A}_1 \hat{Z}_a^{N_0} + \tilde{B}_1 v + \tilde{L} C_0 E^{N_0} + \tilde{L} C_1 E^{N-N_0} + \tilde{L}\zeta \\
&\quad + \tilde{B}_0 \phi_{(l_0,\ldots,l_{n_s-1})}(\mathcal{V}) + \tilde{B}_1 \phi_{l_{n_s}}(v) \\
&= (\tilde{A}_1 + \tilde{B}_1 K)\hat{Z}_a^{N_0} + \tilde{L} C_0 E^{N_0} + \tilde{L} C_1 E^{N-N_0} + \tilde{L}\zeta + \tilde{B}\phi_\ell(\mathcal{V}_a)
\end{aligned}
$$

where

$$
\mathcal{V}_a = \begin{bmatrix} \mathcal{V} \\ v \end{bmatrix}, \quad \tilde{A}_1 = \begin{bmatrix} A_0 & B_0 E_1 \\ 0 & J_{n_s} \end{bmatrix}, \quad \tilde{B} = \begin{bmatrix} \tilde{B}_0 & \tilde{B}_1 \end{bmatrix}, \quad \ell = (l_0, \ldots, l_{n_s}).
$$

Noting finally that $\mathcal{V} = \begin{bmatrix} 0 & I \end{bmatrix} \hat{Z}_a^{N_0}$ and from (3.16f) we infer that $\mathcal{V}_a = \tilde{K} \hat{Z}_a^{N_0}$ where $\tilde{K} = \begin{bmatrix} 0 & I \\ K_z & K_u \end{bmatrix}$. Therefore, we obtain that

$$
\begin{aligned}
\dot{\hat{Z}}_a^{N_0} &= (\tilde{A}_1 + \tilde{B}_1 K)\hat{Z}_a^{N_0} + \tilde{L} C_0 E^{N_0} + \tilde{L} C_1 E^{N-N_0} + \tilde{L}\zeta + \tilde{B}\phi_\ell(\tilde{K}\hat{Z}_a^{N_0}) && \text{(19a)} \\
\dot{E}^{N_0} &= (A_0 - L C_0) E^{N_0} - L C_1 E^{N-N_0} - L\zeta && \text{(19b)} \\
\dot{\hat{Z}}^{N-N_0} &= A_1 \hat{Z}^{N-N_0} + B_1 E_1 \begin{bmatrix} 0 & I \end{bmatrix} \hat{Z}_a^{N_0} + B_1 E_1 \begin{bmatrix} I & 0 \end{bmatrix} \phi_\ell(\tilde{K}\hat{Z}_a^{N_0}) && \text{(19c)} \\
\dot{E}^{N-N_0} &= A_1 E^{N-N_0}. && \text{(19d)}
\end{aligned}
$$

Introducing

$$
X = \operatorname{col}\left(\hat{Z}_a^{N_0}, E^{N_0}, \hat{Z}^{N-N_0}, E^{N-N_0}\right) \tag{20}
$$

we have

$$
\dot{X} = FX + \mathcal{L}\zeta + \mathcal{L}_\phi \phi_\ell(\tilde{K}\hat{Z}_a^{N_0}) \tag{21}
$$

where

$$
F = \begin{bmatrix} \tilde{A}_1 + \tilde{B}_1 K & \tilde{L} C_0 & 0 & \tilde{L} C_1 \\ 0 & A_0 - L C_0 & 0 & -L C_1 \\ B_1 E_1 \begin{bmatrix} 0 & I \end{bmatrix} & 0 & A_1 & 0 \\ 0 & 0 & 0 & A_1 \end{bmatrix}, \quad \mathcal{L} = \begin{bmatrix} \tilde{L} \\ -L \\ 0 \\ 0 \end{bmatrix}, \quad \mathcal{L}_\phi = \begin{bmatrix} \tilde{B} \\ 0 \\ B_1 E_1 \begin{bmatrix} I & 0 \end{bmatrix} \\ 0 \end{bmatrix}.
$$

Remark 5 Due to the structure of the matrix F, it is easy to see that $\mathrm{sp}_\mathbb{C}(F) = \mathrm{sp}_\mathbb{C}(\tilde{A}_1 + \tilde{B}_1 K) \cup \mathrm{sp}_\mathbb{C}(A_0 - L C_0) \cup \mathrm{sp}_\mathbb{C}(A_1)$. Owing to its definition, the matrix A_1 is Hurwitz, consequently the matrix F is Hurwitz if and only if the matrices $\tilde{A}_1 + \tilde{B}_1 K$ and $A_0 - L C_0$ are Hurwitz. Since A_0 is diagonal with simple eigenvalues, the pair (A_0, C_0) satisfies the Kalman condition if and only if $c_n \neq 0$ for all $1 \leq n \leq N_0$. Let us now focus on the pair $(\tilde{A}_1, \tilde{B}_1)$ for which we apply the Hautus test. Let[1] $\lambda \in \mathbb{R}$ and $x = \operatorname{col}(\alpha, \beta)$ with $\alpha \in \mathbb{R}^{N_0}$ and $\beta \in \mathbb{R}^{n_s}$ so that $x^\top \tilde{A}_1 = \lambda x^\top$ and $x^\top \tilde{B}_1 = 0$. This is equivalent to

[1] We restrict the study to real eigenvalues and eigenvectors owing to the fact that the eigenvalues of the matrix $\tilde{A}_1$ are all real.

$$\alpha^\top A_0 = \lambda\alpha^\top, \quad \alpha^\top B_0 E_1 + \beta^\top J_{n_s} = \lambda\beta^\top, \quad \beta^\top F_{n_s} = 0. \tag{22}$$

The latter identity is equivalent to $\beta_{n_s} = 0$. Therefore, recalling that $E_1 = \begin{bmatrix} 1 \; 0 \cdots 0 \end{bmatrix}$, the second identity now reads $\begin{bmatrix} \alpha^\top B_0 \; \beta_1 \; \ldots \; \beta_{n_s-1} \end{bmatrix} = \lambda \begin{bmatrix} \beta_1 \; \ldots \; \beta_{n_s-1} \; 0 \end{bmatrix}$. Hence $\beta = 0$ while $\alpha^\top A_0 = \lambda\alpha^\top$ and $\alpha^\top B_0 = 0$. We deduce that $(\tilde{A}_1, \tilde{B}_1)$ satisfies the Kalman condition if and only if (A_0, B_0) satisfies the Kalman condition. This latter condition holds if and only if $b_n \neq 0$ for all $1 \leq n \leq N_0$.

Remark 6 Since the E^{N-N_0} dynamics described by (19c) is stable (recall that the matrix A_1 is Hurwitz) and decoupled from the rest of the modes, see (19), one can perform the stability analysis in two parts. Indeed, one can first complete the stability analysis as performed in the next subsections based on a truncated model similar to (21) but excluding the E^{N-N_0} dynamics. Then one can deduce the exponential stability of the full closed-loop system by invoking an exponential ISS argument for (19c). Such an approach leads to the derivation of reduced order LMIs compared to the ones presented in the next theorems. For sake of simplicity and conciseness, we focus the presentation on stability conditions derived from the full dynamics (21).

Finally, in preparation for the statement of the main results of this Chapter, we introduce a number of matrices that are useful to relate the different quantities previously introduced. Defining $E = \begin{bmatrix} I \; 0 \; 0 \; 0 \end{bmatrix}$, we have

$$\hat{Z}_a^{N_0} = EX. \tag{23}$$

Introducing $\tilde{E} = \begin{bmatrix} 0_{1\times N_0} \; 1 \; 0_{1\times(n_s-1)} \; 0_{1\times(2N-N_0)} \end{bmatrix}$ and $\tilde{G} = \begin{bmatrix} 1 \; 0_{1\times n_s} \end{bmatrix}$, we also have

$$u = \tilde{E}X, \quad \phi_{l_0}(u) = \tilde{G}\phi_\ell(\tilde{K}\hat{Z}_a^{N_0}),$$

hence

$$u_{\text{sat}} = u + \phi_{l_0}(u) = \tilde{E}X + \tilde{G}\phi_\ell(\tilde{K}\hat{Z}_a^{N_0}). \tag{24}$$

4.3 Main Stability Results

We now study the local exponential stability, while estimating the domain of attraction, of the closed-loop system composed of the plant (6) with measured output (7) and the control law (15). In that case, note that the control design constraints (8) will be structurally achieved due to the introduction of the saturation mechanisms (15c–15e) that are embedded into the control architecture and which constrain the magnitude of the n_s first time derivatives of the control input u.

4.3.1 Stability in L^2 Norm

We start by stating a stability result for PDE trajectories evaluated in L^2 norm.

Theorem 1 *Let $\theta_1, \theta_2 \in [0, \pi/2]$, $p \in C^1([0,1])$ with $p > 0$, $\tilde{q} \in C^0([0,1])$, $b, c \in L^2(0,1)$, and $\ell = (l_0, \dots, l_{n_s}) \in (\mathbb{R}_{>0})^{n_s+1}$. Let $q \in C^0([0,1])$ and $q_c \in \mathbb{R}$ be such that (12) holds. Consider the reaction-diffusion system described by (6) with measured output (7). Let $N_0 \geq 1$ and $\delta > 0$ be given such that $-\lambda_n + q_c < -\delta < 0$ for all $n \geq N_0 + 1$. Assume that $b_n \neq 0$ and $c_n \neq 0$ for all $1 \leq n \leq N_0$. Let $K \in \mathbb{R}^{1\times(N_0+n_s)}$ and $L \in \mathbb{R}^{N_0}$ be such that $\tilde{A}_1 + \tilde{B}_1 K$ and $A_0 - LC_0$ are Hurwitz with eigenvalues that have a real part strictly less than $-\delta < 0$. For a given $N \geq N_0 + 1$, assume that there exist a symmetric positive definite $P \in \mathbb{R}^{(2N+n_s)\times(2N+n_s)}$, $\alpha, \beta, \gamma, \mu, \kappa > 0$, a diagonal positive definite $T \in \mathbb{R}^{(n_s+1)\times(n_s+1)}$, and $C \in \mathbb{R}^{(n_s+1)\times(N_0+n_s)}$ such that*

$$\Theta_1(\kappa) \preceq 0, \quad \Theta_2 \succeq 0, \quad \Theta_3(\kappa) \leq 0 \tag{25}$$

where

$$\Theta_1(\kappa) = \begin{bmatrix} \Theta_{1,1,1}(\kappa) & P\mathcal{L} & -E^\top C^\top T + P\mathcal{L}_\phi \\ * & -\beta & 0 \\ * & * & \alpha\gamma\|\mathcal{R}_N b\|_{L^2}^2 \tilde{G}^\top \tilde{G} - 2T \end{bmatrix}$$

$$\Theta_2 = \begin{bmatrix} P & E^\top(\tilde{K} - C)^\top \\ * & \mu \operatorname{diag}(\ell)^2 \end{bmatrix},$$

$$\Theta_3(\kappa) = 2\gamma\left\{-\lambda_{N+1} + q_c + \kappa + \frac{1}{\alpha}\right\} + \beta\|\mathcal{R}_N c\|_{L^2}^2$$

with $\Theta_{1,1,1}(\kappa) = F^\top P + PF + 2\kappa P + \alpha\gamma\|\mathcal{R}_N b\|_{L^2}^2 \tilde{E}^\top \tilde{E}$. Consider the block representation $P = (P_{i,j})_{1\leq i,j\leq 5}$ with dimensions that are compatible with (17) and (20) and define

$$\mathcal{E}_1 = \left\{ (z, \mathcal{V}) \in L^2(0,1) \times \mathbb{R}^{n_s} : \right.$$
$$\left. \begin{bmatrix} \mathcal{V} \\ \pi_{N_0} z \\ \pi_{N_0+1,N} z \end{bmatrix}^\top \begin{bmatrix} P_{2,2} & P_{2,3} & P_{2,5} \\ P_{3,2} & P_{3,3} & P_{3,5} \\ P_{5,2} & P_{5,3} & P_{5,5} \end{bmatrix} \begin{bmatrix} \mathcal{V} \\ \pi_{N_0} z \\ \pi_{N_0+1,N} z \end{bmatrix} + \gamma\|\mathcal{R}_N z\|_{L^2}^2 \leq \frac{1}{\mu} \right\}. \tag{26}$$

Then, considering the closed-loop system composed of the plant (6) with measured output (7) and the control law (15), there exists $M > 0$ such that for any initial condition $(z_0, \mathcal{V}_0) \in \mathcal{E}_1$ and with a zero initial condition of the observer (i.e., $\hat{z}_n(0) = 0$ for all $1 \leq n \leq N$), the system trajectory satisfies

$$\|z(t,\cdot)\|_{L^2}^2 + \sum_{n=1}^{N} \hat{z}_n(t)^2 + \|\mathcal{V}(t)\|^2 \leq M e^{-2\kappa t}\left(\|z_0\|_{L^2}^2 + \|\mathcal{V}_0\|^2\right) \tag{27}$$

for all $t \geq 0$. Moreover, for any fixed $\kappa \in (0, \delta]$, the constraints (25) are always feasible for N selected large enough.

Proof Let the Lyapunov function candidate be defined by

$$V(X, z) = X^\top P X + \gamma \sum_{n \geq N+1} \langle z, \phi_n \rangle^2, \quad \forall X \in \mathbb{R}^{2N+n_s}, \ \forall z \in L^2(0, 1).$$

The computation of the time derivative of V along the system trajectories (14) and (21) for classical solutions gives

$$\begin{aligned} \dot{V} + 2\kappa V = {} & X^\top (F^\top P + PF + 2\kappa P) X + 2X^\top P\mathcal{L}\zeta + 2X^\top P\mathcal{L}_\phi \phi_\ell(\tilde{K}\hat{Z}_a^{N_0}) \\ & + 2\gamma \sum_{n \geq N+1} (-\lambda_n + q_c + \kappa) z_n^2 + 2\gamma \sum_{n \geq N+1} z_n b_n u_{\mathrm{sat}}. \end{aligned}$$

From (24) and using Young's inequality, we infer that

$$\begin{aligned} & 2 \sum_{n \geq N+1} z_n b_n u_{\mathrm{sat}} \\ & = 2 \sum_{n \geq N+1} z_n b_n \tilde{E} X + 2 \sum_{n \geq N+1} z_n b_n \tilde{G} \phi_\ell(\tilde{K}\hat{Z}_a^{N_0}) \\ & \leq \frac{2}{\alpha} \sum_{n \geq N+1} z_n^2 + \alpha \|\mathcal{R}_N b\|_{L^2}^2 \left\{ X^\top \tilde{E}^\top \tilde{E} X + \phi_\ell(\tilde{K}\hat{Z}_a^{N_0})^\top \tilde{G}^\top \tilde{G} \phi_\ell(\tilde{K}\hat{Z}_a^{N_0}) \right\}. \end{aligned}$$

Moreover, recalling that $\zeta = \sum_{n \geq N+1} c_n z_n$, we infer from Cauchy-Schwarz inequality that

$$\zeta^2 \leq \|\mathcal{R}_N c\|_{L^2}^2 \sum_{n \geq N+1} z_n^2. \tag{28}$$

The two latter estimates imply that

$$\begin{aligned} & \dot{V} + 2\kappa V \\ & \leq \begin{bmatrix} X \\ \zeta \\ \phi_\ell(\tilde{K}\hat{Z}_a^{N_0}) \end{bmatrix}^\top \begin{bmatrix} \Theta_{1,1,1}(\kappa) & P\mathcal{L} & P\mathcal{L}_\phi \\ * & -\beta & 0 \\ * & * & \alpha\gamma \|\mathcal{R}_N b\|_{L^2}^2 \tilde{G}^\top \tilde{G} \end{bmatrix} \begin{bmatrix} X \\ \zeta \\ \phi_\ell(\tilde{K}\hat{Z}_a^{N_0}) \end{bmatrix} + \sum_{n \geq N+1} \Gamma_n z_n^2 \end{aligned}$$

where $\Gamma_n = 2\gamma\left(-\lambda_n + q_c + \kappa + \frac{1}{\alpha}\right) + \beta \|\mathcal{R}_N c\|_{L^2}^2$. Invoking now Lemma 1, we know that

$$\forall \hat{Z}_a^{N_0} \in \mathbb{R}^{N_0+n_s}, \quad |(\tilde{K} - C)\hat{Z}_a^{N_0}| \leq \ell \Rightarrow \phi_\ell(\tilde{K}\hat{Z}_a^{N_0})^\top T(\phi_\ell(\tilde{K}\hat{Z}_a^{N_0}) + C\hat{Z}_a^{N_0}) \leq 0.$$

Owing to (23), we obtain for any $X \in \mathbb{R}^{2N+n_s}$ with $|(\tilde{K}-C)EX| \leq \ell$ that

$$\dot{V} + 2\kappa V \leq \begin{bmatrix} X \\ \zeta \\ \phi_\ell(\tilde{K}\hat{Z}_a^{N_0}) \end{bmatrix}^\top \Theta_1(\kappa) \begin{bmatrix} X \\ \zeta \\ \phi_\ell(\tilde{K}\hat{Z}_a^{N_0}) \end{bmatrix} + \sum_{n \geq N+1} \Gamma_n z_n^2.$$

Since $\Gamma_n \leq \Theta_3(\kappa)$ for all $n \geq N+1$, we infer from (25) that

$$\forall X \in \mathbb{R}^{2N+n_s}, \quad |(\tilde{K}-C)EX| \leq \ell \Rightarrow \dot{V} + 2\kappa V \leq 0.$$

Invoking now the Schur complement, the condition $\Theta_2 \succeq 0$ from (25) implies that $P \succeq \frac{1}{\mu} E^\top (\tilde{K}-C)^\top \mathrm{diag}(\ell)^{-2} (\tilde{K}-C)E$. Hence, for any $X \in \mathbb{R}^{2N+n_s}$ and $z \in L^2(0,1)$ so that $V(X,z) \leq 1/\mu$, we have $\|\mathrm{diag}(\ell)^{-1}(\tilde{K}-C)EX\| \leq 1$ hence $|(\tilde{K}-C)EX| \leq \ell$. Combining this result with the one of the previous paragraph, we deduce that

$$\forall X \in \mathbb{R}^{2N+n_s}, \ \forall z \in L^2(0,1), \quad V(X,z) \leq \frac{1}{\mu} \Rightarrow \dot{V} + 2\kappa V \leq 0.$$

Consider now an initial condition $(z_0, \mathcal{V}_0) \in \mathcal{E}_1$ with $z_0 \in D(\mathcal{A})$. Recalling that the initial condition of the observer is set equal to zero we note that $V(X(0), z_0) \leq 1/\mu$. If $(z_0, \mathcal{V}_0) = 0$, the trajectory is identically zero. Otherwise, for $(z_0, \mathcal{V}_0) \neq 0$ we have that $\dot{V}(X(0), z_0) \leq -2\kappa V(X(0), z_0) < 0$. We infer from a classical contradiction argument that $V(X(t), z(t,\cdot)) \leq 1/\mu$ for all $t \geq 0$. This implies from the above developments that $\dot{V}(X(t), z(t,\cdot)) + 2\kappa V(X(t), z(t,\cdot)) \leq 0$ for all $t \geq 0$ hence $V(X(t), z(t,\cdot)) \leq e^{-2\kappa t} V(X(0), z_0)$. The exponential stability estimate (27) immediately follows for classical solutions. The result for any $(z_0, \mathcal{V}_0) \in \mathcal{E}_1$ is obtained by invoking the concept of mild solutions and by using a classical density argument [36, Theorem 6.1.2].

For any fixed $\kappa \in (0, \delta]$, the feasibility of the constraints (25) for N selected large enough can be obtained using the same arguments as the ones in [19].

4.3.2 Stability in H^1 Norm

We also state a stability result for PDE trajectories evaluated in H^1 norm.

Theorem 2 *Let $\theta_1, \theta_2 \in [0, \pi/2]$, $p \in C^1([0,1])$ with $p > 0$, $\tilde{q} \in C^0([0,1])$, $b, c \in L^2(0,1)$, and $\ell = (l_0, \ldots, l_{n_s}) \in (\mathbb{R}_{>0})^{n_s+1}$. Let $q \in C^0([0,1])$ and $q_c \in \mathbb{R}$ be such that (12) holds with the further constraint $q > 0$. Consider the reaction-diffusion system described by (6) with measured output (7). Let $N_0 \geq 1$ and $\delta > 0$ be given such that $-\lambda_n + q_c < -\delta < 0$ for all $n \geq N_0 + 1$. Assume that $b_n \neq 0$ and $c_n \neq 0$ for all $1 \leq n \leq N_0$. Let $K \in \mathbb{R}^{1 \times (N_0 + n_s)}$ and $L \in \mathbb{R}^{N_0}$ be such that $\tilde{A}_1 + \tilde{B}_1 K$ and $A_0 - LC_0$ are Hurwitz with eigenvalues that have a real part strictly less than $-\delta < 0$. For a given $N \geq N_0 + 1$, assume that there exist a symmetric positive definite $P \in \mathbb{R}^{(2N+n_s)\times(2N+n_s)}$, $\alpha > 1$, $\beta, \gamma, \mu, \kappa > 0$, a diagonal positive definite*

$T \in \mathbb{R}^{(n_s+1)\times(n_s+1)}$, and $C \in \mathbb{R}^{(n_s+1)\times(N_0+n_s)}$ such that

$$\Theta_1(\kappa) \preceq 0, \quad \Theta_2 \succeq 0, \quad \Theta_3(\kappa) \leq 0 \tag{29}$$

where $\Theta_1(\kappa)$ and Θ_2 are defined as in Theorem 1 while

$$\Theta_3(\kappa) = 2\gamma\left\{-\left(1-\frac{1}{\alpha}\right)\lambda_{N+1} + q_c + \kappa\right\} + \frac{\beta\|\mathcal{R}_N c\|_{L^2}^2}{\lambda_{N+1}}.$$

Consider the block representation $P = (P_{i,j})_{1\leq i,j\leq 5}$ with dimensions that are compatible with (17) and (20) and define

$$\mathcal{E}_2 = \left\{(z, \mathcal{V}) \in D(\mathcal{A}) \times \mathbb{R}^{n_s} : \right.$$
$$\left.\begin{bmatrix} \mathcal{V} \\ \pi_{N_0} z \\ \pi_{N_0+1,N} z \end{bmatrix}^\top \begin{bmatrix} P_{2,2} & P_{2,3} & P_{2,5} \\ P_{3,2} & P_{3,3} & P_{3,5} \\ P_{5,2} & P_{5,3} & P_{5,5} \end{bmatrix} \begin{bmatrix} \mathcal{V} \\ \pi_{N_0} z \\ \pi_{N_0+1,N} z \end{bmatrix} + \gamma\|\mathcal{R}_N \mathcal{A}^{1/2} z\|_{L^2}^2 \leq \frac{1}{\mu}\right\}. \tag{30}$$

Then, considering the closed-loop system composed of the plant (6) with measured output (7) and the control law (15), there exists $M > 0$ such that for any initial condition $(z_0, \mathcal{V}_0) \in \mathcal{E}_2$ and with a zero initial condition of the observer (i.e., $\hat{z}_n(0) = 0$ for all $1 \leq n \leq N$), the system trajectory satisfies

$$\|z(t,\cdot)\|_{H^1}^2 + \sum_{n=1}^{N} \hat{z}_n(t)^2 + \|\mathcal{V}(t)\|^2 \leq Me^{-2\kappa t}\left(\|z_0\|_{H^1}^2 + \|\mathcal{V}_0\|^2\right) \tag{31}$$

for all $t \geq 0$. Moreover, for any fixed $\kappa \in (0, \delta]$, the constraints (29) are always feasible for N selected large enough.

Proof Let the Lyapunov function candidate be defined by

$$V(X, z) = X^\top P X + \gamma \sum_{n\geq N+1} \lambda_n \langle z, \phi_n\rangle^2, \quad \forall X \in \mathbb{R}^{2N+n_s}, \ \forall z \in D(\mathcal{A}).$$

The series is connected to the H^1 norm of z via (10) because we assumed that $q > 0$. The computation of the time derivative of V along the system trajectories (14) and (21) for classical solutions gives

$$\dot{V} + 2\kappa V = X^\top(F^\top P + PF + 2\kappa P)X + 2X^\top P\mathcal{L}\zeta + 2X^\top P\mathcal{L}_\phi \phi_\ell(\tilde{K}\hat{Z}_a^{N_0})$$
$$+ 2\gamma \sum_{n\geq N+1} \lambda_n(-\lambda_n + q_c + \kappa) z_n^2 + 2\gamma \sum_{n\geq N+1} \lambda_n z_n b_n u_{\text{sat}}.$$

From (24) and using Young's inequality, we infer that

$$
\begin{aligned}
&2 \sum_{n \geq N+1} \lambda_n z_n b_n u_{\mathrm{sat}} \\
&= 2 \sum_{n \geq N+1} \lambda_n z_n b_n \tilde{E} X + 2 \sum_{n \geq N+1} \lambda_n z_n b_n \tilde{G} \phi_\ell(\tilde{K} \hat{Z}_a^{N_0}) \\
&\leq \frac{2}{\alpha} \sum_{n \geq N+1} \lambda_n^2 z_n^2 + \alpha \|\mathcal{R}_N b\|_{L^2}^2 \left\{ X^\top \tilde{E}^\top \tilde{E} X + \phi_\ell(\tilde{K} \hat{Z}_a^{N_0})^\top \tilde{G}^\top \tilde{G} \phi_\ell(\tilde{K} \hat{Z}_a^{N_0}) \right\}.
\end{aligned}
$$

Combining the above estimate and (28), we obtain that

$$
\begin{aligned}
&\dot{V} + 2\kappa V \\
&\leq \begin{bmatrix} X \\ \zeta \\ \phi_\ell(\tilde{K} \hat{Z}_a^{N_0}) \end{bmatrix}^\top \begin{bmatrix} \Theta_{1,1,1}(\kappa) & P\mathcal{L} & P\mathcal{L}_\phi \\ * & -\beta & 0 \\ * & * & \alpha\gamma \|\mathcal{R}_N b\|_{L^2}^2 \tilde{G}^\top \tilde{G} \end{bmatrix} \begin{bmatrix} X \\ \zeta \\ \phi_\ell(\tilde{K} \hat{Z}_a^{N_0}) \end{bmatrix} + \sum_{n \geq N+1} \lambda_n \Gamma_n z_n^2.
\end{aligned}
$$

where $\Gamma_n = 2\gamma \left\{ -\left(1 - \frac{1}{\alpha}\right) \lambda_n + q_c + \kappa \right\} + \frac{\beta \|\mathcal{R}_N c\|_{L^2}^2}{\lambda_n}$. Using again Lemma 1, we have that

$$
\forall \hat{Z}_a^{N_0} \in \mathbb{R}^{N_0+n_s}, \quad |(\tilde{K} - C)\hat{Z}_a^{N_0}| \leq \ell \Rightarrow \phi_\ell(\tilde{K} \hat{Z}_a^{N_0})^\top T (\phi_\ell(\tilde{K} \hat{Z}_a^{N_0}) + C \hat{Z}_a^{N_0}) \leq 0.
$$

In view of (23), we deduce that, for any $X \in \mathbb{R}^{2N+n_s}$ so that $|(\tilde{K} - C)EX| \leq \ell$,

$$
\dot{V} + 2\kappa V \leq \begin{bmatrix} X \\ \zeta \\ \phi_\ell(\tilde{K} \hat{Z}_a^{N_0}) \end{bmatrix}^\top \Theta_1(\kappa) \begin{bmatrix} X \\ \zeta \\ \phi_\ell(\tilde{K} \hat{Z}_a^{N_0}) \end{bmatrix} + \sum_{n \geq N+1} \lambda_n \Gamma_n z_n^2.
$$

Since $\alpha > 1$, we note that $\Gamma_n \leq \Theta_3(\kappa) \leq 0$ for all $n \geq N+1$. Using in addition the condition $\Theta_1(\kappa) \preceq 0$, we deduce that $\dot{V} + 2\kappa V \leq 0$ for any $X \in \mathbb{R}^{2N+n_s}$ such that $|(K - C)EX| \leq \ell$. The reminder of the proof follows the same arguments that the ones reported in the proof of Theorem 1 while using (10) for the evaluation of the PDE trajectories in H^1 norm.

4.4 *Comments on the Main Results*

4.4.1 Numerical Considerations

The design constraints from Theorems 1 and 2 are nonlinear w.r.t. the decision variables. However, following the same approach that the one reported in [19], these constraints can be specialized into tractable LMI conditions. These latter LMI conditions remain feasible provided the order of the observer is selected to be large enough. We refer the reader to [19] for details.

In the context of Theorem 1 (a similar remark applies in the context of Theorem 2), the ellipsoid $\mathcal{E}_1$ defined by (26) is a subset of the region of attraction of the origin. A possible approach for enlarging $\mathcal{E}_1$ is discussed in [19] (see also [18]) and goes as follows. For a given symmetric positive definite matrix $R \in \mathbb{R}^{(N+n_s+1)\times(N+n_s+1)}$, let $r > 0$ be such that

$$\mathcal{P} \triangleq \begin{bmatrix} P_{2,2} & P_{2,3} & P_{2,5} & 0 \\ P_{3,2} & P_{3,3} & P_{3,5} & 0 \\ P_{5,2} & P_{5,3} & P_{5,5} & 0 \\ 0 & 0 & 0 & \gamma \end{bmatrix} \preceq \frac{r}{\mu} R \tag{32}$$

along with the constraints (25) of Theorem 1. Note that since $R \succ 0$, (32) can always be obtained by selecting $r > 0$ sufficiently large. Under (32) we directly infer that

$$\left\{ (z, \mathcal{V}) \in L^2(0,1) \times \mathbb{R}^{n_s} : \begin{bmatrix} \mathcal{V} \\ \pi_{N_0} z \\ \pi_{N_0+1,N} z \\ \|\mathcal{R}_N z\|_{L^2} \end{bmatrix}^\top R \begin{bmatrix} \mathcal{V} \\ \pi_{N_0} z \\ \pi_{N_0+1,N} z \\ \|\mathcal{R}_N z\|_{L^2} \end{bmatrix} \leq \frac{1}{r} \right\} \subset \mathcal{E}_1 \tag{33}$$

Hence one can try to maximize the size of the ellipsoid $\mathcal{E}_1$ by minimizing the parameter $r > 0$ under the constraints (25).

It is also worth noting that the degrees of freedom available in the choice of the feedback gain K and the observer gain L can be used to enlarge the (estimation of the) domain of attraction. Indeed, fixing K and L as in Theorems 1–2, one can first determine a dimension N of the observer such that the constraints (25) of Theorem 1 or the constraints (29) of Theorem 2 are feasible. Then one can try to maximize the domain of attraction by considering successively: 1) K, L as decision variables and P as fixed; 2) K, L as fixed and P as a decision variable.

4.4.2 Extension to Boundary Measurements

We have discussed in this chapter the local stabilization of the reaction-diffusion PDE (6) under the saturation constraints (8) for a measurement taking the form of a bounded operator described by (7). The extension to unbounded measurement operators such as Dirichlet boundary measurement $y_D(t) = z(t, 0)$ and Neumann boundary measurement $y_N(t) = z_x(t, 0)$ in the case $p \in C^2([0, 1])$ can be achieved by adopting the procedures reported in [28, 29]. This was achieved in [19] in the case of the reaction-diffusion PDE (6) without constraints on the time derivatives of the control input. The extension to the setting of this chapter can be easily obtained by merging the approach reported in [19] with the arguments of the previous sections. Using arguments from [28, 29], the approach can also be extended to boundary control provided the saturation function is sufficiently smooth for well-posedness issues.

Table 1 Results of the estimation of the domain of attraction in L^2 and H^1 norms for different dimensions N of the observer

Dimension of the observer	$N = 3$	$N = 4$	$N = 5$	$N = 6$	$N = 7$
L^2 norm	$r = 4.8310$	$r = 3.7142$	$r = 3.5604$	$r = 3.4689$	$r = 3.4238$
H^1 norm	$r = 4.7536$	$r = 3.5864$	$r = 3.5352$	$r = 3.4695$	$r = 3.4237$

5 Numerical Illustration

We propose in this section a numerical illustration of Theorems 1 and 2. We consider the reaction-diffusion PDE (6) with Dirichlet boundary conditions ($\theta_1 = \theta_2 = 0$), $p = 1, \tilde{q} = -10, b(x) = 1_{[1/10,3/10]}(x)$, and $c(x) = 1_{[7/10,9/10]}(x)$. These parameters yield an open-loop unstable plant.

We consider the case of the saturation of the command input u and its first time derivative ($n_s = 1$) with saturation levels set as $l_0 = l_1 = 1$. The feedback and observer gains are set as $K = \begin{bmatrix} -4.3489 & -1.8304 \end{bmatrix}$ and $L = 31.3729$, respectively.

The constraints (25) of Theorem 1 and (29) of Theorem 2 are found feasible for $N = 3$, ensuring the exponential stability of the system trajectories in L^2 and H^1 norms.

In order to estimate the domain of attraction of the origin, we consider the additional constraint (32) with $R = \text{diag}(10, I, 0.005)$. The results obtained when trying to minimize the value of the parameter $r > 0$ for different dimensions N of the observer are summarized in Table 1.

For numerical simulations, we consider the dimension of the observer $N = 6$ and we set the initial condition of the reaction-diffusion PDE as $z_0(x) = 2.9x(x-1)$ while $\mathcal{V}_0 = 0$. From the above numerical results, $(z_0, \mathcal{V}_0)$ is checked to be an element of $\mathcal{E}_1 \cap \mathcal{E}_2$, hence is in the domain of attraction of the origin for system trajectories evaluated in both L^2 and H^1 norms. As shown in Figs. 1 and 2, the system trajectory converges exponentially to zero in both L^2 and H^1 norms, confirming the theoretical predictions of Theorems 1 and 2. Finally, Fig. 3 shows that this exponential decrease of the system trajectory is obtained despite significant saturations in both position and rate of change of the command input u.

6 Conclusion

This chapter discussed the output feedback stabilization of reaction-diffusion PDEs in the presence of saturations of the command input and its time derivatives. We provide in the process an estimation of the domain of attraction. The presented results extend the approach reported in [19] dealing with the case of a sole input saturation. While the presentation focused on the case of measurements taking the form of a bounded

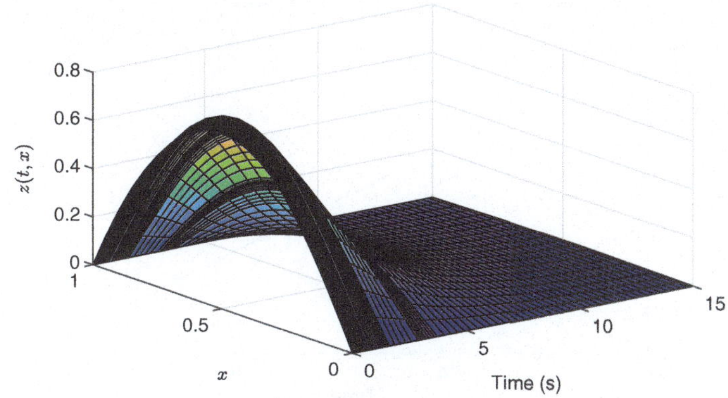

(a) State $z(t,x)$

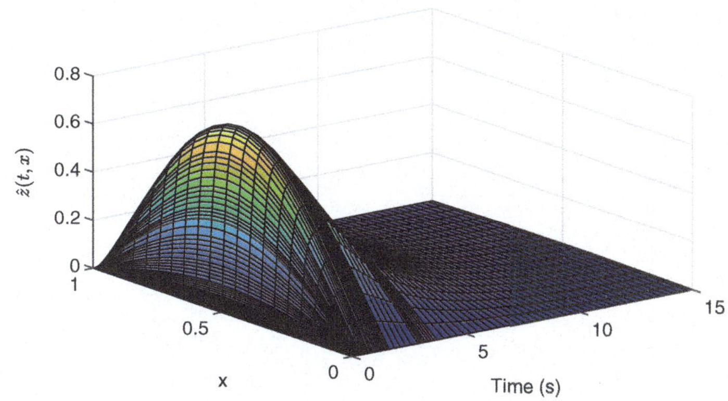

(b) Estimation $\hat{z}(t,x) = \sum_{n=1}^{N} \hat{z}_n(t)\phi_n(x)$

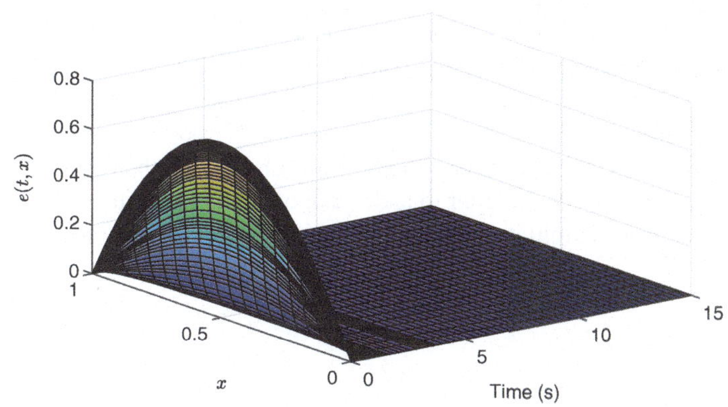

(c) Observation error $z(t,x) - \hat{z}(t,x)$

Fig. 1 Temporal behavior of the closed-loop system

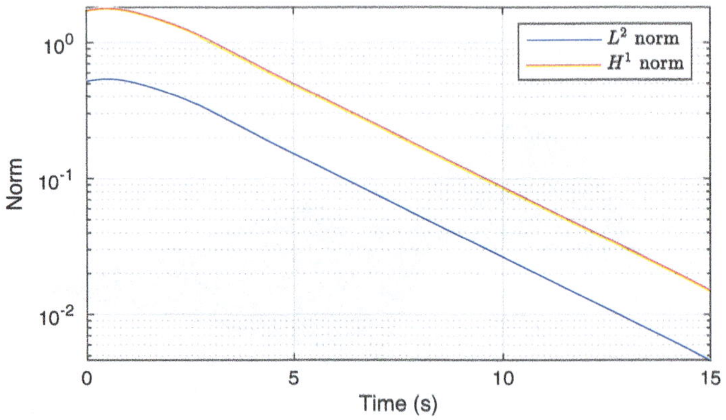

Fig. 2 Temporal behavior of the PDE trajectories in L^2 and H^1 norms

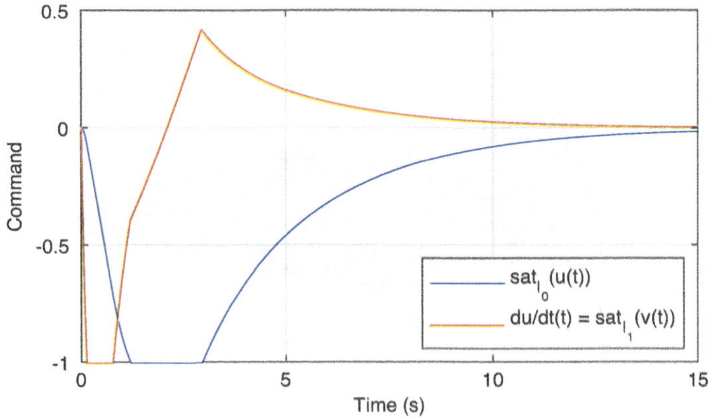

Fig. 3 Temporal behavior of the command input

observation operator, the results presented in this chapter can be combined with the approach described in [19] in order to tackle unbounded observation operators. This includes the cases of Dirchlet and Neumann boundary measurements.

References

1. Bernstein, D.S., Michel, A.N.: A chronological bibliography on saturating actuators (1995)
2. Campo, P.J., Morari, M.: Robust control of processes subject to saturation nonlinearities. Comput. Chem. Eng. **14**(4–5), 343–358 (1990)
3. Tarbouriech, S., Garcia, G., Gomes da Silva Jr, J.M., Queinnec, I.: Stability and stabilization of linear systems with saturating actuators. Springer Science & Business Media (2011)

4. Zaccarian, L., Teel, A.R.: Modern anti-windup synthesis: control augmentation for actuator saturation. Princeton University Press (2011)
5. Slemrod, M.: Feedback stabilization of a linear control system in Hilbert space with an a priori bounded control. Math. Control Signals Syst. **2**(3), 265–285 (1989)
6. Lasiecka, I., Seidman, T.I.: Strong stability of elastic control systems with dissipative saturating feedback. Syst. Control Lett. **48**(3–4), 243–252 (2003)
7. Prieur, C., Tarbouriech, S., Gomes da Silva Jr, J.M.: Wave equation with cone-bounded control laws. IEEE Trans. Autom. Control **61**(11), 3452–3463 (2016)
8. Marx, S., Andrieu, V., Prieur, C.: Cone-bounded feedback laws for m-dissipative operators on Hilbert spaces. Math. Control Signals Syst. **29**(4), 1–32 (2017)
9. Marx, S., Cerpa, E., Prieur, C., Andrieu, V.: Global stabilization of a Korteweg-de Vries equation with saturating distributed control. SIAM J. Control Optim. **55**(3), 1452–1480 (2017)
10. Dubljevic, S., El-Farra, N.H., Mhaskar, P., Christofides, P.D.: Predictive control of parabolic PDEs with state and control constraints. Int. J. Robust Nonlinear Control **16**(16), 749–772 (2006)
11. El-Farra, N.H., Armaou, A., Christofides, P.D.: Analysis and control of parabolic PDE systems with input constraints. Automatica **39**(4), 715–725 (2003)
12. Russell, D.L.: Controllability and stabilizability theory for linear partial differential equations: recent progress and open questions. SIAM Rev. **20**(4), 639–739 (1978)
13. Coron, J.-M., Trélat, E.: Global steady-state controllability of one-dimensional semilinear heat equations. SIAM J. Control Optim. **43**(2), 549–569 (2004)
14. Coron, J.-M., Trélat, E.: Global steady-state stabilization and controllability of 1D semilinear wave equations. Commun. Contemp. Math. **8**(4), 535–567 (2006)
15. Lhachemi, H., Prieur, C., Shorten, R.: An LMI condition for the robustness of constant-delay linear predictor feedback with respect to uncertain time-varying input delays. Automatica **109**, 108551 (2019)
16. Lhachemi, H., Prieur, C.: Feedback stabilization of a class of diagonal infinite-dimensional systems with delay boundary control. IEEE Trans. Autom. Control **66**(1), 105–120 (2021)
17. Lhachemi, H., Shorten, R., Prieur, C.: Exponential input-to-state stabilization of a class of diagonal boundary control systems with delay boundary control. Syst. Control Lett. **138**, 104651 (2020)
18. Mironchenko, A., Prieur, C., Wirth, F.: Local stabilization of an unstable parabolic equation via saturated controls. IEEE Trans. Autom. Control **66**(5), 2162–2176 (2021)
19. Lhachemi, H., Prieur, C.: Local output feedback stabilization of a reaction-diffusion equation with saturated actuation. IEEE Trans. Aut. Control, to appear (2023)
20. Curtain, R.: Finite-dimensional compensator design for parabolic distributed systems with point sensors and boundary input. IEEE Trans. Autom. Control **27**(1), 98–104 (1982)
21. Sakawa, Y.: Feedback stabilization of linear diffusion systems. SIAM J. Control Optim. **21**(5), 667–676 (1983)
22. Balas, M.J.: Finite-dimensional controllers for linear distributed parameter systems: exponential stability using residual mode filters. J. Math. Anal. Appl. **133**(2), 283–296 (1988)
23. Harkort, C., Deutscher, J.: Finite-dimensional observer-based control of linear distributed parameter systems using cascaded output observers. Int. J. Control **84**(1), 107–122 (2011)
24. Sano, H.: Stability-enhancing control of a coupled transport-diffusion system with Dirichlet actuation and Dirichlet measurement. J. Math. Anal. Appl. **388**(2), 1194–1204 (2012)
25. Grüne, L., Meurer, T.: Finite-dimensional output stabilization of linear diffusion-reaction systems–a small-gain approach (2021). arXiv:2104.06102
26. Katz, R., Fridman, E.: Constructive method for finite-dimensional observer-based control of 1-D parabolic PDEs. Automatica **122**, 109285 (2020)
27. Katz, R., Fridman, E.: Delayed finite-dimensional observer-based control of 1-D parabolic PDEs. Automatica **123**, 109364 (2021)
28. Lhachemi, H., Prieur, C.: Finite-dimensional observer-based boundary stabilization of reaction-diffusion equations with a either Dirichlet or Neumann boundary measurement. Automatica, **135**, 109955 (2022)

29. Lhachemi, H., Prieur, C.: Nonlinear boundary output feedback stabilization of reaction diffusion PDEs (2021). arXiv:2105.08418
30. Lhachemi, H., Prieur, C.: Finite-dimensional observer-based PI regulation control of a reaction-diffusion equation. IEEE Trans. Aut. Control, to appear (2022)
31. Lhachemi, H., Prieur, C.: Predictor-based output feedback stabilization of an input delayed parabolic PDE with boundary measurement. Automatica, **137**, 110115 (2022)
32. Lhachemi, H., Prieur, C.: Boundary Output Feedback Stabilization of Reaction-Diffusion PDEs with Delayed Boundary Measurement (2021). arXiv:2106.13637
33. Lhachemi, H., Shorten, R.: Boundary output feedback stabilization of state delayed reaction-diffusion PDEs (2021). arXiv:2105.15056
34. Orlov, Y.: On general properties of eigenvalues and eigenfunctions of a Sturm-Liouville operator: comments on "ISS with respect to boundary disturbances for 1-D parabolic PDEs". IEEE Trans. Autom. Control **62**(11), 5970–5973 (2017)
35. Curtain, R.F., Zwart, H.: An introduction to infinite-dimensional linear systems theory, vol. 21. Springer Science & Business Media (2012)
36. Pazy, A.: Semigroups of linear operators and applications to partial differential equations, vol. 44. Springer Science & Business Media (2012)

Numerical Control of a Semilinear Wave Equation on an Interval

Marcelo Cavalcanti, Valeria Domingos Cavalcanti, Carole Rosier, and Lionel Rosier

Abstract We are concerned with the numerical exact controllability of the semilinear wave equation on the interval $(0, 1)$. We introduce a Picard iterative scheme yielding a sequence of approximated solutions which converges towards a solution of the null controllability problem, provided that the initial data are small enough. The boundary control, which is applied at the endpoint $x = 1$, is taken in the space $H_0^1(0, T)$ for $T = 2$. For the linear part, the control input is obtained by imposing a transparent boundary condition at $x = 1$. Next, we provide several simulations to show the efficiency of the algorithm, using collocation pseudospectral methods on Chebychev grids to discretize the second order derivative in space in the wave equation.

1 Introduction

The issue of the exact controllability of the wave equation is well understood for a long time [4, 24]. Unfortunately, the numerical control of the wave equation turns out to be delicate, as the discrete control may not converge towards the continuous con-

The authors are partially supported by CAPES: n° 88881.520205/2020-01 and MATH AMSUD: 21-MATH-03.

M. Cavalcanti (✉) · V. D. Cavalcanti
Departamento de Matemática, Universidade Estadual de Maringá,
87020-900 Maringá, PR, Brazil
e-mail: mmcavalcanti@uem.br

V. D. Cavalcanti
e-mail: vndcavalcanti@uem.br

C. Rosier · L. Rosier
Laboratoire de Mathématiques Pures et Appliquées J. Liouville, Université du Littoral Côte d'Opale, 699, F-62228 Calais, France
e-mail: Carole.Rosier@univ-littoral.fr

L. Rosier
e-mail: Lionel.Rosier@univ-littoral.fr

J. Auriol et al. (eds.), *Advances in Distributed Parameter Systems*, Advances in Delays and Dynamics 14, https://doi.org/10.1007/978-3-030-94766-8_4

trol of the wave equation. This is due to the emergence of spurious high-frequencies oscillations for the discrete control problem. Several strategies have been proposed to face this problem: multigrid strategy, filtering of the high frequency modes, introduction of viscous terms, etc. We refer to [3, 8, 9, 14, 20–22, 26] for some works about the numerical control of the wave equation, and to [13, 27] for some surveys. See also [6, 11] for the numerics corresponding to the wave equation with damping. We refer also to [5] for the stabilization of hyperbolic systems, and to [1, 10] for the reduction of initial-boundary value problems for hyperbolic PDEs to initial-value problems for integro-differential-difference equations.

In most of the papers concerned with the theoretical or numerical control of the wave equation, the control strategy is based upon the Hilbert Uniqueness Method (HUM) introduced by J.-L. Lions. However, in dimension one there exists another strategy based upon the use of transparent boundary conditions leading to the finite-time stability of the system. This was first noticed by Majda in [19], and next used in [17, 22] to give explicit control inputs for the wave equation in dimension one. The transparent boundary conditions can be as well used for the finite-time stabilization of systems of conservation laws on networks [2, 15, 18, 23]. The idea is to let waves leave the domain in a natural way without bounce (or damping) at the boundary.

We shall see that the trajectory corresponding to the transparent boundary conditions is the same as those obtained by applying homogeneous Dirichlet conditions on a larger domain, thanks to the finite propagation speed of the wave equation. Thus, to design numerically the control, we can replace HUM (unstable) by a numerical method to solve a classical Cauchy problem for the wave equation (stable).

In this chapter, we are concerned with the exact boundary controllability of a semilinear wave equation

$$y_{tt} - y_{xx} = f(y)$$

on the interval $(0, 1)$ from a numerical perspective. By the time reversibility of the wave equation, it is sufficient to restrict ourselves to the null controllability problem. The aim of the chapter is to propose an algorithm at the continuous level which gives a sequence of controls $h^n \to h$ and a sequence of "trajectories" $y^n \to y$ such that, at the limit, y satisfies both

$$\begin{aligned} &y_{tt} - y_{xx} = f(y), \qquad (t, x) \in (0, T) \times (0, 1)\\ &y(t, 0) = 0, \ y(t, 1) = h(t), \qquad t \in (0, T)\\ &(y(0, .), y_t(0, .)) = (\xi^0, \xi^1) \end{aligned}$$

and

$$(y(T, .), y_t(T, .)) = (0, 0).$$

Here, (ξ^0, ξ^0) denotes any small initial data in some appropriate space, and the function f is locally Lipschitz continuous and it vanishes at 0. The state (y, y_t) and the control h will be taken more regular (see [12]) than in the linear case ($f \equiv 0$) in order to give a sense to the nonlinear term $f(y)$.

The control h^n is computed numerically by solving several linear control problems, and the state y^n, obtained by solving numerically an initial boundary-value problem, provides a numerical approximation of the trajectory y.

The chapter is organized as follows. In Sect. 2, we describe our iterative scheme and prove the convergence of both the trajectories and the control inputs in Theorem 1, the main result in the chapter. Section 3 is devoted to the description of the numerical approach used here and Sect. 4 provides several numerical simulations. Some concluding remarks are displayed in Sect. 5. Some technical lemmas needed to prove Theorem 1 are gathered in an appendix.

2 Picard Iteration Scheme

We first propose, at the continuous level, an exponentially convergent iterative scheme providing at the limit both the controlled trajectory and the control input for the semilinear wave equation. Due to obvious storage limitations, it is preferable to keep in the memory of the computer the control input $h(t)$ computed at the previous step than the corresponding trajectory $y(t,x)$. That choice, needed for a practical implementation of the algorithm, results in a scheme for which the convergence turns to be more complicated to establish than for a scheme based on the storage of the trajectory.

To describe the scheme, we need to introduce a few notations. For the sake of simplicity, we write

$$H_x^k = H^k(0,1)\ (k \in \mathbb{Z}), \quad L_x^2 = L^2(0,1), \quad H_t^1 = H^1(0,T)$$

and introduce the spaces

$$\mathcal{H}_0 := H_0^1(0,1) \times L^2(0,1) \subset \mathcal{H} := \{(y,z) \in H_x^1 \times L_x^2;\ y(0) = 0\}$$

endowed with the norm

$$||(y,z)||_{\mathcal{H}} = \left(||y_x||_{L_x^2}^2 + ||z||_{L_x^2}^2\right)^{\frac{1}{2}}.$$

Let $\left(S(t)\right)_{t\in\mathbb{R}}$ denote the unitary group on $\mathcal{H}_0$ generated by the operator $A(y,z) = (z, y_{xx})$ with domain

$$D(A) = (H^2(0,1) \cap H_0^1(0,1)) \times H_0^1(0,1) \subset \mathcal{H}_0.$$

Thus, for given $(\xi^0, \xi^1) \in \mathcal{H}_0$, if $(y(t), z(t)) = S(t)(\xi^0, \xi^1)$, then $z(t) = y_t(t)$ and y solves the system

$$y_{tt} - y_{xx} = 0, \quad (t, x) \in \mathbb{R} \times (0, 1) \tag{1}$$
$$y(t, 0) = y(t, 1) = 0, \quad t \in \mathbb{R} \tag{2}$$
$$(y(0, .), y_t(0, .)) = (\xi^0, \xi^1). \tag{3}$$

Pick $\delta_0 > 0$ and $f \in W^{1,\infty}(-\delta_0, \delta_0)$ such that

$$f(0) = 0, \text{ and } ||f'||_{L^\infty(-\delta_0,\delta_0)} \leq \eta < \eta_0 < \frac{1}{2}. \tag{4}$$

Let finally $\Lambda : \mathcal{H}_0 \to H_0^1(0, T)$ denote a continuous map such that for any given $(\xi^{0,T}, \xi^{1,T}) \in \mathcal{H}_0$, the solution y of the controlled system

$$y_{tt} - y_{xx} = 0, \quad (t, x) \in (0, T) \times (0, 1) \tag{5}$$
$$y(t, 0) = 0, \; y(t, 1) = h(t) := \Lambda(\xi^{0,T}, \xi^{1,T})(t), \quad t \in (0, T) \tag{6}$$
$$(y(0, .), y_t(0, .)) = (0, 0) \tag{7}$$

satisfies

$$(y(T, .), y_t(T, .)) = (\xi^{0,T}, \xi^{1,T}).$$

From well-known results, Λ is well-defined and continuous from $L_x^2 \times H_x^{-1}$ into $L^2(0, T)$ if $T \geq 2$, but Λ is also well-defined and continuous from $\mathcal{H}_0$ into $H_0^1(0, T)$ (see e.g. [12]). (Note that $h(0) = 0$ by (7) and $h(T) = \xi^{0,T}(1) = 0$, for $\xi^{0,T} \in H_0^1(0, 1)$.) Thus, there is some constant $C_\Lambda > 0$ such that

$$||h||_{H_t^1} \leq C_\Lambda ||(\xi^{0,T}, \xi^{1,T})||_{\mathcal{H}} \tag{8}$$

Actually Λ may be constructed explicitly by using d'Alembert's formula, and this is precisely the numerical strategy used here.

We are in a position to describe our Picard iterative scheme. Pick $T = 2$ and let $(\xi^0, \xi^1) \in \mathcal{H}_0$.

1. FIRST ITERATION.

We set $h^1(t) = 0$ for $t \in [0, T]$. Let y_1^1 solve the nonlinear system

$$y_{1,tt}^1 - y_{1,xx}^1 = f(y_1^1), \quad (t, x) \in (0, T) \times (0, 1) \tag{9}$$
$$y_1^1(t, 0) = 0, \; y_1^1(t, 1) = h^1(t) = 0, \quad t \in (0, T) \tag{10}$$
$$(y_1^1(0, .), y_{1t}^1(0, .)) = (\xi^0, \xi^1). \tag{11}$$

Next, we set

$$(\delta h)^1 := -\Lambda(y_1^1(T, .), y_{1t}^1(T, .)), \tag{12}$$

and introduce the solution of the linear system

$$y^1_{2,tt} - y^1_{2,xx} = 0, \quad (t,x) \in (0,T) \times (0,1) \tag{13}$$
$$y^1_2(t,0) = 0, \ \ y^1_2(t,1) = (\delta h)^1(t), \quad t \in (0,T) \tag{14}$$
$$(y^1_2(0,.), y^1_{2t}(0,.)) = (0,0). \tag{15}$$

Set $y^1 := y^1_1 + y^1_2$. Then, by construction, y^1 solves

$$y^1_{tt} - y^1_{xx} = f(y^1_1), \quad (t,x) \in (0,T) \times (0,1) \tag{16}$$
$$y^1(t,0) = 0, \ \ y^1(t,1) = h^1(t) + (\delta h)^1(t), \quad t \in (0,T) \tag{17}$$
$$(y^1(0,.), y^1_t(0,.)) = (\xi^0, \xi^1), \tag{18}$$

and it satisfies

$$(y^1(T,.), y^1_t(T,.)) = (0,0). \tag{19}$$

Thus, $(\delta h)^1$ is a correction term added to h^1 in order that (19) be satisfied.

2. nTH ITERATION ($n \geq 2$).

Assume that $h^{n-1} \in H^1_0(0,T)$ and $(\delta h)^{n-1} \in H^1_0(0,T)$ have been constructed. Set

$$h^n(t) := h^{n-1}(t) + (\delta h)^{n-1}(t), \tag{20}$$

and let y^n_1 denote the solution of the nonlinear system

$$y^n_{1,tt} - y^n_{1,xx} = f(y^n_1), \quad (t,x) \in (0,T) \times (0,1) \tag{21}$$
$$y^n_1(t,0) = 0, \ \ y^n_1(t,1) = h^n(t), \quad t \in (0,T) \tag{22}$$
$$(y^n_1(0,.), y^n_{1t}(0,.)) = (\xi^0, \xi^1). \tag{23}$$

Next, set

$$(\delta h)^n = -\Lambda(y^n_1(T,.), y^n_{1t}(T,.)), \tag{24}$$

and let y^n_2 denote the solution of the linear system

$$y^n_{2,tt} - y^n_{2,xx} = 0, \quad (t,x) \in (0,T) \times (0,1) \tag{25}$$
$$y^n_2(t,0) = 0, \ \ y^n_2(t,1) = (\delta h)^n(t), \quad t \in (0,T) \tag{26}$$
$$(y^n_2(0,.), y^n_{2t}(0,.)) = (0,0). \tag{27}$$

Let

$$y^n := y^n_1 + y^n_2. \tag{28}$$

Then y^n satisfies

$$y^n_{tt} - y^n_{xx} = f(y^n_1), \quad (t,x) \in (0,T) \times (0,1) \tag{29}$$
$$y^n(t,0) = 0, \ y^n(t,1) = h^n(t) + (\delta h)^n(t), \quad t \in (0,T) \tag{30}$$
$$(y^n(0,.), y^n_t(0,.)) = (\xi^0, \xi^1), \tag{31}$$
$$(y^n(T,.), y^n_t(T,.)) = (0,0). \tag{32}$$

We are in a position to state the convergence result for the Picard iterative scheme at the continuous level.

Theorem 1 *Let $T = 2$. Then there exist some constants $\delta \in (0, \delta_0)$, $\varepsilon \in (0,1)$ and $C > 0$ such that for any $(\xi^0, \xi^1) \in \mathcal{H}_0$ with*

$$||(\xi^0, \xi^1)||_{\mathcal{H}} < \delta \tag{33}$$

the sequence $(y^n_1, y^n_2, h^n, (\delta h)^n)_{n\geq 1}$ is well defined in $C([0,T], H^1_x)^2 \times (H^1_t)^2$ and we have

$$||(y^n_1 - y, y^n_{1,t} - y_t)||_{C([0,T],\mathcal{H})} + ||(y^n_2, y^n_{2,t})||_{C([0,T],\mathcal{H})} + ||h^n - h||_{H^1_t} + ||(\delta h)^n||_{H^1_t} \leq C\varepsilon^n, \tag{34}$$

where $(y, y_t, h) \in C([0,T], \mathcal{H}) \times H^1_0(0,T)$ satisfies

$$y_{tt} - y_{xx} = f(y), \quad (t,x) \in (0,T) \times (0,1) \tag{35}$$
$$y(t,0) = 0, \ y(t,1) = h(t), \quad t \in (0,T) \tag{36}$$
$$(y(0,.), y_t(0,.)) = (\xi^0, \xi^1), \tag{37}$$

and

$$(y(T,.), y_t(T,.)) = (0,0). \tag{38}$$

Proof We need two lemmas whose proofs are postponed in appendix.

Lemma 1 (Estimates for the linear system) *There exists a constant $C_1 > 0$ such that for any $(z^0, z^1) \in \mathcal{H}_0$, any $h \in H^1_t$ with $h(0) = 0$, and any $g \in L^1(0,T,L^2_x)$, the solution of the system*

$$z_{tt} - z_{xx} = g, \quad (t,x) \in (0,T) \times (0,1) \tag{39}$$
$$z(t,0) = 0, \ z(t,1) = h(t), \quad t \in (0,T) \tag{40}$$
$$(z(0,.), z_t(0,.)) = (z^0, z^1) \tag{41}$$

fulfills

$$||(z, z_t)||_{C([0,T],\mathcal{H})} \leq ||(z^0, z^1)||_{\mathcal{H}} + ||g||_{L^1(0,T,L^2_x)} + C_1||h||_{H^1_t}. \tag{42}$$

Let $(z, z_t) = W(h)$ denote the solution of (39)–(41) for $h \in H^1(0,T)$ with $h(0) = 0$ and $(z^0, z^1, g) = (0,0,0)$.

Lemma 2 (Estimate for the nonlinear problem) *Let C_1 be as in Lemma 1. Then there exist some constants $\delta_1 \in (0, \delta_0)$, $R > 0$ and $C_2 > 0$ such that for $(z^0, z^1, h) \in \mathcal{H}_0 \times H_t^1$ with $h(0) = 0$ and*

$$||(z^0, z^1)||_{\mathcal{H}} + C_1||h||_{H_t^1} \le \delta_1, \tag{43}$$

the map

$$\Gamma : (z, \tilde{z}) \to S(t)(z^0, z^1) + W(h) + \int_0^t S(t-s)(0, f(z(s)))ds \tag{44}$$

has a fixed-point $(z, \tilde{z})$ in the ball $B(0, R) \subset C([0, T], \mathcal{H})$. Furthermore, $\tilde{z} = z_t$, and z solves the nonlinear system

$$z_{tt} - z_{xx} = f(z), \qquad (t, x) \in (0, T) \times (0, 1) \tag{45}$$

$$z(t, 0) = 0, \ z(t, 1) = h(t), \qquad t \in (0, T) \tag{46}$$

$$(z(0, .), z_t(0, .)) = (z^0, z^1), \tag{47}$$

and we have

$$||(z, z_t)||_{C([0,T],\mathcal{H})} \le C_2 \left(||(z^0, z^1)||_{\mathcal{H}} + C_1||h||_{H_t^1} \right). \tag{48}$$

□

The proof of Theorem 1 is done in two steps. In the first step, we show by induction on n that $||h^n||_{H_t^1}$ is uniformly bounded, and that for some $\varepsilon \in (0, 1)$

$$||(\delta h)^{n+1}||_{H_t^1} \le \varepsilon||(\delta h)^n||_{H_t^1}.$$

In the second step, we combine the above estimates to prove the exponential convergence of all the sequences. Pick any $\delta \in (0, \delta_1)$ and set (for notational convenience) $(\delta h)^0(t) := \sin(\pi t/T)$. Let

$$\varepsilon := \frac{2C_\Lambda C_1}{\pi - 2\eta}\eta, \qquad C_3 = \frac{C_\Lambda C_2}{1 - \varepsilon}. \tag{49}$$

where C_Λ is as in (8). We have $\varepsilon < 1$ and $C_3 > 0$ if $0 < \eta < \eta_0 < \frac{1}{2}$ with η_0 *small enough.*

STEP 1. ESTIMATES OF THE CONTROL INPUTS.

Proposition 1 *Let ε and C_3 be as in (49). Then we have*

$$||h^n||_{H_t^1} \le C_3\delta, \qquad \forall n \ge 1, \tag{50}$$

$$||(\delta h)^n||_{H_t^1} \le \varepsilon ||(\delta h)^{n-1}||_{H_t^1}, \qquad \forall n \ge 1. \tag{51}$$

Proof of Proposition 1. We proceed by induction on $n \ge 1$. Assume first that $n = 1$. Then (50) is obvious, for $h^1 = 0$. To prove (51), we estimate $(\delta h)^1$ in H_t^1 as follows:

$$\begin{aligned} ||(\delta h)^1||_{H_t^1} &= ||\Lambda(y_1^1(T,.), y_{1t}^1(T,.))||_{H_t^1} \\ &\le C_\Lambda ||(y_1^1(T,.), y_{1t}^1(T,.))||_{\mathcal{H}} \\ &\le C_\Lambda C_2 ||(\xi^0, \xi^1)||_{\mathcal{H}} \\ &\le C_\Lambda C_2 \delta \qquad (52) \\ &\le \varepsilon ||(\delta h)^0||_{H_t^1} \end{aligned}$$

if δ is small enough.

Assume that (50) and (51) are valid for any integer $n' < n$, where $n \ge 2$. Let us check that (50) and (51) are also true for n. Since $h^n = \sum_{k=1}^{n-1} (\delta h)^k$, we have that

$$\begin{aligned} ||h^n||_{H_t^1} &\le \sum_{k=1}^{n-1} \varepsilon^{k-1} ||(\delta h)^1||_{H_t^1} \\ &\le \frac{C_\Lambda C_2}{1-\varepsilon}\delta = C_3\delta, \end{aligned}$$

where we used (51) for $k < n$ and (52). Let us proceed to the proof of (51). Let $w^n := y^{n-1} - y_1^n = (y_1^{n-1} + y_2^{n-1}) - y_1^n$. Then w^n satisfies

$$\begin{aligned} &w_{tt}^n - w_{xx}^n = f(y_1^{n-1}) - f(y_1^n), \qquad (t,x) \in (0,T) \times (0,1) \\ &w^n(t,0) = 0, \ w^n(t,1) = 0, \qquad t \in (0,T) \\ &w^n(0,.) = w_t^n(0,.) = 0, \\ &(w^n(T,.), w_t^n(T,.)) = -(y_1^n(T), y_{1t}^n(T)). \qquad (53) \end{aligned}$$

Thus, by (24), (42) and (53), we have

$$\begin{aligned} ||(\delta h)^n||_{H_t^1} &\le C_\Lambda ||(y_1^n(T,.), y_{1t}^n(T,.))||_{\mathcal{H}} \\ &\le C_\Lambda ||(w^n, w_t^n)||_{C([0,T],\mathcal{H})} \\ &\le C_\Lambda ||f(y_1^{n-1}) - f(y_1^n)||_{L^1(0,T,L_x^2)} \\ &\le C_\Lambda \eta ||y_1^{n-1} - y_1^n||_{L^1(0,T,L_x^2)}. \qquad (54) \end{aligned}$$

For the last line, we used (4) and the estimate $\sup(||y_1^{n-1}||_{L_{t,x}^\infty}, ||y_1^n||_{L_{t,x}^\infty}) \le \delta_0$ valid if δ is small enough. Indeed, using (48) and (50) for $n-1$ and n,

$$\begin{aligned}||y_1^n||_{L^\infty((0,T)\times(0,1))} &\le ||(y_1^n)_x||_{L^\infty(0,T,L^2(0,1))}\\ &\le C_2\big(||(\xi^0,\xi^1)||_{\mathcal{H}} + C_1||h^n||_{H_t^1}\big)\\ &\le C_2\big(\delta + C_1C_3\delta\big)\\ &\le \delta_0\end{aligned}$$

and similarly $||y_1^{n-1}||_{L^\infty((0,T)\times(0,1))} \le \delta_0$, for δ small enough. It remains to estimate $||y_1^{n-1} - y_1^n||_{L^1(0,T,L_x^2)}$ in terms of $||(\delta h)^{n-1}||_{H_t^1}$. Let $v^n = y_1^{n-1} - y_1^n$. Then v^n solves

$$\begin{aligned}&v_{tt}^n - v_{xx}^n = f(y_1^{n-1}) - f(y_1^n), \qquad (t,x) \in (0,T)\times(0,1)\\ &v^n(t,0) = 0,\ v^n(t,1) = -(\delta h)^{n-1}(t), \qquad t \in (0,T)\\ &v^n(0,.) = v_t^n(0,.) = 0.\end{aligned} \tag{55}$$

An application of Lemma 1 yields

$$\begin{aligned}||(y_1^{n-1})_x - (y_1^n)_x||_{C([0,T],L_x^2)} &\le ||(y_1^{n-1} - y_1^n, y_{1,t}^{n-1} - y_{1,t}^n)||_{C([0,T],\mathcal{H})}\\ &\le ||f(y_1^{n-1}) - f(y_1^n)||_{L^1(0,T,L_x^2)} + C_1||(\delta h)^{n-1}||_{H_t^1}\\ &\le \eta||y_1^{n-1} - y_1^n||_{L^1(0,T,L_x^2)} + C_1||(\delta h)^{n-1}||_{H_t^1}.\end{aligned} \tag{56}$$

It follows from Poincaré's inequality $\|f\|_{L_x^2} \le \pi^{-1}\|f_x\|_{L_x^2}$ (valid for $f \in H_x^1$ with $f(0) = 0$) that

$$||y_1^{n-1} - y_1^n||_{L^1(0,T,L_x^2)} \le \frac{T}{\pi}||(y_1^{n-1})_x - (y_1^n)_x||_{C([0,T],L_x^2)} \tag{57}$$

which, combined with (56), gives

$$||y_1^{n-1} - y_1^n||_{L^1(0,T,L_x^2)} \le (1 - \frac{T\eta}{\pi})^{-1}\frac{C_1T}{\pi}||(\delta h)^{n-1}||_{H_t^1}. \tag{58}$$

Gathering together (54), (58) and the fact that $T = 2$, we conclude that

$$||(\delta h)^n||_{H_t^1} \le \varepsilon||(\delta h)^{n-1}||_{H_1^t}$$

with ε as in (49). The proof of Proposition 1 is complete.

STEP 2. EXPONENTIAL DECAY

Let us start with the convergence of the sequence of controls $(h^n)_{n\ge1}$. From (51) and (52), we infer that

$$||h^{n+1} - h^n||_{H_t^1} = ||(\delta h)^n||_{H_t^1} \le C_\Lambda C_2\delta\varepsilon^{n-1}, \qquad \forall n \ge 1, \tag{59}$$

so that for $n, p \ge 1$

$$||h^{n+p} - h^n||_{H_t^1} \leq C_\Lambda C_2 \delta (\varepsilon^{n-1} + \cdots + \varepsilon^{n+p-2}) \\ \leq \frac{C_\Lambda C_2 \delta}{1-\varepsilon} \varepsilon^{n-1}. \tag{60}$$

Thus $(h^n)_{n\geq 1}$ is a Cauchy sequence in H_t^1, and its limit $h \in H_t^1$ satisfies

$$||h - h^n||_{H_t^1} \leq \frac{C_\Lambda C_2 \delta}{1-\varepsilon} \varepsilon^{n-1} \tag{61}$$

by letting $p \to \infty$ in (60).

It follows from (25)–(27), (42), (51) and (52) that

$$||(y_2^n, y_{2,t}^n)||_{C([0,T],\mathcal{H})} \leq C_1 ||(\delta h)^n||_{H_t^1} \leq C_1 \varepsilon^{n-1} ||(\delta h)^1||_{H_t^1} \\ \leq C_\Lambda C_1 C_2 \delta \varepsilon^{n-1} =: C_4 \delta \varepsilon^{n-1},$$

and hence $(y_2^n, y_{2,t}^n) \to (0,0)$ in $C([0,T],\mathcal{H})$ as $n \to \infty$. From (55), (58) and (59), we infer that

$$||(y_1^{n+1} - y_1^n, y_{1,t}^{n+1} - y_{1,t}^n)||_{C([0,T],\mathcal{H})} \leq \big(C_1 + \eta(1 - \frac{T\eta}{\pi})^{-1} \frac{C_1 T}{\pi}\big) ||(\delta h)^n||_{H_t^1} \leq C_5 \delta \varepsilon^{n-1}$$

with $C_5 = \big(C_1 + \eta(1 - \frac{T\eta}{\pi})^{-1} \frac{C_1 T}{\pi}\big) C_\Lambda C_2$. This yields

$$||(y_1^{n+p} - y_1^n, y_{1,t}^{n+p} - y_{1,t}^n)||_{C([0,T],\mathcal{H})} \leq \frac{C_5 \delta}{1-\varepsilon} \varepsilon^{n-1},$$

so that the sequence $(y_1^n, y_{1,t}^n)$ converges towards some function (y, y_t) in $C([0,T],\mathcal{H})$ with $y(t,0) = 0$ and

$$||(y - y_1^n, y_t - y_{1,t}^n)||_{C([0,T],\mathcal{H})} \leq \frac{C_5 \delta}{1-\varepsilon} \varepsilon^{n-1}.$$

Thus (34) is proved for some constant $C > 0$.

It remains to check that the pair (y, h) satisfies (35)–(38). Equations (35)–(37) follow from (21)–(23) and (34), while (38) follows from (28), (32) and (34).

3 Numerical Schemes

Let us describe the numerical schemes used to compute y_1^n, y_2^n and $(\delta h)^n$. To simplify the notations, we remove the index n (corresponding to the nth Picard iteration) in y_1^n, y_2^n and $(\delta h)^n$. We focus on the computation of y_1 and h, as y_2 is not needed to approximate y.

3.1 Chebychev Collocation Spectral Method

We have to solve numerically several times the linear (or semilinear) wave equation on a bounded domain. After an obvious change of spatial variables, we can assume that $x \in (-1, 1)$, incorporating a coefficient c in the wave operator. Consider the Cauchy problem for the semilinear wave equation:

$$u_{tt} - c^2 u_{xx} = g(t, x, u), \quad (t, x) \in (0, T) \times (-1, 1), \tag{62}$$

$$u(t, -1) = h_{-1}(t), \quad u(t, 1) = h_1(t), \quad t \in (0, T), \tag{63}$$

$$u(0, x) = u^0(x), \quad u_t(0, x) = u^1(x), \quad x \in (-1, 1), \tag{64}$$

where g, h_{-1}, h_1, u^0 and u^1 are given functions. In this chapter, we shall use the Chebychev collocation spectral method to solve (62)–(64), as described e.g. in [25]. Let $N \in \mathbb{N}^*$ be given. We introduce the Chebychev points $x_j = \cos(\theta_j)$, where $\theta_j = (j-1)\pi/N$ for $1 \le j \le N+1$. Note that $x_1 = 1$ and $x_{N+1} = -1$. Any smooth function $v = v(x)$ defined on $[-1, 1]$ is conveniently approximated by the Lagrange polynomial $p \in \mathbb{R}[x]$ of degree at most N defined by $p(x_i) = v(x_i)$ for $1 \le j \le N+1$. It turns that a good approximation of du/dx can be obtained by using the Chebychev differentiation matrix $D = (D_{ij})_{1 \le i,j \le N+1}$ defined as

$$D_{jj} = \sum_{\substack{k=1 \\ k \ne j}}^{N+1} (x_j - x_k)^{-1}, \quad 1 \le j \le N+1,$$

$$D_{ij} = \frac{1}{a_j} \sum_{\substack{k=1 \\ k \ne i,j}}^{N+1} (x_i - x_k), \quad 1 \le i \ne j \le N+1,$$

where $a_j = \sum_{\substack{k=1 \\ k \ne j}}^{N+1} (x_j - x_k)$ for $1 \le j \le N+1$. Indeed, introducing the vector $U = (U_j)_{1 \le j \le N+1}$ with $U_j = u(x_j)$ $(1 \le j \le N+1)$ and setting $\tilde{U} = DU$, $\tilde{U} = (\tilde{U}_j)_{1 \le j \le N+1}$, then we notice that the Lagrange polynomial q of degree $\le N$ defined by $q(x_j) = \tilde{U}_j$ for $1 \le j \le N+1$ provides a good approximation of du/dx; namely, we have for some constants $C > 0$ and $K > 1$

$$|\frac{du}{dx}(x_j) - \tilde{U}_j| \le CK^{-N}, \quad \forall N \ge 1, \ \forall j \in \{1, \ldots, N+1\}.$$

(See e.g. [7, 25].)

Let $M \in \mathbb{N}^*$ and $\Delta t = T/M$, where $T = 2$. For $1 \le j \le N+1$, $0 \le m \le M$, let U_j^m stand for the numerical approximation of $u(m\Delta t, x_j)$. To solve (62)–(64) numerically, we use a leap-frog formula in t and Chebychev spectral differentiation in x to get

$$(\Delta t)^{-2}(U_j^{m+1} - 2U_j^m + U_j^{m-1}) - c^2(D^2 U)_j = g(m\Delta t, x_j, U_j^m),$$
$$2 \le j \le N,\ 0 \le m \le M-1,$$
$$U_1^{m+1} = h_1((m+1)\Delta t),\ U_{N+1}^{m+1} = h_{-1}((m+1)\Delta t),\quad 0 \le m \le M-1,$$
$$U_j^0 = u^0(x_j),\ U_j^{-1} = u^0(x_j) - (\Delta t)u^1(x_j),\quad 2 \le j \le N.$$

3.2 Numerical Scheme for the Linear Control Problem

To compute $(\delta h)^n$ as in (24), we have to solve numerically the following linear control problem: for given $(y_1(T,.), y_{1,t}(T,.)) \in \mathcal{H}_0$, find $h \in H_0^1(0,T)$ such that the solution y_2 of the system

$$y_{2,tt} - y_{2,xx} = 0,\quad (t,x) \in (0,T) \times (0,1) \tag{65}$$
$$y_2(t,0) = 0,\ y_2(t,1) = h(t),\quad t \in (0,T) \tag{66}$$
$$y_2(0,x) = y_{2,t}(0,x) = 0,\quad x \in (0,1) \tag{67}$$

satisfies

$$y_2(T,x) = -y_1(T,x),\ y_{2,t}(T,x) = -y_{1,t}(T,x),\quad x \in (0,1). \tag{68}$$

The change of unknown function $z(t,x) = y_2(T-t,x)$ transforms the exact controllability problem (65)–(68) into the following null controllability problem

$$z_{tt} - z_{xx} = 0,\quad (t,x) \in (0,T) \times (0,1) \tag{69}$$
$$z(t,0) = 0,\ z(t,1) = h(T-t),\quad t \in (0,T) \tag{70}$$
$$z(0,x) = z^0(x) = -y_1(T,x),\ z_t(0,x) = z^1(x) = y_{1,t}(T,x),\quad x \in (0,1) \tag{71}$$
$$z(T,x) = z_t(T,x) = 0,\quad x \in (0,1). \tag{72}$$

Instead of considering a discretization of the control problem (69)–(72) and solving a finite-dimensional control problem, we design a control for the problem (69)–(72) based upon the finite-time stabilization method widely used for systems of conservation laws (see e.g. [2, 15, 18, 23]). It is well known that the transparent boundary condition $z_x = -z_t$ taken at $x = 1$ (replacing the condition $z(t,1) = h(T-t)$), which is a control in feedback form, leads to the finite-time stability of (69)–(72). The idea is to let the waves leave the domain (0, 1) in a free way, as they do naturally when the spatial domain is $\mathbb{R}$ instead of (0, 1). To achieve such a natural control, we proceed as follows. We extend z^0 and z^1 as odd functions with compact support. Namely, we set

$$z^0(x) = \begin{cases} -y_1(T,x) & \text{if } 0 < x < 1, \\ y_1(T,-x) & \text{if } -1 < x < 0, \\ 0 & \text{if } x \in (-\infty,1) \cup (1,+\infty), \end{cases} \tag{73}$$

and

$$z^1(x) = \begin{cases} y_{1,t}(T,x) & \text{if } 0 < x < 1, \\ -y_{1,t}(T,-x) & \text{if } -1 < x < 0, \\ 0 & \text{if } x \in (-\infty, 1) \cup (1, +\infty). \end{cases} \tag{74}$$

Let $z = z(t, x)$ solve now the system

$$z_{tt} - z_{xx} = 0, \quad 0 < t < T, \ x \in \mathbb{R} \tag{75}$$

$$z(0, x) = z^0(x), \ z_t(0, x) = z^1(x), \quad x \in \mathbb{R}. \tag{76}$$

Since $y_1(T, .) \in H_0^1(0, 1)$ and $y_{1,t}(T, .) \in L^2(0, 1)$, $(z^0, z^1) \in H^1(\mathbb{R}) \times L^2(\mathbb{R})$ and $(z, z_t) \in C([0, T], H^1(\mathbb{R}) \times L^2(\mathbb{R}))$. Furthermore, z is odd in x, for z^0 and z^1 are odd in x. Thus $z(0, t) = 0$ for all $t \in (0, T)$. Finally, it follows from d'Alembert's formula

$$z(t, x) = \frac{1}{2}[z^0(x + t) + z^0(x - t)] + \frac{1}{2}\int_{x-t}^{x+t} z^1(s)ds \tag{77}$$

and from the choice of z^0, z^1, and $T = 2$ that

$$z(T, x) = z_t(T, x) = 0, \qquad \text{for } -1 < x < 1. \tag{78}$$

We infer from (77) that

$$\text{supp}\, z(t, .) \subset [-1 - t, 1 + t] \subset [-3, 3] \quad \text{for } 0 \le t \le T = 2.$$

Thus, we can actually limit the computations to the spatial domain $[-3, 3]$ by imposing homogenous Dirichlet boundary conditions at $x = \pm 3$, and use e.g. the scheme described in Subsection 3.1. The control $h(t) = z(T - t, 1)$ is numerically obtained by computing $z(m\Delta t, 1)$ for $0 \le m \le M$. Note that 1 may not be a Chebychev point $3x_j$ for some $j \in [1, N + 1]$. Assuming that $1 \in [3x_{j+1}, 3x_j]$, then a linear interpolation between $z(m\Delta t, 3x_j)$, $z(m\Delta t, 3x_{j+1})$ gives an approximation of $z(m\Delta t, 1)$.

Another way to compute numerically $h(t) = z(T - t, 1)$ is to use (77) for $x = 1$, which gives for $t > 0$

$$z(t, 1) = \frac{1}{2}z^0(1 - t) + \frac{1}{2}\int_{1-t}^{1} z^1(s)ds. \tag{79}$$

(This expression was already derived in [22].) The integral term in (79) can be estimated by using the trapezoid rule.

A third way to compute numerically $h(t) = z(T - t, 1)$ is to go back to the transparent boundary condition $z_x(t, 1) = -z_t(t, 1)$ and to introduce the Riemann invariants $s := z_t + z_x$ and $d := z_t - z_x$. Set $d^0 := z^1 - z_x^0$. Then the transport equation $d_t + d_x = 0$ is integrated as $d(t, x) = d^0(x - t)$. On the other hand $z_t = (s + d)/2$ and $s(t, 1) = 0$, for $z_x(t, 1) = -z_t(t, 1)$. It follows that

$$z(t,1) = \int_0^t z_t(s,1)ds = \frac{1}{2}\int_0^t d(s,1)ds = \frac{1}{2}\int_0^t d^0(1-s)ds \tag{80}$$

which is nothing but (79). Numerically, $h(T - m\Delta t) = z(m\Delta t, 1)$ is approximated by $\frac{\Delta t}{4}\sum_{n=0}^{m-1}[d^0(1-n\Delta t) + d^0(1-(n+1)\Delta t)]$. The main avantage of this way to compute the control in the linear problem is that there is no need to solve the wave equation for $(t,x) \in (0,T) \times (-3,3)$. What is needed is the computation of d^0, first on the Chebychev points $(x_j)_{1\le j\le N+1}$ (using the Chebychev differentiation matrix to evaluate z_x^0), and next on the fine grid $(1-n\Delta t)_{0\le n\le M}$ by using the associated Lagrange polynomial.

4 Numerical Simulations

4.1 Linear Problem

We first test the boundary control of the linear wave equation by using the numerical integration of the linear wave equation (75)–(76) supplemented with the boundary conditions $z(t,\pm 3) = 0$. We take as initial data $z^0(x) = \sin(\pi x)$ and $z^1(x) = 0$ for $x \in (0,1)$ and we extend z^0 and z^1 to $\mathbb{R}$ as in (73)–(74). We use the numerical scheme in Subsection 3.1 with $N = 200$ and $\Delta t = 0.1/N^2$. The simulation is displayed in Fig. 1. As expected, the solution z vanishes at $T = 2$ for $x \in (-1,1)$. If we let $e(z^0)$ denote the sup-norm of the vector of the values of z^0 at the Chebychev points inside $(-1,1)$, which gives an estimate of $\|z^0\|_{L^\infty(-1,1)}$, then we obtain as relative error

$$\frac{e(z(T,.))}{e(z^0)} \approx 2.5\ 10^{-3}.$$

4.2 Nonlinear Control Problem

From now on, we compute numerically the control h in the linear control problem (69)–(72) by using a discretization of (80).

4.2.1 The General Case

We are concerned with the numerical simulation of the control problem (35)–(11), where f is as in (4). We shall take $f(u) = Cu^2$, where $C \in \mathbb{R}$ is some parameter. We shall see that the parameter C has to take its values in a bounded interval for the control problem to be numerically solved. We take as initial data

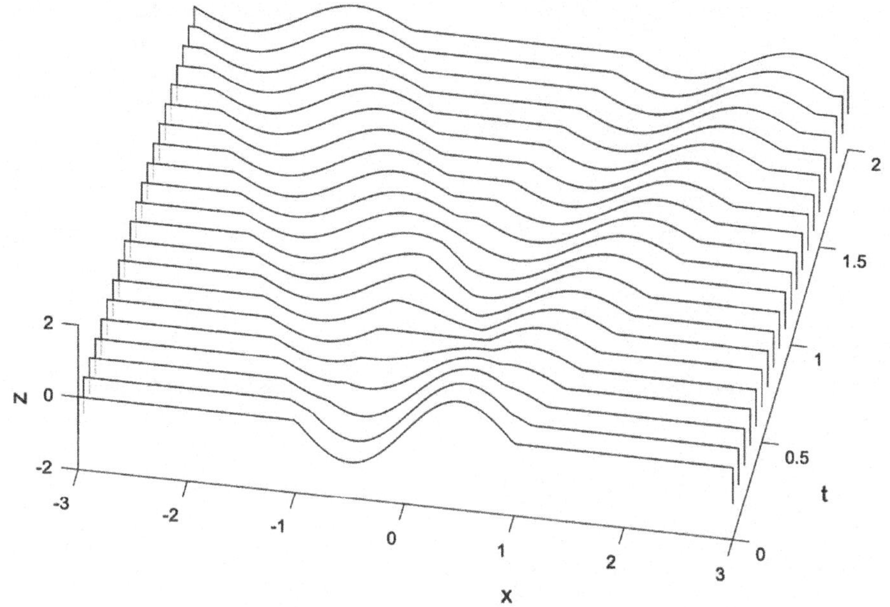

Fig. 1 Free evolution of the linear wave equation

$$\xi^0(x) = 1.5\sin(3\pi x), \quad \xi^1(x) = x^2 \quad \text{for } x \in (0, 1).$$

For $C = 4$, we apply our Picard iterative scheme with $n = 5$ iterations, $N = 200$, $\Delta t = 1/N^2$. The simulation is displayed in Fig. 2 (bottom). The relative error is found to be $\frac{e(z(T,.))}{e(z^0)} \approx 6.9\,10^{-3}$. Another simulation is performed with $C = 0$ (linear case) and is also displayed in Fig. 2 (top). The presence of the nonlinear term $4u^2$ in the PDE leads to a visible change of the control input.

4.2.2 The Case of an Odd Function f

When f is odd, then the solution of the system

$$\begin{aligned} y_{tt} - y_{xx} &= f(y), \quad t \in (0, T), \ x \in (-1, 1), \\ y(t, \pm 1) &= \pm h(t), \quad t \in (0, T), \\ y(0, x) = \mathrm{sign}(x)\xi^0(|x|), \ y_t(t, 0) &= \mathrm{sign}(x)\xi^1(|x|), \ x \in (-1, 1) \end{aligned}$$

is *odd*, and hence it vanishes for $x = 0$. This allows to do the computations on the interval $(-1, 1)$.

We first test our numerical scheme with $f(u) = Cu^3$. It is numerically proved to work well for $C \in [-2\,,\ 1.8]$. For $C = 2$, we have a blow-up, while for $C \leq -3$ we

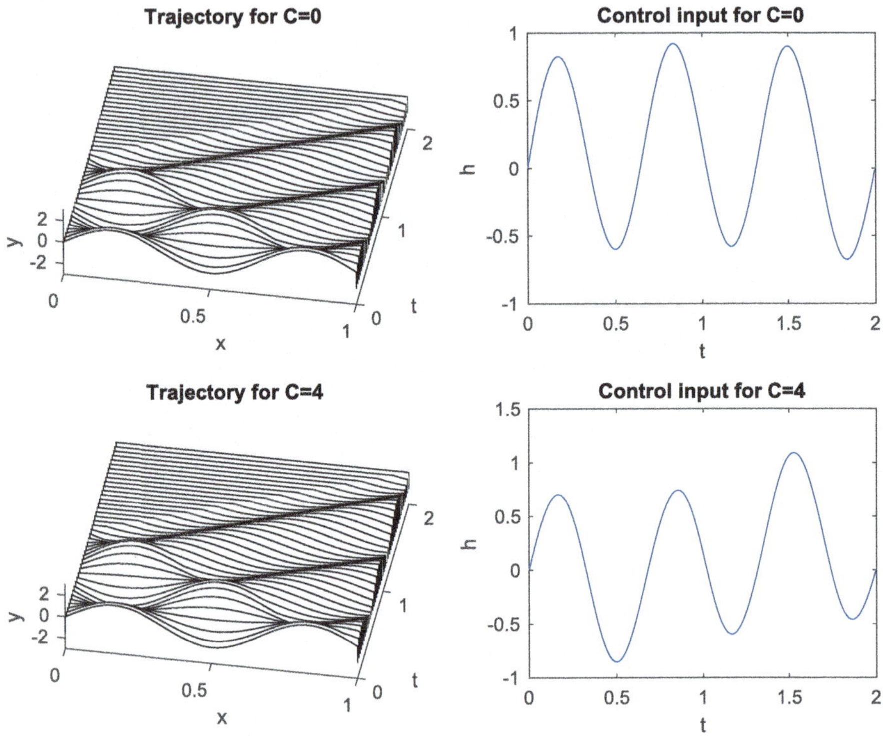

Fig. 2 Control of the nonlinear wave equation: $f(u) = Cu^2$

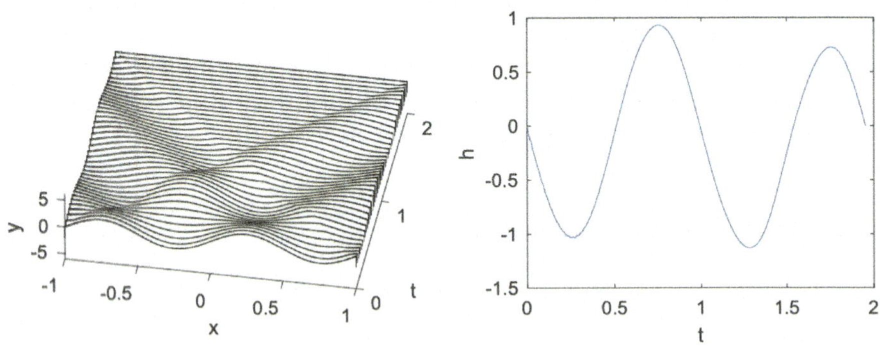

Fig. 3 Control of the semilinear wave equation: $f(u) = 1.5\,u^3$

never reach the null state at $T = 2$. This suggests that the Lipschitz condition (4) cannot be replaced by the passivity-like sector condition $f(y)y > 0$. Note that, even for small values of C, the best result is achieved for a "small" number of iterations: the relative error reaches its minimal value and next increases as the number of iterations tends to infinity. We take $\xi^0(x) = 2\sin(2\pi x)$ and $\xi^1(x) = 0$ as initial data,

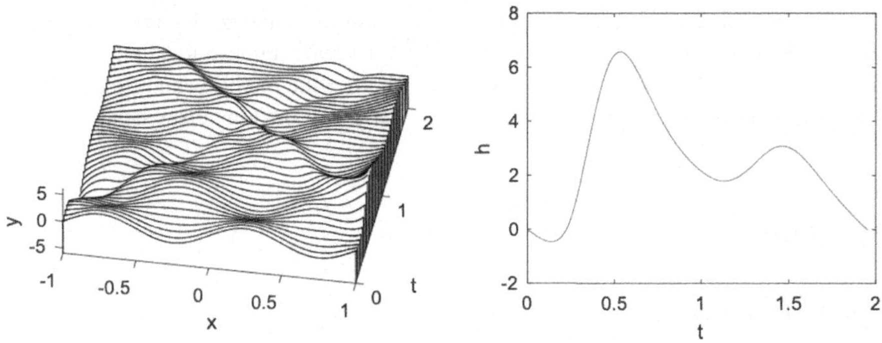

Fig. 4 Control of the semilinear wave equation: $f(u) = -4u^3$

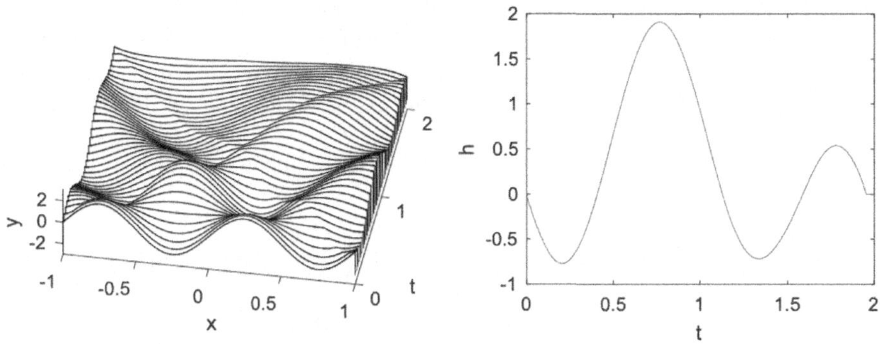

Fig. 5 Control of the semilinear wave equation: $f(u) = 6 \sin u$

and $N = 200$ and $\Delta t = 4/N^2$ as parameters. For $C = 1.5$ we obtain as relative error $5.7\ 10^{-3}$ with 5 iterations (see Fig. 3). For $C = -4$, the "best" relative error is 0.54 and it is obtained with 4 iterations (see Fig. 4). Finally, with the choice $f(u) = C \sin u$ (which is both Lipschitz continuous and bounded), we obtain with $C = 6$ as relative error 0.26 with 3 iterations (see Fig. 5). This suggests that our algorithm based on a perturbative argument does not work for large (Lipschitz continuous and bounded) perturbations.

5 Conclusion

An iterative algorithm was proposed to derive a boundary control for the null controllability of a semilinear wave equation in dimension one. Designing the control of the linear part by using a transparent boundary condition and using a collocation pseudospectral method on a Chebychev grid to compute the second order derivative in x in the wave equation, we provided several numerical simulations showing the efficiency of the method, at least for *small* nonlinearities and *small* initial data. Of course, further investigations are required to provide a numerical analysis of the algorithm. It would be also interesting to see whether the method can be extended to the wave equation in dimension two or three. This will be done elsewhere.

Appendix

Proof of Lemma 1

The estimate for $h \equiv 0$ follows from classical semigroup theory, $\big(S(t)\big)_{t\in\mathbb{R}}$ being a unitary group on $\mathcal{H}_0$. Assume now that $h \neq 0$ and that $(z^0, z^1, g) = (0, 0, 0)$. Since $||(z, \tilde{z})||_{\mathcal{H}}$ and $||z||_{H^1_x} + ||\tilde{z}||_{L^2_x}$ are equivalent norms on $\mathcal{H}$, it is sufficient to prove

$$||z||_{C([0,T],H^1_x)} + ||z_t||_{C([0,T],L^2_x)} \leq C||h||_{H^1_t}.$$

(Here, C denotes some constant that may vary from line to line.) It is well-known that

$$||z||_{C([0,T],L^2_x)} + ||z_t||_{C([0,T],H^{-1}_x)} \leq C\left(||z^0||_{L^2_x} + ||z^1||_{H^{-1}_x} + ||h||_{L^2_t}\right) \tag{81}$$

where z denotes the solution by transposition of (39)–(41) with $(z^0, z^1, g) = (0, 0, 0)$. Introduce $w := z_t$. Then w solves the system

$$\begin{aligned}
&w_{tt} - w_{xx} = 0, \qquad (t, x) \in (0, T) \times (0, 1)\\
&w(t, 0) = 0,\ \ w(t, 1) = h'(t), \qquad t \in (0, T)\\
&(w(0, .), w_t(0, .)) = (z^1, z^0_{xx}) = (0, 0).
\end{aligned}$$

Since $h \in H^1_t$, we infer from (81) that

$$||w||_{C([0,T],L^2_x)} + ||w_t||_{C([0,T],H^{-1}_x)} \leq C||h||_{H^1_t}, \tag{82}$$

so that the estimate for $||z_t||_{C([0,T],L^2_x)}$ is established. For $||z||_{C([0,T],H^1_x)}$, it is sufficient to notice that $z_{xx} = z_{tt} = w_t$ and to combine

$$||z_{xx}||_{C([0,T],H^{-1}_x)} \leq C||h||_{H^1_t}$$

with the elliptic estimate

$$||z(t, .)||_{H^1_x} \leq C\left(||z_{xx}(t, .)||_{H^{-1}_x} + |z(t, 0)| + |z(t, 1)|\right) \leq C(||z_{xx}(t, .)||_{H^{-1}_x} + |h(t)|).$$

Proof of Lemma 2

For $(z, \tilde{z}) \in C([0, T], \mathcal{H})$, let

$$|||(z, \tilde{z})||| = \max_{t \in [0,T]} ||(z(t), \tilde{z}(t))||_{\mathcal{H}} = \max_{t \in [0,T]} \left(||z_x(t)||^2_{L^2_x} + ||\tilde{z}(t)||^2_{L^2_x}\right)^{\frac{1}{2}}.$$

We shall apply the contraction mapping principle in the ball $B(0, R) \subset C([0, T], \mathcal{H})$, where $R > 0$ will be chosen later on. Note that

$$||z||_{L^\infty_x(0,1)} \leq ||z_x||_{L^2_x}$$

if $z \in H^1_x$ with $z(0) = 0$. Assume that $z^0 \in H^1_0(0, 1)$, and that

$$||(z^0, z^1)||_{\mathcal{H}} + C_1||h||_{H^1_t} =: \delta \leq \delta_0. \tag{83}$$

$$R < \delta_0. \tag{84}$$

Note that, for $(z, \tilde{z}) \in B(0, R)$,

$$||z(t)||_{L^\infty_x(0,1)} \leq ||z_x(t)||_{L^2_x} \leq |||(z, \tilde{z})||| < \delta_0 \qquad \forall t \in [0, T].$$

Pick two pairs $(z_1, \tilde{z}_1)$ and $(z_2, \tilde{z}_2)$ in $B(0, R)$. Then by Lemma 1

$$\begin{aligned} |||\Gamma(z_1, \tilde{z}_1) - \Gamma(z_2, \tilde{z}_2)||| &= |||\int_0^t S(t-s)(0, f(z_1(s)) - f(z_2(s)))ds||| \\ &\leq ||f(z_1) - f(z_2)||_{L^1(0,T,L^2_x)} \\ &\leq \eta||z_1 - z_2||_{L^1(0,T,L^2_x)} \\ &\leq T\eta|||(z_1 - z_2, \tilde{z}_1 - \tilde{z}_2)|||. \end{aligned} \tag{85}$$

Note that Γ contracts in $B(0, R)$, for $T = 2$ and η as in (4). On the other hand, for any $(z, \tilde{z}) \in B(0, R)$, we have by Lemma 1 and (85)

$$\begin{aligned} |||\Gamma(z, \tilde{z})||| &\leq |||S(t)(z^0, z^1) + W(h)||| + |||\int_0^t S(t-s)(0, f(z(s)))ds||| \\ &\leq \delta + 2\eta|||(z, \tilde{z}))|||. \end{aligned} \tag{86}$$

Thus $\Gamma(B(0, R)) \subset B(0, R)$ if

$$\delta + 2\eta R \leq R.$$

Pick

$$R = \frac{\delta}{1 - 2\eta}.$$

with

$$0 < \delta < \delta_1 = (1 - 2\eta)\delta_0 < \delta_0.$$

Then Γ has a unique fixed point $(z, \tilde{z})$ in $B(0, R)$ by the contraction mapping principle. Clearly, $\tilde{z} = z_t$ and (48) holds with $C_2 = (1 - 2\eta)^{-1}$.

References

1. Abolinya, V.È., Myshkis, A.D.: Mixed problems for quasi-linear hyperbolic systems in the plane. (Russian) Mat. Sb. (N.S.) **50**(92), 423–442 (1960)
2. Alabau-Boussouira, F., Perrollaz, V., Rosier, L.: Finite-time stabilization of a network of strings. Math. Control Relat. Fields **5**(4), 721–742 (2015)
3. Asch, M., Lebeau, G.: Geometrical aspects of exact boundary controllability for the wave equation- a numerical study. ESAIM: COCV **3**, 163–212 (1998)
4. Bardos, C., Lebeau, G., Rauch, J.: Sharp sufficient conditions for the observation, control and stabilization of waves from the boundary. SIAM J. Control Optim. **30**(5), 1024–1065 (1992)
5. Bastin, G., Coron, J.-M.: Stability and Boundary Stabilization of 1-D Hyperbolic Systems. In: Progress in Nonlinear Differential Equations and their Applications, 88. Subseries in Control, Birkhaüser/Springer (2016)
6. Cavalcanti, M.M., Corrêa, W.J., Rosier, C., Dias Silva, F.R.: General decay rate estimates and numerical analysis for a transmission problem with locally distributed nonlinear damping. Comput. Math. Appl. **73**(10), 2293–2318 (2017)
7. Canuto, C., Hussaini, M.Y., Quarteroni, A., Zang, T.A.: Spectral methods in Fluid Dynamics, Springer Series in Computational Physics, Springer Berlin Heidelberg (1988)
8. Castro, C., Micu, S.: Boundary controllability of a linear semi-discrete 1-D wave equation derived from a mixed finite element method. Numer. Math. **102**(3), 413–462 (2006)
9. Castro, C., Micu, S., Münch, A.: Numerical approximation of the boundary control for the wave equation with mixed finite elements in a square. IMA J. Numer. Anal. **28**(1), 186–214 (2008)
10. Cooke, K.L., Krumme, D.W.: Differential-difference equations and nonlinear initial-boundary value problems for linear hyperbolic partial differential equations. J. Math. Anal. Appl. **24**, 372–387 (1968)
11. Domingos Cavalcanti, V.N., Rodrigues, J.H., Rosier, C.: Numerical analysis for the wave equation with locally nonlinear distributed damping. J. Comput. Appl. Math. **301**, 144–160 (2016)
12. Ervedoza, S., Zuazua, E.: A systematic method for building smooth controls for smooth data. Discrete Contin. Dyn. Syst., Ser. B **14**(4), 1375–1401 (2010)
13. Ervedoza, S., Zuazua, E.: Numerical approximation of exact controls for waves. Springer Briefs in Mathematics. Springer, New York (2013)
14. Glowinski, R., Li, C., Lions, J.L.: A numerical approach to the exact boundary controllability of the wave equation. Japan J. Appl. Math. **3**, 1–76 (1990)
15. Gugat, M., Perrollaz, V., Rosier, L.: Boundary stabilization of quasilinear hyperbolic systems of balance laws: exponential decay for small source terms. J. Evol. Equ. **18**(3), 1471–1500 (2018)
16. Ignat, L., Zuazua, E.: Convergence of a multi-grid method for the control of waves. J. Eur. Math. Soc. **11**, 351–391 (2009)
17. Komornik, V.: Exact Controllability and Stabilization. Research in Applied Mathematics, Masson, Paris, The Multiplier Method. RAM (1994)
18. Leugering, G., Schmidt, E.J.P.G.: On the modelling and stabilization of flows in networks of open canals. SIAM J. Control Optim. **41**, 164–180 (2002)

19. Majda, A.: Disappearing solutions for the dissipative wave equation. Indiana Univ. Math. J. **24**, 1119–1133 (1975)
20. Micu, S.: Uniform boundary controllability of a semidiscrete 1-D wave equation with vanishing viscosity. SIAM J. Control Optim. **47**(6), 2857–2885 (2008)
21. Münch, A.: A uniformly controllable and implicit scheme for the 1-D wave equation, M2AN Math. Model. Numer. Anl. **39**(2), 377–418 (2005)
22. Pedregal, P., Periago, F., Villena, J.: A numerical method of local energy decay for the boundary controllability of time-reversible distributed parameter systems. Stud. Appl. Math. **121**(1), 27–47 (2008)
23. Perrollaz, V., Rosier, L.: Finite-time stabilization of 2×2 hyperbolic systems on tree-shaped networks. SIAM J. Control Optim. **52**(1), 143–163 (2014)
24. Russell, D.L.: Controllability and stabilization theory for linear partial differential equations: recent progress and open questions. SIAM Rev. **20**(4), 639–737 (1978)
25. Trefethen, L.N.: Spectral Methods in Matlab, SIAM Philadelphia (2000)
26. Zuazua, E.: Boundary observability for the finite difference space semi-discretizations of the 2-D wave equation in the square. J. Math. Pures Appl. **78**, 523–563 (1999)
27. Zuazua, E.: Propagation, observation, control and numerical approximation of waves. SIAM Rev. **47**(2), 197–243 (2005)

Stabilization and Tracking of Interconnected Systems

Insight into the Stability Analysis of the Reaction-Diffusion Equation Interconnected with a Finite-Dimensional System Taking Support on Legendre Orthogonal Basis

Mathieu Bajodek, Alexandre Seuret, and Frédéric Gouaisbaut

Abstract This chapter deals with the stability analysis of the reaction-diffusion subject to dynamic boundary conditions. More particularly, the objective is to propose a linear matrix inequality criterion which ensures the stability of such infinite-dimensional system. By the use of Fourier-Legendre series, the Lyapunov functional is split into an augmented finite-dimensional state including within it the first Fourier-Legendre coefficients and the residual part. A link between this modeling and Padé approximation is briefly highlighted. Then, from Bessel and Wirtinger inequalities applied to the Fourier-Legendre remainder and using its orthogonality properties, a sufficient condition of stability expressed in terms of linear matrix inequalities is obtained. This efficient and scalable stability condition is finally performed on examples.

1 Introduction

The wide class of infinite-dimensional systems [7, 8, 11] generates issues when considering stability or stabilization problems. In most of the cases, when the eigenvalue decomposition of the infinite-dimensional operator cannot be given analytically, the stability properties of these systems remain unknown and require a dedicated study. For instance, the stability analysis of a system coupled with a partial differential equation is a tough task, which has been infrequently studied. Most of recent studies bypass the problem by focusing on the stabilization of such infinite-dimensional systems thanks to the design of infinite-dimensional control laws [1, 23].

M. Bajodek (✉) · A. Seuret · F. Gouaisbaut
LAAS-CNRS, Equipe MAC (E49), 7 avenue du colonel Roche, 31400 Toulouse, France
e-mail: mathieu.bajodek@laas.fr

A. Seuret
e-mail: alexandre.seuret@laas.fr

F. Gouaisbaut
e-mail: frederic.gouaisbaut@laas.fr

J. Auriol et al. (eds.), *Advances in Distributed Parameter Systems*, Advances in Delays and Dynamics 14, https://doi.org/10.1007/978-3-030-94766-8_5

In the field of numerical analysis, researchers approximate the solutions from spacial discretization [18] or the tau method [13]. Under dissipativity conditions [14], convergence properties of the numerical schemes emerge. However, they never deal with the stability of the original system. Therefore, to rule on the stability of interconnected ordinary-partial differential equations, input-to-state [16, 20, 21] or Lyapunov [22, 27] approaches are privileged. For generic cases, quadratic constraints and complete Lyapunov functionals lead to criteria solved by semi-definite programming [9, 10, 26]. To refine, adjust and better understand these generic results, works have been pursued on the model and the inequalities involved in the sufficient condition of stability. For instance, for transport [24], heat [6] or wave [5] equations, an augmented system enriched by Legendre polynomials coefficients of the partial-differential part can be built. Applying Bessel-Legendre inequality, hierarchical stability results are expressed in terms of linear matrix inequalities and dedicated to each equation. Concerning the reaction-diffusion equation, the work is slightly modified since the Wirtinger inequality has to be invoked. In this chapter, the objective is to emphasize the structure of the linear matrix inequality by enlightening the Fourier-Legendre series. This contribution can be seen as an extension of [3] with other boundary conditions, and which provides a deeper understanding of the underlying structure of the conditions.

To sum up, a numerical approach to state on the stability of an ordinary differential equation interconnected with a reaction-diffusion equation is proposed. The novelty of this contribution comes from the model transformation which eases the expression and understanding of the stability analysis. Indeed, by defining signals derived from Fourier-Legendre remainder, the Bessel-Legendre and the first Wirtinger inequalities can be rewritten in a nice formalism. Around these signals, the dynamical model can be constructed and related to Padé $(n-1|n)$ approximant. It includes the extended finite-dimensional based on the first Fourier-Legendre polynomials coefficients. Taking a quadratic Lyapunov function based on this augmented system, a tight linear matrix inequality condition of stability is obtained. Numerical results are finally performed to show the effectiveness of the approximation and of the sufficient stability criterion.

Notations: In this paper, the set of natural numbers, real numbers, real positive numbers, matrices of size $n \times m$ and of symmetric positive definite matrices of size n are respectively denoted $\mathbb{N}$, $\mathbb{R}$ $\mathbb{R}_{\geq 0}$, $\mathbb{R}^{n\times m}$ and $\mathbb{S}^n_+$, respectively. The identity matrix of dimension n is denoted by I_n and $\text{diag}(d_0, \ldots, d_n)$ stands for the diagonal matrix whose coefficients are $(d_0, \ldots, d_n)$. For any matrix M, $M^{i,j}$ refers to the coefficient located on the ith row and j^{th} column. For any square matrix M, $\mathcal{H}(M) = M + M^\top$, $M \succ 0$ means that M is symmetric positive definite and $\sigma(M)$ denotes the spectrum of M. Furthermore, if M is symmetric, $\underline{\sigma}(M)$ and $\bar{\sigma}(M)$ stands for its lower and larger eigenvalues, respectively. For any analytic function G_1 and G_2, $G_1(s) = \underset{s\to\lambda}{O}\left(G_2(s)\right)$ means that the ratio $\frac{G_1}{G_2}(s)$ is finite when s tends to $\lambda \in \mathbb{R}$. Moreover, the space of square integrable functions $\mathcal{L}^2(a, b; \mathbb{R})$ is associated to the scalar product $\langle z_1 | z_2 \rangle_{a,b} = \int_a^b z_1(\theta) z_2(\theta) \mathrm{d}\theta$ and the induced norm $\|z\|^2_{a,b} = \int_a^b z^2(\theta)\mathrm{d}\theta$. Set $\mathcal{H}^1(a, b; \mathbb{R})$ stands for the set of functions z, such that z and $\partial_\theta z$ are in $\mathcal{L}^2(a,b;\mathbb{R})$. With a light abuse of notations, the notation for inner product $\langle z_1|z_2\rangle_{a,b}$ will be used when z_1 and z_2 are vector functions, and will be consired component-wise. In addition, when $a=-1$ and $b=1$, the index $_{a,b}$ is omitted.

2 Presentation of the System

2.1 Interconnected System

Consider the following system composed of a linear ordinary differential equation interconnected with a reaction-diffusion partial differential equation with cross-type boundary conditions

$$\begin{cases} \dot{x}(t) = Ax(t) + Bz(t,1), \quad \forall t \in \mathbb{R}_{\geq 0}, & \text{(1a)} \\ \partial_t z(t,\theta) = (\delta\partial_{\theta\theta} + \lambda)z(t,\theta), \quad \forall (t,\theta) \in \mathbb{R}_{\geq 0} \times (0,1), & \text{(1b)} \\ \begin{bmatrix} z(t,0) \\ \partial_\theta z(t,1) \end{bmatrix} = \begin{bmatrix} 0 \\ Cx(t) \end{bmatrix}, \quad \forall t \in \mathbb{R}_{\geq 0}, & \text{(1c)} \\ (x(0), z(0,\cdot)) = (x_0, z_0). & \text{(1d)} \end{cases}$$

Coefficients $\delta > 0$, $\lambda \in \mathbb{R}$, matrix $A \in \mathbb{R}^{n_x \times n_x}$ and vectors $B, C^\top \in \mathbb{R}^{n_x}$ are supposed to be constant and known.

Remark 1 In this chapter, cross-type boundary conditions have been selected in order to make more readable and understandable the presented results. However, our approach can be easily extended to Dirichlet, Neumann or Robin boundary conditions [3].

Remark 2 It is worth noticing that the stability analysis of system (1) is not straighforward. Indeed, under some transformations, the system operator can be given by $\mathbb{A} = \begin{bmatrix} \delta\partial_{\theta\theta}+\lambda-\theta CB\gamma_1 & -\theta CA \\ A & B\gamma_1 \end{bmatrix}$ where γ_1 is the trace operator at $\theta = 1$. Since the ordinary and partial differential equations are fully interconnected, an eigenbasis decomposition is no longer possible.

2.2 Existence and Uniqueness of Solutions

As a first step and before studying stability of such class of system, it is important to show that system (1) is well-posed despite non trivial boundary conditions on the partial differential equation part [19]. This is formulated in the next proposition.

Proposition 1 *Assuming that* (x_0, z_0) *belongs to* $\mathbb{R}^{n_x} \times \mathcal{L}^2(0,1;\mathbb{R})$*, system (1) admits a continuous and unique solution* $(x,z) \in \mathbb{R}^{n_x} \times \mathcal{L}^2(0,1;\mathbb{R})$.

Proof Let us define the energy of system (1) by $\mathcal{E}(t) = x^T(t)x(t) + (2\delta)^{-1}\|z(t)\|_{0,1}^2$. Computing its derivatives with respect to the time, several calculations lead to

$$\begin{aligned}\tfrac{\mathrm{d}}{\mathrm{d}t}\mathcal{E}(t) = x^T(t)(A^T + A)x(t) + 2x^T(t)Bz(t,1) + \tfrac{\lambda}{\delta}\|z(t)\|_{0,1}^2 \\ -\|\partial_\theta z(t)\|_{0,1}^2 + x^T(t)C^T z(t,1).\end{aligned}$$

Getting rid of the cross terms by application of Young's inequality, one obtains

$$\frac{\mathrm{d}}{\mathrm{d}t}\mathcal{E}(t) \leq x^T(t)\left(\mathcal{H}(A) + 4BB^T + C^T C\right)x(t) + \frac{\lambda}{\delta}\|z(t)\|_{0,1}^2 - \|\partial_\theta z(t)\|_{0,1}^2 + \frac{1}{2}z^2(t,1).$$

However, from Jensen's inequality, one gets to

$$\|\partial_\theta z(t)\|_{0,1}^2 \geq (z(t,1) - z(t,0))^2 = z^2(t,1), \tag{2}$$

which simplifies the previous inequality on $\frac{\mathrm{d}}{\mathrm{d}t}\mathcal{E}(t)$, leading to

$$\tfrac{\mathrm{d}}{\mathrm{d}t}\mathcal{E}(t) \leq K\mathcal{E}(t) - \tfrac{1}{2}\|\partial_\theta z(t)\|_{0,1}^2 \leq K\mathcal{E}(t),$$

where K is a generic constant depending on $A, B, C, \lambda, \epsilon$.

From Grönwall's inequality, there exists a unique solution (x,z) in the continuous space from $\mathbb{R}_{\geq 0}$ to $\mathbb{R}^{n_x} \times \mathcal{L}^2(a,b;\mathbb{R})$.

Remark 3 In Sect. 5, to deal with stability of system (1), a Lyapunov functional is chosen to be equivalent to the energy $\mathcal{E}(t)$ in the sense of the $\mathbb{R}^{n_x} \times \mathcal{L}^2(0,1;\mathbb{R})$. In parallel with the calculations carried out here, the positive term in $\|z\|$ is upper bounded by Wirtinger inequality and Jensen's inequality (2) is improved thanks to Bessel-Legendre inequality.

2.3 Equilibrium Point

As a second step, it is also important to characterize the equilibrium of (1). More particularly, one has to understand under which conditions, system (1) admits a unique equilibrium. This is formulated below.

Proposition 2 *System (1) admits a unique equilibrium* $(0,0)$ *if and only if matrix* Ω *given by (3) has a full rank.*

$$\Omega = \begin{cases} \begin{bmatrix} A & B\sinh(\tilde{\lambda}) \\ C & -\tilde{\lambda}\cosh(\tilde{\lambda}) \end{bmatrix}, & \text{if } \lambda < 0, \\ \begin{bmatrix} A & B \\ C & -1 \end{bmatrix}, & \text{if } \lambda = 0, \\ \begin{bmatrix} A & B\sin(\tilde{\lambda}) \\ C & -\tilde{\lambda}\cos(\tilde{\lambda}) \end{bmatrix}, & \text{if } \lambda > 0. \end{cases} \tag{3}$$

Proof Let $(\bar{x}, \bar{z})$ be an equilibrium of system (1), meaning that the following relations hold.

$$\begin{cases} A\bar{x} + B\bar{z}(1) = 0, & \text{(4a)} \\ (\delta\partial_{\theta\theta} + \lambda)\bar{z}(\theta) = 0, & \text{(4b)} \\ \bar{z}(0) = 0, & \text{(4c)} \\ \partial_\theta \bar{z}(1) = C\bar{x}. & \text{(4d)} \end{cases}$$

Integrating the differential Eq. (4b) and recalling the boundary condition (4c), one obtains

$$\bar{z}(\theta) = \begin{cases} \gamma\sinh(\tilde{\lambda}\theta), & \text{if } \lambda < 0, \\ \gamma\theta, & \text{if } \lambda = 0, \\ \gamma\sin(\tilde{\lambda}\theta), & \text{if } \lambda > 0, \end{cases}$$

where γ in $\mathbb{R}$ to be fixed and $\tilde{\lambda} = \sqrt{|\lambda|/\delta}$. By re-injecting this expression into (4a) and (4d), it yields $\Omega\left[\begin{smallmatrix}\bar{x}\\ \gamma\end{smallmatrix}\right] = 0$ with Ω given by (3). Hence, system (4) admits a unique solution leading to the equilibrium $(\bar{x}, \bar{z}) = (0, 0)$ if and only if $\det(\Omega) \neq 0$ which is equivalent to have Ω a full matrix.

2.4 Expected Results

Under the conditions stated above, the aim of this paper is to state accurately on the stability of the equilibrium point (0, 0) using the linear matrix inequality framework. More specifically the following results will be obtained:

- The main result is the proposition of a scalable stability analysis for this coupled ordinary-partial differential equations based on Lyapunov arguments. The analysis is performed thanks to an accurate Lyapunov functional, which is build using Legendre polynomials. The calculations developments are highly inspired by [3] and have been simplified thanks to the consideration of easier boundary condition applied to the reaction-diffusion equation.
- Contrary to a previous study provided in [6], the analysis will be performed through the introduction of the remainder of Legendre-Fourier series, allowing to simplify some technical aspects. Thanks to this remainder, it is possible to rewrite the Wirtinger and the Bessel-Legendre inequalities in a simpler manner compared to the formulation presented in [6].

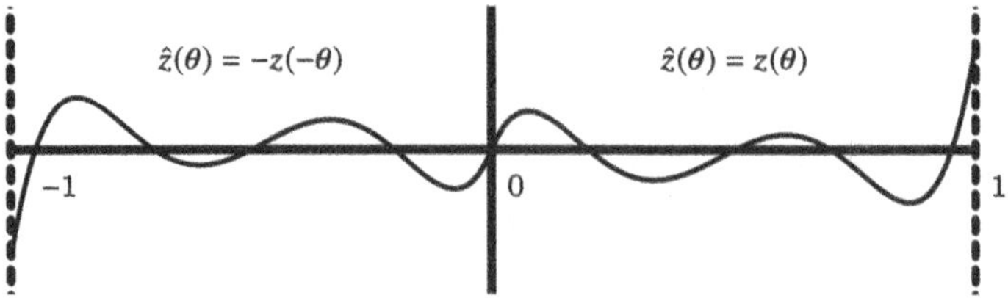

Fig. 1 Function z extended on the interval $[-1, 1]$

- Moreover, thanks to the introduction of an augmented system, a necessary condition for the stability condition emerges: a certain matrix $\mathbf{A}_n$ has to be stable. In Sect. 4, as a subsidiary result, it is shown that this matrix is the Redheffer product between a realization of the Padé $(n-1|n)$ approximant of the reaction-diffusion transfer function and the finite-dimensional system $\left[\begin{smallmatrix} A & B \\ C & 0 \end{smallmatrix}\right]$.

3 Fourier-Legendre Remainder and Inequalities

Based on the nature of the boundary conditions of the reaction-diffusion (1c), several approximations could be investigated. Here, the use of Fourier series is not possible because of nonhomogeneous boundary conditions. Indeed, by using the eigenfunctions of operator $\partial_{\theta\theta}$ subject to cross-type boundary conditions, the interconnection with the ordinary would not be taken into consideration in the stability analysis. Encouraged by the promising results on time-delay systems [2], the Legendre-Fourier approximation is taken under consideration.

We consider then a function z in $\mathcal{L}^2(0, 1; \mathbb{R})$ such as $z(0) = 0$. To describe this function z, it is possible to work on a function $\hat{z}$ constructed by imparity on an extended interval $[-1, 1]$ as shown in Fig. 1. Then, a complete characterization can be done using only odd polynomials on this extended interval $[-1, 1]$. After recalling the basics of Legendre polynomials and some of their properties, this section provides the definition of Fourier-Legendre coefficients and remainder on the interval $[-1, 1]$. In a last step, several relevant inequalities are presented.

3.1 Fourier-Legendre Coefficients and Remainder

Legendre polynomials, denoted as l_k for any positive integer k, are given by

$$l_k(\theta) = \sum_{i=0}^{k} (-1)^i \frac{(k+i)!}{(i!)^2(k-i)!} \left(\frac{1-\theta}{2}\right)^i, \quad \forall \theta \in [-1, 1]. \tag{5}$$

The orthogonal family $\{l_k\}_{k\in\mathbb{N}}$ spans $\mathcal{L}^2(-1, 1; \mathbb{R})$. For writing comfort, let introduce the notation ℓ_n for any $n \in \mathbb{N}$, which gathers the n first odd Legendre polynomials in vector formulation, that is

$$\ell_n(\theta) = \left[l_1(\theta)\ l_3(\theta)\ \dots\ l_{2n-1}(\theta)\right]^\top \in \mathbb{R}^n. \tag{6}$$

Recall also some important properties of Legendre polynomials [12], that will be useful along the paper, that are

$$\langle \ell_n | \ell_n \rangle = \mathcal{I}_n^{-1}, \tag{7a}$$
$$\ell_n''(\theta) = \mathcal{L}_n \ell_n(\theta), \quad \forall \theta \in [-1, 1], \tag{7b}$$
$$\ell_n(-\theta) = -\ell_n(\theta), \quad \forall \theta \in [0, 1], \tag{7c}$$

where matrices $\mathcal{L}_n$ and $\mathcal{I}_n$ are square matrices of dimension n, given by

$$\begin{aligned} \mathcal{I}_n &= \frac{1}{2}\mathrm{diag}(3, \dots, 4n-1), \\ \mathcal{L}_n^{i,j} &= (4j-1)\sum_{p=j}^{i-1}(4p+1), \quad \forall i, j \in \{1, \dots, n\}. \end{aligned} \tag{8}$$

Functions ℓ_n can be easily evaluated at $\theta \in \{0, 1\}$ and is given by

$$\ell_n(1) = [\,1\ 1\ \dots\ 1\,]^\top \in \mathbb{R}^n, \quad \ell_n(0) = [\,0\ 0\ \dots\ 0\,]^\top \in \mathbb{R}^n, \tag{9a}$$
$$\ell_n'(1) = [\,1\ 6\ \dots\ n(2n-1)\,]^\top \in \mathbb{R}^n. \tag{9b}$$

Remark 4 From definition 5, note that the first two odd polynomials are given by $l_1(\theta) = \theta$ and $l_3(\theta) = -\frac{3}{2}\theta + \frac{5}{2}\theta^3$ and satisfy the properties given above.

As stated in the previous section, we will define here the main features of this paper in the following definition.

Definition 1 For any function $z \in \mathcal{L}^2(0, 1; \mathbb{R})$ such that $z(0) = 0$ and for integer n in $\mathbb{N}$, we define

- the extended state $\hat{z}$ on the extended interval $[-1, 1]$ (see Fig. 1) by

$$\hat{z}(\theta) = \begin{cases} z(\theta), & \forall \theta \in [0, 1], \\ -z(-\theta), & \forall \theta \in [-1, 0], \end{cases} \tag{10}$$

- the n first non null Fourier-Legendre coefficients of $\hat{z}$ as

$$\zeta_n = \begin{bmatrix} \int_{-1}^{1} l_1(\theta)\hat{z}(\theta)\mathrm{d}\theta \\ \int_{-1}^{1} l_3(\theta)\hat{z}(\theta)\mathrm{d}\theta \\ \vdots \\ \int_{-1}^{1} l_{2n-1}(\theta)\hat{z}(\theta)\mathrm{d}\theta \end{bmatrix} = \langle \ell_n | \hat{z} \rangle \in \mathbb{R}^n, \tag{11}$$

- the associated Fourier-Legendre remainder of $\hat{z}$ at order n as

$$w_n(\theta) = \hat{z}(\theta) - \ell_n^\top(\theta)\mathcal{I}_n\zeta_n, \quad \forall\theta \in [-1, 1]. \tag{12}$$

It is worth noting that w_n is in $\mathcal{L}^2(-1, 1; \mathbb{R})$ and that the norm and scalar product used in the sequel are considered in the interval $[-1, 1]$. The main interest for introducing this remainder is stated in the two following lemmas.

Lemma 1 *For any n in $\mathbb{N}$, the Fourier-Legendre remainder w_n given by (12) verifies*

- $\langle l_k | w_n \rangle = 0$, *for any integer $k \in \{0, \dots, 2n\}$, which means in other words that w_n is orthogonal to the $2n + 1$ first Legendre polynomials,*
- $w_n(0) = 0$, *which guarantee that the boundary condition at $\theta = 0$ satisfied by z is maintained for w_n.*

Proof By symmetry of all even Legendre polynomials, one obtains directly that $\langle l_{2k} | w_n \rangle = 0$, for all positive integers k. Moreover, thanks to the orthogonality (7a) of the Legendre polynomials, re-injecting the definition of w_n into $\langle \ell_n | w_n \rangle$ yields

$$\langle \ell_n | w_n \rangle = \langle \ell_n | \hat{z} \rangle - \langle \ell_n | \ell_n \rangle \mathcal{I}_n \langle \ell_n | \hat{z} \rangle = \langle \ell_n | \hat{z} \rangle - \langle \ell_n | \hat{z} \rangle = 0,$$

which leads to the orthogonality on the n first odd Legendre polynomials. Lastly, thanks to the value at $\theta = 0$ given by (9a), we have

$$w_n(0) = z(0) - \ell_n^\top(0)\mathcal{I}_n\zeta_n = \hat{z}(0) = 0.$$

which concludes the proof.

In the next subsections, Bessel and Wirtinger inequalities are rewritten in an adequate manner to be consistent with the Fourier-Legendre remainder introduced above. The forthcoming Lyapunov analysis could then be pursued on an augmented system which takes aside the Fourier-Legendre remainder w_n of the signal $\hat{z}$.

3.2 Bessel Inequalities

Let us first recall the Bessel-Legendre inequality.

Lemma 2 *For any function $\hat{z} \in \mathcal{L}^2(-1, 1; \mathbb{R})$ and for any integer n in $\mathbb{N}$, the following inequality holds:*

$$\|\hat{z}\|^2 \geq \sum_{k=0}^{n} \left(\frac{2k+1}{2}\right) \langle l_k | \hat{z}^2 \rangle . \tag{13}$$

Proof The proof can be found in [3, 6]. It is directly derived from the positivity of the $\mathcal{L}^2(-1, 1; \mathbb{R})$ norm of the Fourier-Legendre remainder of the function $\hat{z}$.

The application of Lemma 2 on Fourier-Legendre remainder yields $\|w_n\|^2 \geq 0$, since the remainder is orthogonal to the first Legendre polynomials, i.e. $\langle \ell_n | w_n \rangle = 0$. This means that the information encapsulated in the Bessel inequality is already included in the remainder.

The interest of using the remainder is related to the formulation of this inequality when it is applied to its derivatives $\|w_n'\|$ that is presented in the next lemma.

Lemma 3 *For any function z in $\mathcal{H}^1(0, 1; \mathbb{R})$ such that $z(0) = 0$ and any integer n in $\mathbb{N}$, the derivatives of the remainder w_n given by (12) verifies*

$$\|w_n'\|^2 \geq \kappa_n^B w_n^2(1), \quad \textit{with} \quad \kappa_n^B = 2(n+1)(2n+1). \tag{14}$$

Proof Following Definition 1, we define the extended signal $\hat{z}$ by (10) and we call its Fourier-Legendre remainder w_n. Thanks to the Bessel-Legendre inequality at order $2n+1$, the following inequality holds

$$\|w_n'\|^2 \geq \sum_{k=0}^{2n+1} \left(\frac{2k+1}{2}\right) \langle l_k | w_n' \rangle^2 . \tag{15}$$

In addition, performing an integration by parts yields

$$\langle l_k | w_n' \rangle = \left(l_k(1) + l_k(-1)\right) w_n(1) - \langle l_k' | w_n \rangle - \left(1 - (-1)^k\right) w_n(1) - \langle l_k' | w_n \rangle .$$

Then, we recall that w_n is the remainder of the Fourier-Legendre series, which is consequently orthogonal to the $2n+1$ first Legendre polynomials. Therefore, the last term of the previous equality is zero so that

$$\|w_n'\|^2 \geq \left(\sum_{p=0}^{n} \frac{4p+1}{2} \right) (2 w_n(1))^2 .$$

To complete the proof, we have

$$\sum_{k=0}^{n} (4p+1) = 4\frac{n(n+1)}{2} + (n+1) = (n+1)(2n+1),$$

which yields the result.

Remark 5 It is worth noticing that the error made in this new Bessel-inequality is equal to the $\mathcal{L}^2(-1, 1; \mathbb{R})$ norm of the Fourier-Legendre remainder of $\hat{z}'$. Then, this error converges to zero since $\hat{z}'$ belongs to $\mathcal{L}^2(0, 1; \mathbb{R})$.

It is important to note that the previous lemma allows to express a lower bound on the derivative with respect to θ of the Fourier-Legendre series, which only depends on the evaluation of w_n at the boundary $\theta = 1$. This bound does not depends on the first Fourier-Legendre coefficients, since we have chosen to consider the remainder only, which is orthogonal to the $2n + 1$ first Legendre polynomial. This will simplify many technical calculations in the next developments.

3.3 Modified Wirtinger's Inequality

In the literature [15], Wirtinger's inequalities refer to inequalities which estimate the integral of the derivative function with the help of the integral of the function. These inequalities have been widely used in the context of analysis, control and observation of time-delay and reaction-diffusion systems [25]. In this paper, one proposes to use Wirtinger's inequality of the first type, stated as follows.

Lemma 4 *For any function $\tilde{z}$ in $\mathcal{H}^1(-1, 1; \mathbb{R})$ such that $\tilde{z}(-1) = \tilde{z}(1)$ and that $\langle \tilde{z} | 1 \rangle = 0$, the following inequality holds:*

$$\|\tilde{z}'\| \geq \pi \|\tilde{z}\|. \tag{16}$$

Proof The proof comes along the arguments provided in [15]. Relying on the Fourier series of $\tilde{z}$ given by $\sum_{k=0}^{\infty} c_k e^{jk\pi\theta}$ with $c_k = \left\langle \tilde{z}(\theta) \middle| e^{-jk\pi\theta} \right\rangle$, it follows $\|\tilde{z}\|^2 = \sum_{k=0}^{\infty} c_k^2$. Moreover, taking also the Fourier series of $\tilde{z}'(\theta)$ as $\sum_{k=0}^{\infty} d_k e^{jk\pi\theta}$ with $d_k = \left\langle \tilde{z}'(\theta) \middle| e^{jk\pi\theta} \right\rangle$, we also have $\|\tilde{z}'\|^2 = \sum_{k=0}^{\infty} d_k^2$. The result is then obtained by integration by parts, noticing that $d_k = jk\pi c_k$.

The next lemma is an application of the previous Wirtinger inequality to the Fourier-Legendre remainder w_n given by (12).

Lemma 5 *For any function z in $\mathcal{H}^1(0, 1; \mathbb{R})$ such as $z(0) = 0$ and for any n in $\mathbb{N}$, the Fourier-Legendre remainder w_n given by (12) verifies*

$$\|w_n'\|^2 - \pi^2 \|w_n\|^2 \geq \kappa_n^W w_n^2(1), \quad \text{with} \quad \kappa_n^W = 2n(2n-1) + \frac{2\pi^2}{4n-1}. \tag{17}$$

Proof In order to apply Lemma 4, let us introduce function $\tilde{w}_n$, defined by

$$\tilde{w}_n(\theta) = w_n(\theta) - l_{2n-1}(\theta)w_n(1), \quad \forall \theta \in [-1, 1], \tag{18}$$

where w_n is given by (12). First, one has to verify the assumptions of the Wirtinger inequality of the first kind, that is $\tilde{w}_n(-1) = \tilde{w}_n(1) = 0$ and $\langle \tilde{w}_n | 1 \rangle = 0$. Recalling that $l_{2k-1}(1) = 1$ and $l_{2k-1}(-1) = -1$ for any $k \in \{1, \dots, n\}$ (see (7c), (9a)), we have

$$\begin{cases} \tilde{w}_n(1) = w_n(1) - w_n(1) = 0, \\ \tilde{w}_n(-1) = w_n(-1) + w_n(1) = -w_n(1) + w_n(1) = 0. \end{cases}$$

Furthermore, one directly obtains $\langle \tilde{w}_n | 1 \rangle = 0$ from the orthogonality of the Fourier-Legendre remainder w_n with $l_0(\theta) = 1$ (see Lemma 1). Therefore, Wirtinger's first inequality (16) states that, under the equality and integral conditions, $\|\tilde{w}_n'\| \geq \pi \|\tilde{w}_n\|$ holds. It remains to compute $\|\tilde{w}_n'\|$ and $\|\tilde{w}_n\|$. On the first hand,

$$\begin{aligned} \|\tilde{w}_n'\|^2 &= \int_{-1}^{1} \left(w_n'(\theta) - l_{2n-1}'(\theta)w_n(1)\right)^2 \mathrm{d}\theta, \\ &= \|w_n'\|^2 - 2\left\langle l_{2n-1}' \middle| w_n' \right\rangle w_n(1) + \left\langle l_{2n-1}' \middle| l_{2n-1}' \right\rangle w_n^2(1). \end{aligned}$$

Integration by parts yield

$$\begin{aligned} \|\tilde{w}_n'\|^2 &= \|w_n'\|^2 - 2\left(-\left\langle l_{2n-1}'' \middle| w_n \right\rangle + [l_{2n-1}' w_n]_{-1}^{1}\right) w_n(1) \\ &\quad + \left(-\left\langle l_{2n-1}'' \middle| l_{2n-1} \right\rangle + [l_{2n-1}' l_{2n-1}]_{-1}^{1}\right) w_n^2(1), \\ &= \|w_n'\|^2 + 2\left\langle l_{2n-1}'' \middle| w_n \right\rangle w_n(1) - \left\langle l_{2n-1}'' \middle| l_{2n-1} \right\rangle w_n^2(1) + l_{2n-1}'(1) w_n^2(1). \end{aligned}$$

By orthogonality, we can get rid of the second and third terms. Moreover, knowing

$$l_{2n-1}'(1) = \sum_{k=0}^{2(n-1)} (1 - (-1)^k)(2k+1)l_k(1) = \sum_{p=0}^{n-1} 2(4p+1),$$

we obtain

$$\|\tilde{w}_n'\|^2 = \|w_n'\|^2 - \sum_{p=0}^{n-1} 2(4p+1) w_n^2(1).$$

The last term of the previous expression has already been computed in (15), and we know that

$$\|\tilde{w}_n'\|^2 = \|w_n'\|^2 - 2n(2n-1)w_n^2(1), \tag{19}$$

On the other hand, the norm of $\tilde{w}_n$ can be computed as follows. Due to the orthogonality of the Legendre polynomials provided by Lemma 1, we have

$$\|\tilde{w}_n\|^2 = \|w_n\|^2 - 2\,\langle l_{2n-1}|w_n\rangle\, w_n(1) + \langle l_{2n-1}|l_{2n-1}\rangle\, w_n^2(1)$$
$$= \|w_n\|^2 + \frac{2}{4n-1} w_n^2(1). \tag{20}$$

Thus, the proof is concluded by merging the two expressions given in (19) and (20) into $\|\tilde{w}_n'\| \geq \pi\|\tilde{w}_n\|$.

The previous lemma extends the Wirtinger inequality $\|z'\|_{0,1} \leq \frac{\pi}{2}\|z\|_{0,1}$, which holds in the situation that z belongs to $\mathcal{H}_1(0, 1; \mathbb{R})$and that $z(0) = 0$. The main advantages of using the Fourier-Legendre remainder appears in the simple formulation of the lower bound in (17). It is important to stress that the orthogonality condition $\langle \ell_n|w_n\rangle = 0$ drastically simplifies the expression and the calculations.

It is worth mentioning that both constants κ_n^B and $\kappa_n^W = \kappa_{n-1}^B + \frac{2\pi^2}{4n-1}$ increase in a quadratic manner with respect to n. This information will not be used in this paper but could be a key point in the proof of the convergence of the sufficient stability criterion developed in the last section.

4 Modeling of an Augmented System

In this section, one proposes the design of an augmented model issued from the interconnected system (1). The introduction of such model will allow to perform an assurate stability test.

4.1 Transfer Function of the Reaction-Diffusion Equation

Take the reaction-diffusion equation with cross-type boundary conditions:

$$\begin{cases} \partial_t z(t,\theta) = (\delta\partial_{\theta\theta} + \lambda)z(t,\theta), & \forall (t,\theta) \in \mathbb{R}_{\geq 0} \times (0,1), \quad (21a) \\ \begin{bmatrix} z(t,0) \\ \partial_\theta z(t,1) \end{bmatrix} = \begin{bmatrix} 0 \\ e(t) \end{bmatrix}, & \forall t \in \mathbb{R}_{\geq 0}, \quad (21b) \end{cases}$$

where $e(t)$ is an input signal.

In the following, system (21) is transformed in Laplace domain where $s \in \mathbb{C}$ is the Laplace variable and where variables in capital letters represents the Laplace transform associated with variable in non-capital letters. As an infinite-dimensional system, the irrational transfer function between $E(s) = \frac{\partial}{\partial\theta}Z(s, 1)$ and $Z(s, \theta)$ can be expressed as in [17] by

$$G(s,\theta) = \frac{\mathrm{sh}\left(\theta\sqrt{\frac{s-\lambda}{\delta}}\right)}{\sqrt{\frac{s-\lambda}{\delta}}\mathrm{ch}\left(\sqrt{\frac{s-\lambda}{\delta}}\right)}, \quad \forall (s,\theta) \in \mathbb{C} \times [-1, 1]. \tag{22}$$

The main issue is to approximate this infinite-dimensional part in a specific manner and to highlight the link with Legendre approximation methods.

As $G(s, \theta)$ is an odd analytical function with respect to θ, it is possible to define Fourier-Legendre remainder $\tilde{G}_n$ done by truncation at order $2n - 1$ of the Legendre-Fourier series of G on Legendre polynomials coefficients which is, for all $s \in \mathbb{C}$,

$$\tilde{G}_n(s,\theta) = G(s,\theta) - \sum_{k=0}^{n-1} \frac{4k+3}{2} l_{2k+1}(\theta) \langle l_{2k+1} | G(s) \rangle, \quad \forall \theta \in [-1, 1]. \tag{23}$$

Remark 6 This remainder (23) is well-defined on the segment $[-1, 1]$ with respect to θ by the fact that $G(s)$ belongs to $C^\infty([-1, 1], \mathbb{C})$ for any s in $\mathbb{C}$ and $G_n(s)$ converges uniformly on any compact subset of $\mathbb{C}$ [28].

Then, error (23) can be rewritten by the following remained series

$$\tilde{G}_n(s,\theta) = \sum_{k=n}^{\infty} \frac{4k+3}{2} l_{2k+1}(\theta) \int_{-1}^{1} G(s,\theta) l_{2k+1}(\theta) d\theta, \quad \forall \theta \in [-1, 1], \tag{24}$$

and an estimation of its value around $s = \lambda$ can be proposed.

Lemma 6 *Error (23) is $\underset{s \to \lambda}{O}(s^n)$, for all $\theta \in [-1, 1]$.*

Proof Lemma 6 means that $\tilde{G}_n(s, \theta)$ has its $n - 1$ first derivatives with respect to s evaluated at $s = \lambda$ equal to zero and this is proven herafter. From the Maclaurin development of G around $s = \lambda$ with an infinite radius of convergence, for all $\theta \in [-1, 1]$,

$$G(s,\theta) = \frac{1}{\mathrm{ch}\left(\sqrt{\frac{s-\lambda}{\delta}}\right)} \sum_{k=n}^{\infty} \frac{\theta^{2k+1}}{(2k+1)!} \left(\frac{s-\lambda}{\delta}\right)^k.$$

The p successive derivations of G with respect to s evaluated at $s = \lambda$ give

$$\frac{\partial^p}{\partial s^p} G(\lambda,\theta) = \sum_{k=0}^{p} \frac{k!\theta^{2k+1}}{(2k+1)!}.$$

Regarding (24), one obtains zero for $p < n$ because θ^{2p} can be decomposed on the $2p + 1$ first Legendre polynomials.

This error is close to zero for s near λ and allows us to expect that it leads to accurate models, which satisfy the definition of the Padé approximants [4].

4.2 Realization of the Padé Approximant of the Transfer Function Taking Support on the First Legendre Polynomials

In this part, with the measurement $y(t) = z(t, 1)$, the irrational transfer function $H(s)$ of the reaction-diffusion equation from $E(s)$ to $Y(s)$

$$H(s) = G(s, 1) = \frac{\mathrm{th}\left(\sqrt{\frac{s-\lambda}{\delta}}\right)}{\sqrt{\frac{s-\lambda}{\delta}}}, \quad \forall s \in \mathbb{C}, \tag{25}$$

is considered. We show the relation between the modeling inspired by tau methods [14] on Legendre polynomials basis and the $(n-1|n)$ Padé approximated transfer function H_n of the irrational transfer function $H(s)$. Particular state form representation of this nice Padé approximated model is provided. Indeed, the states of this realization are Fourier-Legendre polynomials coefficients of the distributed state $z(t, \theta)$ under the following definition.

Definition 2 For any signal $z(t) \in \mathcal{L}^2(0, 1; \mathbb{R})$ satisfying (21) and for integer n in $\mathbb{N}$, we define,

- the extended state $z(t)$ on the extended interval $[-1, 1]$ by

$$\hat{z}(t, \theta) = \begin{cases} z(t, \theta), & \forall (t, \theta) \in \mathbb{R}_{\geq 0} \times [0, 1], \\ -z(t, -\theta), & \forall (t, \theta) \in \mathbb{R}_{\geq 0} \times [-1, 0], \end{cases} \tag{26}$$

- the n first non null Fourier-Legendre coefficients of $\hat{z}$ as

$$\zeta_n(t) = \begin{bmatrix} \int_{-1}^{1} l_1(\theta)\hat{z}(t,\theta)\mathrm{d}\theta \\ \vdots \\ \int_{-1}^{1} l_{2n-1}(\theta)\hat{z}(t,\theta)\mathrm{d}\theta \end{bmatrix} = \langle \ell_n | \hat{z}(t) \rangle \in \mathbb{R}^n, \quad \forall t \in \mathbb{R}_{\geq 0}, \tag{27}$$

- the associated Fourier-Legendre remainder of $\hat{z}(t)$ at order n as

$$w_n(t, \theta) = \hat{z}(t, \theta) - \ell_n^\top(\theta) \mathcal{I}_n \zeta_n(t), \quad \forall (t, \theta) \in \mathbb{R}_{\geq 0} \times [-1, 1]. \tag{28}$$

Then, we build a finite-dimensional approximation of system (21) taking support on Legendre polynomials coefficients and certify that the approximated model matches with Padé approximants, thanks to Lemmas 6 and 7.

Proposition 3 *System* $\left(\begin{smallmatrix} \mathcal{A}_n & \mathcal{B}_n \\ C_n & 0 \end{smallmatrix}\right)$ *is a realization of the* $(n-1|n)$ *Padé approximant of the transfer function* $H(s)$ *given by (25) with*

$$\begin{aligned} &\mathcal{A}_n = \lambda I_n + \delta \mathcal{L}_n + \mathcal{B}_n^* C_n, \\ &\mathcal{B}_n = 2\delta \ell_n(1), \quad \mathcal{B}_n^* = -2\delta \ell_n'(1), \quad C_n = \ell_n(1)^T \mathcal{I}_n, \quad C_n^* = -\ell_n'^{\top}(1)\mathcal{I}_n, \end{aligned} \tag{29}$$

where $\mathcal{L}_n, \mathcal{I}_n$ *are given in (8) and* $\ell_n(1), \ell_n'(1)$ *recalled in (9).*

Proof By two successive integrations by parts, the dynamics of the n coefficients $\zeta_n(t)$ defined by (27) lead to

$$\begin{aligned} \tfrac{\mathrm{d}}{\mathrm{d}t}\zeta_n(t) &= \delta \left\langle \ell_n \middle| \partial_{\theta\theta}\hat{z}(t)\right\rangle + \lambda \zeta_n(t), \\ &= -\delta \left\langle \ell_n' \middle| \partial_{\theta}\hat{z}(t)\right\rangle + \delta \left[\ell_n(\theta)\partial_\theta \hat{z}(t,\theta)\right]_{-1}^{1} + \lambda \zeta_n(t), \\ &= \delta \left\langle \ell_n'' \middle| z(t)\right\rangle + \lambda \zeta_n(t) + \delta \left[\ell_n(\theta)\tfrac{\partial}{\partial\theta}\hat{z}(t,\theta)\right]_{-1}^{1} - \delta\left[\ell_n'(\theta)\hat{z}(t,\theta)\right]_{-1}^{1}, \\ &= (\lambda I_n + \delta \mathcal{L}_n)\zeta_n(t) + 2\delta \ell_n(1) e(t) - 2\delta \ell_n'(1) z(t,1). \end{aligned}$$

The signal $y(t) = z(t,1)$ can be approximated by its the truncated Fourier-Legendre series at order n and called $w_n(t,1)$. That implies

$$\left\{ \begin{aligned} &\frac{\mathrm{d}}{\mathrm{d}t}\zeta_n(t) = \mathcal{A}_n \zeta_n(t) + \mathcal{B}_n e(t) + \mathcal{B}_n^* w_n(t,1), \\ &y(t) = z(t,1) = C_n \zeta_n(t) + w_n(t,1). \end{aligned} \right. \tag{30}$$

In Laplace domain, the error $\mathcal{W}_n(s,1) = \tilde{H}_n(s)U(s)$ with

$$\tilde{H}_n(s) = \tilde{G}_n(s,1) = H(s) - C_n \left\langle \ell_n | G(s) \right\rangle. \tag{31}$$

It is the remainder of the truncated Fourier-Legendre series of G evaluated at $\theta = 1$. System (30) leads to

$$\left\{ \begin{aligned} &\mathcal{Z}_n(s) = (sI_n - \mathcal{A}_n)^{-1}(\mathcal{B}_n + \mathcal{B}_n^* \tilde{H}_n(s))U(s), \\ &Y(s) = C_n \mathcal{Z}_n(s) + \tilde{H}_n(s)U(s). \end{aligned} \right.$$

The transfer function $H(s)$ from $E(s)$ to $Y(s)$ is

$$H(s) = C_n(sI_n - \mathcal{A}_n)^{-1}\mathcal{B}_n + \left(1 + C_n(sI_n - \mathcal{A}_n)^{-1}\mathcal{B}_n^*\right)\tilde{H}_n(s).$$

Thanks to Lemma 6, we already know that $\tilde{H}_n(s) = \underset{s\to\lambda}{O}(s^n)$. Then, the application of Lemma 7 given in Appendix with vectors $u = -\mathcal{B}_n^*$, $v^T = C_n$ and matrix $L = -\delta\mathcal{L}_n$ ensures that

$$H(s) - C_n(sI_n - \mathcal{A}_n)^{-1}\mathcal{B}_n = \underset{s\to\lambda}{O}(s^{2n}).$$

According to the definition of Padé approximations given in [4], the rational function $H_n(s) = C_n(sI_n - \mathcal{A}_n)^{-1}\mathcal{B}_n$ is a $(n-1|n)$ Padé approximant of $H(s)$ at $s = \lambda$.

This proposition motivates the construction of an augmented model for system (1) which put aside this nice Padé realization of the reaction-diffusion part.

4.3 Modeling of the Interconnected System

Consider now (x, z), solution to system (1) and, resuming Definition 2, denote the first Fourier-Legendre coefficients by ζ_n and the Fourier-Legendre remainder w_n, for any n in $\mathbb{N}$. The objective of this section is to rewrite system (1) by exhibiting a finite-dimensional part composed of the $\xi_n = \left[\begin{smallmatrix}\zeta_n \\ x\end{smallmatrix}\right]$ and an infinite-dimensional part represented by the Fourier-Legendre remainder w_n. This is formulated in the following proposition.

Proposition 4 *If (x, z) is a solution of system (1), then $(\xi_n = \left[\begin{smallmatrix}\zeta_n \\ x\end{smallmatrix}\right], w_n)$ defined by (27), (28) verifies the following dynamics*

$$\begin{cases} \dot{\xi}_n(t) = \mathbf{A}_n\xi_n(t) + \mathbf{B}_n w_n(t, 1), & \text{(32a)} \\ \partial_t w_n(t,\theta) = (\delta\partial_{\theta\theta} + \lambda)w_n(t,\theta) - \ell_n^\top(\theta)\mathcal{I}_n\big(\mathbf{E}_n\xi_n(t) + \mathbf{F}_n w_n(t,1)\big), & \text{(32b)} \\ \left[\begin{smallmatrix} w_n(t,0) \\ \partial_\theta w_n(t,1)\end{smallmatrix}\right] = \left[\begin{smallmatrix} 0 \\ \mathbf{C}_n\xi_n(t)\end{smallmatrix}\right], & \text{(32c)} \\ (\xi_n(0), w_n(0)) = \left(\left[\begin{smallmatrix} x_0 \\ \zeta_{0,n}\end{smallmatrix}\right], w_{0,n}\right), & \text{(32d)} \end{cases}$$

where the matrices that defined this model are given by

$$\mathbf{A}_n = \begin{bmatrix} \mathcal{A}_n & \mathcal{B}_n C \\ BC_n & A \end{bmatrix}, \ \mathbf{B}_n = \begin{bmatrix} \mathcal{B}_n^* \\ B \end{bmatrix}, \ \mathbf{C}_n = \begin{bmatrix} C_n^* & C \end{bmatrix}, \ \mathbf{E}_n = \mathcal{B}_n\mathbf{C}_n, \ \mathbf{F}_n = \mathcal{B}_n^*, \quad (33)$$

with matrices $\mathcal{A}_n$, $\mathcal{B}_n$, $\mathcal{B}_n^*$, C_n and C_n^* are given by (29).

Proof The proof is split into three parts referring to each equation in (32).
Proof of (32a): According to the proof Proposition 3, we already know that $\zeta_n = \langle \ell_n | \hat{z}\rangle$ satisfies (30). Then, adding the dynamics of the ordinary differential equation given by (1a), the dynamics of the finite dimensional state are

$$\underbrace{\frac{\mathrm{d}}{\mathrm{d}t}\begin{bmatrix}\zeta_n(t) \\ x(t)\end{bmatrix}}_{\frac{\mathrm{d}}{\mathrm{d}t}\xi_n(t)} = \underbrace{\begin{bmatrix} \mathcal{A}_n & \mathcal{B}_n C \\ BC_n & A \end{bmatrix}}_{\mathbf{A}_n}\underbrace{\begin{bmatrix}\zeta_n(t) \\ x(t)\end{bmatrix}}_{\xi_n(t)} + \underbrace{\begin{bmatrix}\mathcal{B}_n^* \\ B\end{bmatrix}}_{\mathbf{B}_n} w_n(t, 1), \quad (34)$$

which corresponds to the first Eq. (32a).
Proof of (32b): Let us now focus in the second Eq. (32b), which refers to the partial

differential equation resulting from the changes of variable z to w_n. To do so, differentiating with respect to time of the Fourier-Legendre remainder w_n given in (28) yields $\partial_t w_n(t,\theta) = \partial_t z(t,\theta) - \ell_n^\top(\theta)\mathcal{I}_n \frac{\mathrm{d}}{\mathrm{d}t}\zeta_n(t)$. From one side, we need to express $\partial_t z$ using the new system of coordinates, that is reflected in

$$\begin{aligned}\partial_t z(t,\theta) &= \delta\partial_{\theta\theta} z(t,\theta) + \lambda z(t,\theta)\\ &= (\delta\partial_{\theta\theta} + \lambda)w_n(t,\theta) + \left(\delta\partial_{\theta\theta}\ell_n^\top(\theta) + \lambda\ell_n^\top(\theta)\right)\mathcal{I}_n\zeta_n(t).\end{aligned}$$

Applying the differentiation rules of the Legendre polynomials in (7b), the previous expression resumes to

$$\partial_t z(t,\theta) = (\delta\partial_{\theta\theta} + \lambda)w_n(t,\theta) + \ell_n^\top(\theta)\left(\delta\mathcal{L}_n^\top + \lambda I_n\right)\mathcal{I}_n\zeta_n(t). \tag{35}$$

On the other side, the expression of $\frac{\mathrm{d}}{\mathrm{d}t}\zeta_n(t)$ given by (30) leads to

$$\frac{\mathrm{d}}{\mathrm{d}t}\zeta_n(t) = \mathcal{I}_n^{-1}(\delta\mathcal{L}_n^\top + \lambda I_n)\mathcal{I}_n\zeta_n(t) + \underbrace{\mathcal{B}_n[\,C_n^*\ C\,]}_{\mathbf{E}_n}\begin{bmatrix}\zeta_n(t)\\ x(t)\end{bmatrix} + \underbrace{\mathcal{B}_n^*}_{\mathbf{F}_n} w_n(t,1), \tag{36}$$

and calcultation details can be found in [3]. Thus, collecting (35), (36) and simplifying the term in $\delta\mathcal{L}_n^\top + \lambda I_n$, the partial differential equation verified by w_n is obtained. Proof of (32b): Finally, the first boundary condition is already verified according to Lemma 1 and the second boundary condition is directly obtained by derivation of (28) and evaluation at $\theta = 1$.

Remark 7 Matrix $\mathbf{A}_n$ given by (33) is equal to

$$\mathbf{A}_n = \left(\begin{smallmatrix}\mathcal{A}_n & \mathcal{B}_n\\ C_n & 0\end{smallmatrix}\right) \otimes \left(\begin{smallmatrix}A & B\\ C & 0\end{smallmatrix}\right) = H_n(s) \otimes \left(\begin{smallmatrix}A & B\\ C & 0\end{smallmatrix}\right), \tag{37}$$

where $\otimes$ denotes the Redheffer product and $H_n(s)$ the Padé $(n-1|n)$ approximant of $H(s)$ given by (25) at $s = \lambda$.

In the new formulation, the reaction-diffusion equation, which characterizes the dynamics of w_n, is similar to the one of the original system. The only difference relies on the last term in (32b). Even though, it seems at a first sight more complicated, it will appear in the next developments that this new term has no impact on the complexity of the analysis. This is due to the orthogonality of this new term with the Fourier-Legendre remainder, w_n.

5 Stability Analysis

This last section is dedicated to the construction of a numerical tractable stability criterion for system (1), based on Lemmas 3 and 5 and highly related to the properties of the augmented model (32).

Theorem 1 *For a given integer n in $\mathbb{N}$ and some λ, δ satisfying $\frac{\lambda}{\delta} < \pi^2$, assume that there exists $\mathbf{P}_n$ in $\mathbb{S}_+^{n_x+n+1}$ such that matrix*

$$\Xi_n = \begin{cases} \begin{bmatrix} \mathcal{H}(\mathbf{P}_n\mathbf{A}_n) & \mathbf{P}_n\mathbf{B}_n + \mathbf{C}_n^\top \\ * & -\kappa_n^B \end{bmatrix}, & \text{if } \lambda \leq 0, \\ \begin{bmatrix} \mathcal{H}(\mathbf{P}_n\mathbf{A}_n) & \mathbf{P}_n\mathbf{B}_n + \mathbf{C}_n^\top \\ * & -\left(1 - \frac{\lambda}{\delta\pi^2}\right)\kappa_n^B - \frac{\lambda}{\delta\pi^2}\kappa_n^W \end{bmatrix}, & \text{if } \lambda > 0. \end{cases} \tag{38}$$

is negative definite, where constants κ_n^B, κ_n^W are defined respectively by (14), (17) and matrices $\mathbf{A}_n$, $\mathbf{B}_n$, $\mathbf{C}_n$ and $\mathbf{D}_n$ are defined by (33). Then, under the conditions prescribed by Propositions 1 and 2, the equilibrium $(0, 0)$ of system (1) is globally exponentially stable, in the sense of the $\mathbb{R}^{n_x} \times \mathcal{L}^2(0, 1; \mathbb{R})$ norm.

Proof Consider the Lyapunov functional given by

$$\mathcal{V}_n(x, z) = \underbrace{\xi_n^\top \mathbf{P}_n \xi_n}_{\mathcal{V}_n^a(x,z)} + \underbrace{(2\delta)^{-1}\|w_n\|^2}_{\mathcal{V}_n^b(x,z)}, \tag{39}$$

with $\mathbf{P}_n$ in $\mathbb{S}_+^{n_x+n+1}$. It is a Lyapunov candidate functional for both systems (1), (32). Differentiation of $V_n^a(t) = \mathcal{V}_n^a(x(t), z(t))$ along the trajectories of the system (1) using the dynamics given by (32) yields

$$\frac{\mathrm{d}}{\mathrm{d}t} V_n^a(t) = \xi_n^\top(t)\mathcal{H}(\mathbf{P}_n\mathbf{A}_n)\xi_n(t) + 2\xi_n^\top(t)\mathbf{P}_n\mathbf{B}_n w_n(t, 1).$$

Secondly, from the dynamics of the Fourier-Legendre remainder in (32), denoting $V_n^b(t) = \mathcal{V}_n^b(x(t), z(t))$, we find

$$\frac{\mathrm{d}}{\mathrm{d}t} V_n^b(t) = \int_{-1}^{1} \partial_{\theta\theta} w_n(t, \theta) w_n(t, \theta)\mathrm{d}\theta + \frac{\lambda}{\delta}\int_{-1}^{1} w_n^2(t, \theta)\mathrm{d}\theta.$$

Using an integration by parts, this expression is decomposed in $\mathcal{L}^2(-1, 1; \mathbb{R})$ norms of signals w_n and $\partial_\theta w_n$ as

$$\begin{aligned} \frac{\mathrm{d}}{\mathrm{d}t} V_n^b(t) &= \frac{\lambda}{\delta}\|w_n(t)\|^2 - \|\partial_\theta w_n(t)\|^2 + 2\partial_\theta w_n(t, 1) w_n(t, 1), \\ &= \frac{\lambda}{\delta}\|w_n(t)\|^2 - \|\partial_\theta w_n(t)\|^2 + 2\xi_n^\top(t)\mathbf{C}_n^\top w_n(t, 1). \end{aligned}$$

Thereafter, the proof is split into two cases. If $\lambda \leq 0$, the proof is a straightforward application of Bessel-Legendre inequality given by (14). Indeed, we have

$$\frac{\mathrm{d}}{\mathrm{d}t} V_n(t) \leq -\frac{|\lambda|}{\delta}\|w_n(t)\|^2 + \begin{bmatrix} \xi_n(t) \\ w_n(t,1) \end{bmatrix}^\top \Xi_n \begin{bmatrix} \xi_n(t) \\ w_n(t,1) \end{bmatrix},$$

Assuming $\Xi_n \prec 0$, application of Lyapunov theorem leads to Theorem 1.
For the case $0 < \lambda < \delta\pi^2$, we apply first the adapted Wirtinger inequality (17) on the Fourier-Legendre remainder w_n to obtain

$$\frac{\mathrm{d}}{\mathrm{d}t}V_n^b(t) \leq -\epsilon\pi^2\|w_n(t)\|^2 - \left(1 - \frac{\lambda}{\delta\pi^2} - \epsilon\right)\|\partial_\theta w_n(t)\|^2 + 2\xi_n^\top(t)\mathbf{C}_n^\top w_n(t,1) - \left(\frac{\lambda}{\delta\pi^2} + \epsilon\right)\kappa_n^W w_n^2(t,1),$$

for any sufficiently small $\epsilon > 0$ such that $(1 - \frac{\lambda}{\delta\pi^2} - \epsilon) > 0$. Then, the Bessel-Legendre inequality given by (14) yields

$$\frac{\mathrm{d}}{\mathrm{d}t}V_n^b(t) \leq -\epsilon\pi^2\|w_n\|^2 + 2\xi_n^\top(t)\mathbf{C}_n^\top w_n(t,1) - \left((1 - \frac{\lambda}{\delta\pi^2} - \epsilon)\kappa_n^B + (\frac{\lambda}{\delta\pi^2} + \epsilon)\kappa_n^W\right)w_n^2(t,1).$$

Merging with $\frac{\mathrm{d}}{\mathrm{d}t}V_n^a(t)$ gives

$$\frac{\mathrm{d}}{\mathrm{d}t}V_n(t) \leq -\epsilon\pi^2\|w_n(t)\|^2 + \begin{bmatrix}\xi_n(t)\\ w_n(t,1)\end{bmatrix}^\top \Xi_n \begin{bmatrix}\xi_n(t)\\ w_n(t,1)\end{bmatrix} + \epsilon\kappa_n^B w_n^2(t,1).$$

If the linear matrix inequality $\Xi_n \prec 0$ holds, it is possible to take a small enough $\epsilon > 0$ such that $-|\bar{\sigma}(\Xi_n)| + \epsilon\kappa_n^B < 0$. Then, there exists $\rho > 0$ such that the derivatives $\frac{\mathrm{d}}{\mathrm{d}t}V_n(t)$ satisfies $\frac{\mathrm{d}}{\mathrm{d}t}V_n(t) \leq -\rho V_n(t)$. The proof is concluded by application of Lyapunov theorem on the exponential stability at the equilibrium point.

Remark 8 Notice that $\mathbf{A}_n$ Hurwitz is a necessary condition for the feasibility of linear matrix inequality $\Xi_n \prec 0$, where Ξ_n defined by (38). This is promising to the matter of fact that $\mathbf{A}_n$ is an approximated model for the original system (1).

Remark 9 Compared to generalized linear matrix inequality formulations based on sum of squares [10], the result is condensed, more appropriate to the application and numerical burden are improved. This optimization is due to the transformations made to obtain our linear matrix inequality, highly correlated to the system under study. Nevertheless, to the best of our knowledge, we are not aware of stability condition addressing this particular class of systems and this is the reason why no further comparison will be presented.

The procedure to implement Theorem 1 can be decomposed as follows:

- Pick an order $n \in \mathbb{N}$ and some coefficients λ, δ satisfying $\frac{\lambda}{\delta} < \pi$,
- Compute matrices $\mathbf{A}_n$, $\mathbf{B}_n$, $\mathbf{C}_n$ given by (33) which depend on system parameters,
- Check the feasibility of linear matrix inequality (38) with *Yalmip* tools,
- In the case that the problem is feasible, the proposed system is stable.

Remark 10 To reduce computational efforts, a good strategy could be to test initially low orders n and increase it gradually. In practice, we stop at order $n = 10$ for which the feasability problem is too long to solve numerically. Moreover, we are not able to prove that such criteria is hierarchic and converges towards all stability areas.

6 Application to Numerical Examples

6.1 On Padé Approximations

Without loss of generalization, let take $\lambda = 0$ and $\delta = 1$. The approximated model $\left(\begin{smallmatrix} \mathcal{A}_n & \mathcal{B}_n \\ \mathcal{C}_n & 0 \end{smallmatrix}\right)$ leads to a transfer function H_n. For low orders, the corresponding matrices and approximated transfer functions have been stored in Table 1.

By comparison with the Padé approximant of $H(s) = G(s, 1) = \frac{\mathrm{sh}(\sqrt{s})}{\sqrt{s}\mathrm{ch}(\sqrt{s})}$ at $s = 0$ computed by Matlab *pade* tool, the same transfer function H_n are recognized.

Then, according to Remark 7, the approximated model $\mathbf{A}_n$ will be equivalent to the star product of the finite-dimensional model with this Padé $(n-1|n)$ transfer function of the reaction-diffusion equation.

6.2 On Stability Results

In this subsection, a standard example is chosen. The stability properties with respect to λ are investigated with the method proposed in this chapter.

Example 1 Let system (1) with $A = -1$, B taking values in $[-1, 2]$, $C = 1$ and $\delta = 1$.

The feasibility of linear matrix inequality $\Xi_n \prec 0$ with Ξ_n given by (38) is determined with feasp function and tested for $n = 2$. The maximal bound λ_{max} which ensures the stability of the system is collected in Table 2 for several values of B.

Table 1 Padé $(n-1|n)$ approximants of transfer function $H(s)$

Order n	1	2	3	4
$\mathcal{A}_n$	-3	$\begin{bmatrix} -3 & -7 \\ -3 & -42 \end{bmatrix}$	$\begin{bmatrix} -3 & -7 & -11 \\ -3 & -42 & -66 \\ -3 & -42 & -165 \end{bmatrix}$	$\begin{bmatrix} -3 & -7 & -11 & -15 \\ -3 & -42 & -66 & -90 \\ -3 & -42 & -165 & -225 \\ -3 & -42 & -165 & -420 \end{bmatrix}$
$\mathcal{B}_n$	2	$[2\ 2]^\top$	$[2\ 2\ 2]^\top$	$[2\ 2\ 2\ 2]^\top$
$\mathcal{C}_n$	$\frac{3}{2}$	$\left[\frac{3}{2}\ \frac{7}{2}\right]$	$\left[\frac{3}{2}\ \frac{7}{2}\ \frac{11}{2}\right]$	$\left[\frac{3}{2}\ \frac{7}{2}\ \frac{11}{2}\ \frac{15}{2}\right]$
$H_n(s)$	$\frac{3}{s+3}$	$5\frac{2s+21}{s^2+45s+105}$	$21\frac{s^2+60s+495}{s^3+210s^2+4725s+10395}$	$9\frac{4s^3+770s^2+30030s+225225}{s^4+630s^3+51975s+2027025}$

Table 2 Maximal allowable reaction coefficient $\lambda_{\max}$ guaranteed by Theorem 1

Order n	1	2	3	4	Expected
$B = 2$	−3.751	−3.668	−3.668	−3.668	−3.668
$B = 0$	2.461	2.467	2.467	2.467	$(\frac{\pi}{2})^2 = 2.467$
$B = 0.1$	2.258	2.263	2.263	2.263	2.263
$B = -1$	3.176	3.672	3.685	3.685	3.685

This result is compared with the expected bounds coming from the stability of an approximated model at order 100 (see tau methods [13]).

According to Table 2, we can see that an upper bound for parameter λ has been found to guarantee exponential stability of the origin of system (1). Based on cases $B < 0$, it is important to note that the classical upper bound $(\frac{\pi}{2})^2$ can be exceeded thanks to the stabilizable properties of A, which is Hurwitz. It is also worth noticing that the proposed test seem to be hierarchical and to converge toward the expected bound of the region of stability in a fast manner (from order $n \simeq 3$). At last, the required order n seems to be larger as λ_{max} increases (i.e. B decreases).

7 Conclusion

A Lyapunov approach is applied to derive a new sufficient condition of stability for a linear finite-dimensional system coupled with a reaction diffusion process in the linear matrix framework. Our work focused on the rewriting of Bessel and Wirtinger inequalities involved in the Lyapunov analysis of the reaction-diffusion equation and signals linked with the associated Legendre remainder. Then, a dynamical model has been constructed to enforce these signals and the extended state composed of the first Legendre polynomials coefficients. Finally, taking a classical quadratic Lyapunov functional fitted to this new modeling, a condition expressed as a linear matrix inequality is directly obtained.

Based on the structure of the result, robust approach could be pursued to construct finite-dimensional controllers stabilizing the original infinite-dimensional system. Observer design by early-lumping might also be investigated.

Appendix

Derived from the matrix inversion and determinant lemmas, a usefull lemma can also be formulated.

Lemma 7 *For any $u \in \mathbb{R}^n$ with a non-zero first component, $v \in \mathbb{R}^n$ not equal to the zero vector and $L \in \mathbb{R}^{n\times n}$ a strictly lower triangular matrix with non-zero under diagonal coefficients (i.e. $L_{i+1,i} \neq 0\ \forall i \in \{1,\dots n-1\}$), one obtains*

$$1 - v^T((s-\lambda)I_n + L + uv^T)^{-1}u = \underset{s\to\lambda}{O}(s^n). \tag{40}$$

Proof The matrix inversion lemma applied to vectors $u \in \mathbb{R}^n$, $v \in \mathbb{R}^n$ and non singular matrix $M = sI_n + L \in \mathbb{R}^{n\times n}$ gives, for any $s \in \mathbb{C}\backslash\{0\}$,

$$1 - v^T((s-\lambda)I_n + L + uv^T)^{-1}u = (1 + v^T((s-\lambda)I_n + L)^{-1}u)^{-1},$$

and the matrix determinant lemma leads to

$$1 - v^T((s-\lambda)I_n + L + uv^T)^{-1}u = \frac{\det((s-\lambda)I_n + L)}{\det((s-\lambda)I_n + L + uv^T)}.$$

Then, since L is strictly lower triangular, we have

$$\det((s-\lambda)I_n + L) = \det((s-\lambda)I_n) = (s-\lambda)^n.$$

and, because L has non-zero under diagonal coefficients and under the hypothesis done on vectors u, v, matrix $L + uv^T$ has full rank which means $\det(L + uv^T) \neq 0$. That yields the result for s tending to λ.

References

1. Aarsnes, U., Vazquez, R., Di Meglio, F., Krstic, M.: Delay robust control design of under-actuated PDE-ODE-PDE systems. In: 2019 American Control Conference (ACC), pp. 3200–3205. IEEE (2019)
2. Bajodek, M., Seuret, A., Gouaisbaut, F.: Insight into stability analysis of time-delay systems using legendre polynomials. IFAC-PapersOnLine **53**(2), 4816–4821 (2020). 21st IFAC World Congress
3. Bajodek, M., Seuret, A., Gouaisbaut, F.: Stability analysis of an ordinary differential equation interconnected with the reaction-diffusion equation. Submitted (2021)
4. Baker, G., Graves-Morris, P.: Padé Approximants, 2nd edn. Encyclopedia of Mathematics and its Applications $n°59$ (1996)
5. Barreau, M., Seuret, A., Gouaisbaut, F., Baudouin, L.: Lyapunov stability analysis of a string equation coupled with an ordinary differential system. IEEE Trans. Autom. Control **63**(11), 3850–3857 (2018)
6. Baudouin, L., Seuret, A., Gouaisbaut, F.: Stability analysis of a system coupled to a heat equation. Automatica 195–202 (2019)
7. Curtain, R., Morris, K.: Transfer functions of distributed parameter systems: a tutorial. Automatica **45**(5), 1101–1116 (2009)
8. Curtain, R., Zwart, H.: An Introduction to Infinite-Dimensional Linear Systems Theory. Springer (1995)

9. Das, A., Shivakumar, S., Peet, M., Weiland, S.: Robust analysis of uncertain ODE-PDE systems using PI multipliers, PIEs and LPIs. In: 2020 59th IEEE Conference on Decision and Control (CDC), pp. 634–639. IEEE (2020)
10. Das, A., Shivakumar, S., Weiland, S., Peet, M.: Representation and stability analysis of PDE-ODE coupled systems (2018). ArXiv:1812.07186
11. Evans, L.: Partial Differential Equations. American Mathematical Society (1997)
12. Gautschi, W.: Orthogonal polynomials, quadrature, and approximation: computational methods and software (in Matlab). Lect. Notes Math. **1883**, 1–77 (2006)
13. Gottlieb, D., Orszag, S.: Numerical analysis of spectral methods. Society for Industrial and Applied Mathematics, Philadelphia (1977)
14. Ito, K., Teglas, R.: Legendre-Tau approximations for functional differential equations. SIAM J. Control Optim. **24**(4), 737–759 (1986)
15. Kammler, W.: A first course in Fourier analysis. Cambridge University Press (2007)
16. Karafyllis, I., Krstic, M.: Input-to-State Stability for PDEs. Springer (2018)
17. Krstic, M.: Delay Compensation for Nonlinear, Adaptive, and PDE Systems. Springer (2009)
18. Lasieka, I., Triggiani, R.: Control Theory for Partial Differential Equations: Continuous and Approximation Theories. Cambridge University Press (2000)
19. Lhachemi, H., Shorten, R.: Boundary feedback stabilization of a reaction-diffusion equation with Robin boundary conditions and state-delay. Automatica **116**, 108931 (2020)
20. Logemann, H.: Circle criteria, small-gain conditions and internal stability for infinite-dimensional systems. Automatica **27**(4), 677–690 (1991)
21. Mironchenko, A., Prieur, C.: Input-to-state stability of infinite-dimensional systems: recent results and open questions. SIAM Rev., Soc. Ind. Appl. Math. **62**(3), 529–614 (2020)
22. Prieur, C., Girard, A., Witrant, E.: Stability of switched linear hyperbolic systems by Lyapunov techniques. IEEE Trans. Autom. Control **59**(8), 2196–2202 (2014)
23. Prieur, C., Trelat, E.: Feedback stabilization of a 1-D linear reaction-diffusion equation with delay boundary control. IEEE Trans. Autom. Control **64**(4), 1415–1425 (2019)
24. Safi, M., Baudouin, L., Seuret, A.: Tractable sufficient stability conditions for a system coupling linear transport and differential equations. Syst. Control Lett. **110**, 1–8 (2017)
25. Selivanov, A., Fridman, F.: Boundary observers for a reaction-diffusion system under time-delayed and sampled-data measurements. IEEE Trans. Autom. Control **64**(8), 3385–3390 (2019)
26. Valmorbida, G., Ahmadi, M., Papachristodoulou, A.: Stability analysis for a class of partial differential equations via semidefinite programming. IEEE Trans. Autom. Control **61**(6), 1649–1654 (2016)
27. Wang, Z., Zhang, H., Li, P.: An LMI approach to stability analysis of reaction-diffusion Cohen-Grossberg neural networks concerning Dirichlet boundary conditions and distributed delays. IEEE Trans. Syst. Man Cybern. **40**(6), 1596–1606 (2010)
28. Webb, M., Coppé, V., Huybrechs, D.: Pointwise and uniform convergence of Fourier extensions. Constr. Approx. **52**, 139–175 (2019)

Tracking Control for 2 × 2 Linear Heterodirectional Hyperbolic PDEs that Are Bidirectionally Coupled with Nonlinear ODEs

Abdurrahman Irscheid, Nicole Gehring, Joachim Deutscher, and Joachim Rudolph

Abstract A solution-based approach to trajectory tracking control is presented for 2×2 linear heterodirectional hyperbolic PDEs that are actuated at one boundary and bidirectionally coupled with nonlinear ODEs at the other one. The control strategy relies on the solution of the PDE subsystem, significantly simplified by a preliminary backstepping transformation, to essentially reduce the PDE-ODE system to the nonlinear boundary ODEs with a delayed input. A tracking controller, e.g. a flatness-based feedback, is assumed to exist for this reduced system in the case without an input delay. To control the delay system based on this feedback, a prediction of the ODE state is required, which follows from the solution of a general nonlinear Volterra integro-differential equation. An explicit solution is found in the special case of linear ODEs, which, with an appropriate choice of design parameters, is used for showing that the presented solution-based controller for the PDE-ODE system is equivalent to known backstepping-based stabilizing controllers. The performance of the proposed tracking controller is demonstrated in numerical simulations.

The work of the second author has been supported by the FFG, Contract No. 854184: "Pro^{2}Future".

A. Irscheid (✉) · J. Rudolph
Chair of Systems Theory and Control Engineering, Saarland University, Campus A5 1, 66123 Saarbrücken, Germany
e-mail: a.irscheid@lsr.uni-saarland.de

J. Rudolph
e-mail: j.rudolph@lsr.uni-saarland.de

N. Gehring
Institute of Automatic Control and Control Systems Technology, Johannes Kepler University Linz, Altenberger Straße 69, 4040 Linz, Austria
e-mail: nicole.gehring@jku.at

J. Deutscher
Institute for Measurement, Control and Microtechnology, Ulm University, Albert-Einstein-Allee 41, 89081 Ulm, Germany
e-mail: joachim.deutscher@uni-ulm.de

J. Auriol et al. (eds.), *Advances in Distributed Parameter Systems*, Advances in Delays and Dynamics 14, https://doi.org/10.1007/978-3-030-94766-8_6

1 Introduction

The backstepping-based stabilization of 2×2 linear hyperbolic systems has been discussed thoroughly in [22, 23]. Therein, the plant is mapped into a finite-time stable target system by transforming the state into so-called backstepping coordinates. Similar controllers have been designed in the case of bidirectional coupling at the unactuated boundary with linear ODEs. In [7], the PDE-ODE system is mapped into a cascade of stable subsystems by subsequent backstepping transformations, whereas the delay-robust approach in [2] dispenses with the cascading of the target system, yet still maps the PDE into decoupled transport equations. The case of general linear heterodirectional hyperbolic PDEs is found in [13] and [8, 9] for both the case without and with ODEs, respectively. Another method relying on special state transformations is the flatness-based tracking control discussed in [24] (see also [12]), which uses a (differential) parameterization of the input and canonical forms.

In the case of nonlinear ODEs at the unactuated boundary, [3, 5] combine backstepping and prediction-based control. The former designs a stabilizing controller for an oil-drilling system, described by a wave equation that is bidirectionally coupled with a nonlinear ODE, using a combination of backstepping coordinates and predictors. The same techniques are applied in [5] for a cascade of a wave equation with a nonlinear ODE. Overall, predictors play an important role for the control of PDE-ODE systems with nonlinearities. This is particularly the case for nonlinear PDEs as discussed in [4, 21], where prediction problems arise by considering the solution of quasilinear and semilinear PDEs, respectively.

The present work also uses predictors to design a solution-based tracking controller for 2×2 linear heterodirectional hyperbolic PDEs with spatially-varying coefficients that are bidirectionally coupled with nonlinear ODEs. For that, a tracking controller is assumed to be known for the nonlinear ODE boundary system. This controller could originate from passivity-based or port-Hamiltonian methods, sliding mode control, as well as high-gain or flatness-based approaches, to name but a few. From that, a tracking controller is derived for the PDE-ODE system. It is based on a reduced representation involving only the nonlinear ODEs with delayed actuation, which is obtained by a preliminary backstepping transformation and the method of characteristics. The proposed controller for the overall PDE-ODE system requires future values of the PDE and ODE states. In fact, these predictions are determined by solving a general nonlinear Volterra integro-differential equation. In the end, this approach allows for a modular design of the tracking controller for the PDE-ODE system on the basis of known controllers for the PDE and ODE subsystems. Additionally, this design strategy enables one to easily interchange the chosen controller for the nonlinear ODE boundary system. Similar concepts are introduced in [15] for constant transport velocities and linear ODEs, with a brief sketch for the case of nonlinear ODEs. The equivalence of the design to known backstepping-based stabilizing controllers in the linear case, as claimed in [15], is proven in this work.

This contribution is organized as follows. The proposed solution-based tracking controller is derived in Sect. 3, preceded by the problem formulation in Sect. 2. It is shown that the prediction of the ODE state, necessary for the resulting feedback, is found as the solution of a nonlinear integro-differential equation. Simulation results in Sect. 4 highlight the performance of the designed controller. Finally, the special case of linear PDE-ODE systems is considered in Sect. 5. An explicit solution to the prediction problem is found and the equivalence to the known stabilizing backstepping controller for linear PDE-ODE systems in [7] is shown.

2 Problem Formulation

2.1 System Class

The coupled PDE-ODE system

$$\dot{\boldsymbol{w}}(t) = \boldsymbol{f}(\boldsymbol{w}(t), x_1(0,t)) \tag{1a}$$

$$x_2(0,t) = q_0 x_1(0,t) + c(\boldsymbol{w}(t)) \tag{1b}$$

$$\partial_t \boldsymbol{x}(z,t) = \Lambda(z)\partial_z \boldsymbol{x}(z,t) + A(z)\boldsymbol{x}(z,t) \tag{1c}$$

$$x_1(1,t) = q_1 x_2(1,t) + u(t) \tag{1d}$$

under study consists of a nonlinear ODE subsystem with the lumped state $\boldsymbol{w}(t) \in \mathbb{R}^n$ and a 2×2 linear heterodirectional hyperbolic PDE subsystem with the distributed state $\boldsymbol{x}(z,t) \in \mathbb{R}^2$ defined for $z \in [0,1]$ and $t \geq 0$. The subsystems are bidirectionally coupled by means of the vector field $\boldsymbol{f}$ in (1a) and the function c in (1b), that is locally Lipschitz continuous on any domain of interest. Furthermore, $\Lambda(z) = \operatorname{diag}(\lambda_1(z), -\lambda_2(z))$ with continuously differentiable transport speeds $\lambda_1(z), \lambda_2(z) > 0$ for $z \in [0,1]$. This implies that the corresponding state components $x_1(z,t) = \boldsymbol{e}_1^{\mathrm{T}}\boldsymbol{x}(z,t) \in \mathbb{R}$ and $x_2(z,t) = \boldsymbol{e}_2^{\mathrm{T}}\boldsymbol{x}(z,t) \in \mathbb{R}$, with unit vectors $\boldsymbol{e}_1 = (1,0)^{\mathrm{T}}$ and $\boldsymbol{e}_2 = (0,1)^{\mathrm{T}}$, describe the transport in the negative and positive z-direction, respectively (see Fig. 1). The transport delays between the two boundaries, where x_1 and x_2 are coupled via $q_0, q_1 \in \mathbb{R}$, are denoted by $\phi_1(1)$ and $\phi_2(1)$ with

$$\phi_i(z) = \int_0^z \frac{1}{\lambda_i(\zeta)} \,\mathrm{d}\zeta > 0, \qquad i = 1, 2. \tag{2}$$

The in-domain coupling matrix $A(z) = [a_{i,j}(z)]$ with $a_{i,j} \in C^1([0,1])$ is assumed to satisfy $a_{i,i} = 0$ for $i = 1, 2$. The PDE-ODE system (1) with the input $u(t) \in \mathbb{R}$ is completed by initial conditions (ICs) $\boldsymbol{w}(0) = \boldsymbol{w}_0 \in \mathbb{R}^n$ and piecewise continuous $\boldsymbol{x}(z,0) = \boldsymbol{x}_0(z) \in \mathbb{R}^2$.

Remark 1 Heterodirectional and strictly hyperbolic PDEs may always be written with the assumed structures for $\Lambda(z)$ and $A(z)$ by transforming the PDE state $\boldsymbol{x}(z,t)$ [13]. These structural assumptions therefore pose no restriction. □

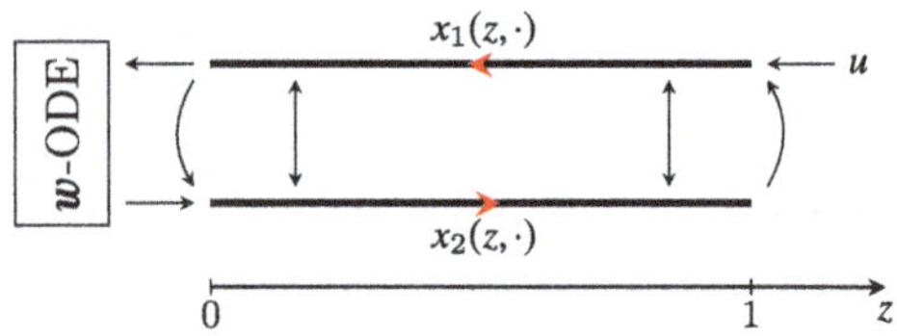

Fig. 1 Schematic overview of the PDE-ODE system (1)

2.2 Objective

A controller for the nonlinear PDE-ODE system (1), tracking reference trajectories $\boldsymbol{w}_\mathrm{r}(t)$ and $\boldsymbol{x}_\mathrm{r}(z,t)$ for $\boldsymbol{w}(t)$ and $\boldsymbol{x}(z,t)$, respectively, is designed in such a way that the origin of the closed-loop system is (locally) asymptotically stable. The proposed method relies on the solution of the PDE-ODE system, and is therefore called *solution-based control design*.

2.3 Assumptions About the ODE Subsystem

The following assumptions are imposed for the control design. Define the fictitious system

$$\dot{\boldsymbol{\omega}}(t) = \boldsymbol{f}(\boldsymbol{\omega}(t), \nu(t)) \tag{3}$$

for $t \geq t_0 \geq 0$ with state $\boldsymbol{\omega}(t) \in \mathbb{R}^n$, input $\nu(t) \in \mathbb{R}$ and IC $\boldsymbol{\omega}(t_0) = \boldsymbol{\omega}_0 \in \mathbb{R}^n$. Note that this system is structurally identical to the nonlinear ODE (1a).

Assumption 1 System (3) is forward complete. □

Forward completeness (see [1]) is a typical assumption for the control of nonlinear ODEs with actuation delay, see e.g. [3]. By that, $\boldsymbol{f}$ is Lipschitz continuous and the existence and uniqueness of a solution $\boldsymbol{\omega}(t)$ is implied for all ICs and every (measurable locally essentially) bounded ν. Assumption 1 is sufficient for predicting the solution $\boldsymbol{w}(\tau)$ for $\tau > t$ of the ODE subsystem (1a), with the current state $\boldsymbol{w}(t)$ as the IC, as it excludes finite escape time. This is an essential part of the control design in Sect. 3.

Assumption 2 There exists a tracking controller

$$\nu(t) = \kappa(t, \boldsymbol{\omega}(t)) \tag{4}$$

for (3), such that, in the closed-loop system, the solution $\boldsymbol{\omega}(t) = \boldsymbol{\omega}_\mathrm{r}(t)$ with a corresponding reference $\boldsymbol{\omega}_\mathrm{r}$ is asymptotically stable, at least locally for all $\boldsymbol{\omega}(t_0)$ and $\boldsymbol{\omega}_\mathrm{r}(t_0)$ in a subset of $\mathbb{R}^n$. □

The derivation of a tracking controller can be done using any appropriate nonlinear control design method. As the stability properties of the overall closed-loop system follow from the specific tracking controller for the nonlinear ODEs, the results of this contribution similarly hold for exponential (or any other kind of) stability. Note that the explicit time dependence of κ in (4) represents its dependence on the reference trajectory $\boldsymbol{\omega}_{\mathrm{r}}(t)$ and its derivatives.

Assumption 3 Sufficiently smooth reference trajectories $\boldsymbol{w}_{\mathrm{r}}(t)$ and $x_{1,\mathrm{r}}(0,t)$ for $\boldsymbol{w}(t)$ and $x_1(0,t)$, respectively, are known and satisfy (1a). □

Obviously, this assumption is only invoked for the tracking problem, i.e., it is not required for stabilization. By inserting $\boldsymbol{w}_{\mathrm{r}}(t)$ and $x_{1,\mathrm{r}}(0,t)$, the boundary condition (1b) immediately gives the reference trajectory $x_{2,\mathrm{r}}(0,t)$ for $x_2(0,t)$. The reference trajectory $\boldsymbol{x}_{\mathrm{r}}(z,t)$ for the entire PDE state $\boldsymbol{x}(z,t)$ is obtained from the solution of the PDE (1c) with known boundary data $\boldsymbol{x}_{\mathrm{r}}(0,t)$. This, in turn, provides the reference trajectory $u_{\mathrm{r}}(t)$ for the input $u(t)$ by means of (1d). The latter constitutes the feed-forward controller for the PDE-ODE system (1). Due to the hyperbolic character of the PDE (1c) with finite transport velocities λ_1 and λ_2, the feed-forward controller $u_{\mathrm{r}}(t)$ depends on the boundary data $\boldsymbol{x}_{\mathrm{r}}(0,t)$ on the interval $[t-\phi_2(1), t+\phi_1(1)]$, i.e., it involves (lumped and distributed) delays and predictions. In conclusion, the reference trajectories $(\boldsymbol{w}_{\mathrm{r}}(t), \boldsymbol{x}_{\mathrm{r}}(z,t), u_{\mathrm{r}}(t))$ satisfy

$$\dot{\boldsymbol{w}}_{\mathrm{r}}(t) = \boldsymbol{f}(\boldsymbol{w}_{\mathrm{r}}(t), x_{1,\mathrm{r}}(0,t)) \tag{5a}$$
$$x_{2,\mathrm{r}}(0,t) = q_0 x_{1,\mathrm{r}}(0,t) + c(\boldsymbol{w}_{\mathrm{r}}(t)) \tag{5b}$$
$$\partial_t \boldsymbol{x}_{\mathrm{r}}(z,t) = \Lambda(z)\partial_z \boldsymbol{x}_{\mathrm{r}}(z,t) + A(z)\boldsymbol{x}_{\mathrm{r}}(z,t) \tag{5c}$$
$$x_{1,\mathrm{r}}(1,t) = q_1 x_{2,\mathrm{r}}(1,t) + u_{\mathrm{r}}(t) \tag{5d}$$

for $z \in [0,1]$ and $t \geq 0$.

This feed-forward control design is well-known in the framework of flatness for hyperbolic PDE systems with dynamic boundary conditions (see [16, 19, 24]). In particular, if the ODE subsystem (1a) is differentially flat, there exists a sufficiently smooth function φ such that $y(t) = \varphi(\boldsymbol{w}(t)) \in \mathbb{R}$ is a flat output (cf. e.g. [20]). Furthermore, the flatness of (1a) implies the existence of differential parameterizations

$$\boldsymbol{w}(t) = \boldsymbol{\theta}\left(y(t), \dot{y}(t), \ldots, y^{(n-1)}(t)\right) \tag{6a}$$
$$x_1(0,t) = \boldsymbol{\rho}\left(y(t), \dot{y}(t), \ldots, y^{(n)}(t)\right) \tag{6b}$$

with sufficiently smooth functions $\boldsymbol{\theta}$ and $\boldsymbol{\rho}$. In fact, this property allows the choice of an arbitrary reference trajectory $y_{\mathrm{r}}(t)$ for the flat output $y(t)$. From that, the corresponding reference trajectories $\boldsymbol{w}_{\mathrm{r}}(t)$ and $x_{1,\mathrm{r}}(0,t)$ for $\boldsymbol{w}(t)$ and $x_1(0,t)$, respectively, follow from inserting $y_{\mathrm{r}}(t)$ and its derivatives in (6). Therefore, Assumption 3 is obviously met for flat ODEs (1a).

Remark 2 Note that the reference trajectory $y_{\mathrm{r}}(t)$ is usually chosen to be n times continuously differentiable in order to guarantee continuity of $u_{\mathrm{r}}(t)$. For instance, a transition between two stationary regimes $\tilde{y}_1, \tilde{y}_2 \in \mathbb{R}$ with finite transition time $T > 0$ may be achieved by planning a polynomial function. The transition starts at time $t_1 \geq \phi_1(1)$ in order to account for the actuation delay. The reference trajectory transitions into the second stationary regime at time $t_2 = t_1 + T$. In particular, the reference trajectory is chosen as

$$y_r(t) = \begin{cases} \tilde{y}_1, & t \leq t_1 \\ p(t), & t_1 < t < t_2 \\ \tilde{y}_2, & t \geq t_2 \end{cases} \tag{7}$$

with a polynomial $p(t) = \sum_{i=0}^{2\mu+1} \alpha_i t^i$ of degree $2\mu + 1$. The coefficients α_i follow from the conditions $p(t_j) = \tilde{y}_j$ and $p^{(i)}(t_j) = 0, i = 1, \ldots, \mu$ for $j = 1, 2$ that guarantee μ-times continuous differentiability of $y_r(t)$. By choosing $\mu \geq n$, the feedforward controller u_r is continuous. □

3 Solution-Based Control Design

One of the main ideas of the solution-based control approach is to reduce the problem to controlling nonlinear ODEs with delayed actuation, as discussed in the sequel, by taking advantage of the PDE's solution. The reduction is simplified significantly by a preliminary backstepping transformation of the PDE state $\boldsymbol{x}$ to remove the in-domain coupling in (1c). This allows for a straightforward solution of the PDE subsystem via the method of characteristics to determine the action of the input $u(t)$ at $z = 1$ on the ODE subsystem at $z = 0$. As a result of the finite transport velocities, a nonlinear delay system is obtained, which captures the dynamics of the PDE-ODE system (1) that are essential for control purposes. Therefore, this delay system is denoted as the reduced system. Building upon that, a tracking controller is designed for the reduced system. The implementation of this controller requires predicted values of the ODE state that are uniquely determined by the current ODE and PDE states. Finally, tracking the reference trajectories $\boldsymbol{w}_r(t)$ and $\boldsymbol{x}_r(z, t)$ for the PDE-ODE system (1) is shown to be implied by the tracking controller designed for the reduced system.

3.1 Preliminary Backstepping Transformation

In order to remove the in-domain coupling due to $A(z)$ in (1c), consider the Volterra integral transformation

$$\tilde{\boldsymbol{x}}(z, t) = \boldsymbol{x}(z, t) - \int_0^z K(z, \zeta)\boldsymbol{x}(\zeta, t)\, \mathrm{d}\zeta =: \mathcal{T}[\boldsymbol{x}(\cdot, t)](z), \tag{8}$$

which is well-known in the backstepping framework [17, 23]. Let $K(z,\zeta) \in \mathbb{R}^{2\times 2}$ be the solution of the kernel equations

$$\Lambda(z)\partial_z K(z,\zeta) + \partial_\zeta [K(z,\zeta)\Lambda(\zeta)] = K(z,\zeta)A(\zeta), \qquad 0 < \zeta < z < 1 \tag{9a}$$

$$K(z,z)\Lambda(z) - \Lambda(z)K(z,z) = A(z) \tag{9b}$$

$$K(z,0)\Lambda(0)[\boldsymbol{e}_1 + \boldsymbol{e}_2 q_0] = \big(0 \; g_2(z)\big)^{\mathrm{T}} =: \boldsymbol{g}(z) \tag{9c}$$

with an arbitrary but continuous function g_2 and define

$$\boldsymbol{h}(z) := \big(h_1(z) \; h_2(z)\big)^{\mathrm{T}} = K(z,0)\Lambda(0)\boldsymbol{e}_2. \tag{10}$$

Then, applying (8) to (1) leads to the transformed system

$$\dot{\boldsymbol{w}}(t) = \boldsymbol{f}(\boldsymbol{w}(t), \tilde{x}_1(0,t)) \tag{11a}$$

$$\tilde{x}_2(0,t) = q_0\tilde{x}_1(0,t) + c(\boldsymbol{w}(t)) \tag{11b}$$

$$\partial_t \tilde{\boldsymbol{x}}(z,t) = \Lambda(z)\partial_z\tilde{\boldsymbol{x}}(z,t) + \boldsymbol{g}(z)\tilde{x}_1(0,t) + \boldsymbol{h}(z)c(\boldsymbol{w}(t)) \tag{11c}$$

$$\tilde{x}_1(1,t) = \tilde{u}(t), \tag{11d}$$

wherein the state feedback[1]

$$u(t) = \tilde{u}(t) - q_1 x_2(1,t) + \int_0^1 \boldsymbol{e}_1^{\mathrm{T}} K(1,\zeta)\boldsymbol{x}(\zeta,t)\,\mathrm{d}\zeta \tag{12}$$

with the new input $\tilde{u}(t) \in \mathbb{R}$ is used. The structure of the PDE-ODE system (11) is depicted in Fig. 2.

Remark 3 The kernel equations (9) are equivalent to those solved for the backstepping control of linear heterodirectional hyperbolic PDEs without an ODE boundary system [23]. Note that in the case of $q_0 = 0$, the second component of (9c) is merely the definition of the function g_2, and the kernel equations (9) have to be complemented by the additional boundary condition

$$\boldsymbol{e}_2^{\mathrm{T}} K(z,0)\boldsymbol{e}_2 = \bar{g}_2(z), \tag{13}$$

with an arbitrary but continuous function $\bar{g}_2$. The kernel $K_{\mathrm{I}}(z,\zeta) \in \mathbb{R}^{2\times 2}$ in the inverse transformation

$$\boldsymbol{x}(z,t) = \tilde{\boldsymbol{x}}(z,t) + \int_0^z K_{\mathrm{I}}(z,\zeta)\tilde{\boldsymbol{x}}(\zeta,t)\,\mathrm{d}\zeta \tag{14}$$

---

[1] The exact compensation of $q_1 x_2(1,t)$ in (12) is known to cause robustness issues [2]. This is not the focus here and will be the subject of future research.

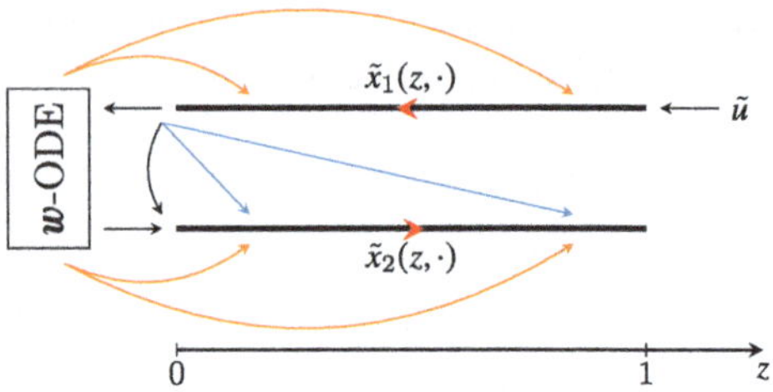

Fig. 2 Schematic overview of the PDE-ODE system (11)

can be obtained from the reciprocity relation

$$K_\mathrm{I}(z,\zeta) = K(z,\zeta) + \int_\zeta^z K(z,\zeta')K_\mathrm{I}(\zeta',\zeta)\,\mathrm{d}\zeta'. \tag{15}$$

See, e.g., [23] for details on the existence and uniqueness of the piecewise $C^1([0,1])$ solutions $K(z,\zeta)$ and $K_\mathrm{I}(z,\zeta)$. Therein, a systematic method for solving the kernel equations can be found as well. □

Up until now, this procedure is identical to the one in [7]. However, the next step will differ from the usual PDE backstepping framework for PDE-ODE systems.

3.2 *Tracking Control*

By the preliminary backstepping transformation (8) and the specific choice of the state feedback (12), the component $\tilde{x}_2$ does not affect the transport of $\tilde{x}_1$, which implies that (11b)–(11d) is a cascade of two transport equations (see Fig. 2). Further inspection of (11c) and considering the definition of $\boldsymbol{g}$ in (9c) reveals that $\tilde{x}_1$ satisfies the inhomogeneous transport equation

$$\partial_t \tilde{x}_1(z,t) = \lambda_1(z)\partial_z \tilde{x}_1(z,t) + h_1(z)c(\boldsymbol{w}(t)). \tag{16}$$

As a consequence, the control input $\tilde{u}(t) = \tilde{x}_1(1,t)$ at $z=1$ affects the ODE at $z=0$ only after a delay $\phi_1(1)$ (see (2)). This relation can be described in the framework of delay systems. For that, the transport equation (16) is solved to find $\tilde{x}_1(0, t+\phi_1(1))$, which acts on the ODE (11a) at time $t+\phi_1(1)$.

The method of characteristics yields the solution

$$\tilde{x}_1(z,t+s) = \tilde{x}_1(\psi_1(\phi_1(z)+s),t) + \int_0^s h_1(\psi_1(\phi_1(z)+s-\sigma))c(\boldsymbol{w}(t+\sigma))\,\mathrm{d}\sigma \tag{17}$$

of (16) for $s \in [0, \phi_1(1) - \phi_1(z)]$ and $z \in [0, 1]$, wherein ψ_1 is the inverse function of ϕ_1, i.e. $\psi_1(\phi_1(z)) = z$. The relation between $\tilde{u}(t)$ and $\tilde{x}_1(0, t + \phi_1(1))$ may be found by considering the solution (17) for $z = 0$ and $s = \phi_1(1)$ and inserting (11d):

$$\tilde{x}_1(0, t + \phi_1(1)) = \tilde{u}(t) + \int_0^{\phi_1(1)} h_1(\psi_1(\phi_1(1) - \sigma))c(\boldsymbol{w}(t + \sigma))\,\mathrm{d}\sigma. \tag{18}$$

For further simplification, using

$$\tilde{u}(t) = \nu(t) - \int_0^{\phi_1(1)} h_1(\psi_1(\phi_1(1) - \sigma))c(\boldsymbol{w}(t + \sigma))\,\mathrm{d}\sigma \tag{19}$$

in (18) leads to

$$\tilde{x}_1(0, t + \phi_1(1)) = \nu(t) \tag{20}$$

for $t \geq 0$. By means of this new input $\nu(t)$, it is possible to control the ODE subsystem (11a) evaluated at $t + \phi_1(1)$. This observation forms the basis for the subsequent considerations.

Applying the time shift $t \mapsto t + \phi_1(1)$ to (11a) and taking (20) into account yields the nonlinear ODE

$$\dot{\boldsymbol{w}}(t + \phi_1(1)) = \boldsymbol{f}(\boldsymbol{w}(t + \phi_1(1)), \nu(t)), \qquad t > 0, \tag{21}$$

that clearly shows the delayed actuation. This is even more apparent from the description

$$\dot{\boldsymbol{w}}(t) = \boldsymbol{f}(\boldsymbol{w}(t), \nu(t - \phi_1(1))), \qquad t > \phi_1(1) \tag{22}$$

as a delay system. In fact, (21) is a reduced representation of (11) (or equivalently (1)) in the sense that it excludes the $\tilde{x}_2$-subsystem. It only incorporates the part of the system relevant for control, i.e. the ODE dynamics (11a) and the transport equation (16) with the boundary condition (11d). This equivalence is also apparent from the IC $[0, \phi_1(1)] \ni s \mapsto \boldsymbol{w}(s)$ for (22) being uniquely determined by $\boldsymbol{w}(0)$ and $\tilde{x}_1(z, 0)$, which dictate the behavior of $\boldsymbol{w}(t)$ for $t \in [0, \phi_1(1)]$.

In light of Assumptions 1 and 2, the control design for (21) is straightforward. This is a consequence of the reduced system (21) and the fictitious system (3) only differing in the actuation delay. In spite of this difference, forward completeness is preserved (see Assumption 1). By that, tracking control of (21) is implied by

$$\nu(t) = \kappa(t + \phi_1(1), \boldsymbol{w}(t + \phi_1(1))). \tag{23}$$

Inserting (21) into (23), the (local) asymptotic stability of the resulting closed loop is guaranteed (see Assumption 2), where the control action affects the ODE state $\boldsymbol{w}(t)$ only for $t \geq \phi_1(1)$.

Whereas the ODE state $\boldsymbol{w}(t)$ and the PDE state $\tilde{\boldsymbol{x}}(z, t) = \mathcal{T}[\boldsymbol{x}(\cdot, t)](z)$ for $z \in [0, 1]$ are available for feedback, knowledge of the future values $\boldsymbol{w}(t + s)$ for

$s \in [0, \phi_1(1)]$, as required in (19) and (23), seems counter-intuitive. This problem is circumvented by predicting $\boldsymbol{w}(t+s)$ based on the available information at time t.

3.3 Prediction of the ODE State

The future values $\boldsymbol{w}(t+s)$ of the ODE state with $s \in [0, \phi_1(1)]$, necessary for evaluating (19) and (23), can be determined from the solution of the ODE subsystem (11a). The assumed forward completeness of (11a) allows for the prediction of $\boldsymbol{w}$ provided that $\tilde{x}_1(0, t+s)$ is known for $s \in [0, \phi_1(1)]$. In fact, the latter can be expressed in terms of the known PDE state $\tilde{x}_1(z, t)$ by considering the solution (17) on the boundary $z = 0$, i.e.

$$\tilde{x}_1(0, t+s) = \tilde{x}_1(\psi_1(s), t) + \int_0^s h_1(\psi_1(s-\sigma))c(\boldsymbol{w}(t+\sigma))\,\mathrm{d}\sigma. \tag{24}$$

Note that $\tilde{x}_1(\psi_1(s), t)$ for $s \in [0, \phi_1(1)]$ corresponds to $\tilde{x}_1(z, t)$ for $z \in [0, 1]$.

In (24), the expression for $\tilde{x}_1(0, t+s)$ in turn depends on distributed predictions of $\boldsymbol{w}$. This does not hinder the prediction, but renders it more involved. In particular, the prediction problem

$$\tfrac{\mathrm{d}}{\mathrm{d}s}\boldsymbol{w}(t+s) = \boldsymbol{f}(\boldsymbol{w}(t+s), \tilde{x}_1(0, t+s)) \tag{25a}$$

$$\tilde{x}_1(0, t+s) = \tilde{x}_1(\psi_1(s), t) + \int_0^s h_1(\psi_1(s-\sigma))c(\boldsymbol{w}(t+\sigma))\,\mathrm{d}\sigma \tag{25b}$$

for $s \in [0, \phi_1(1)]$ follows from (11a) and (24) and, thus, requires solving these coupled equations. At each time t, (25) has to be solved on the basis of the available states $\boldsymbol{w}(t)$ and $\tilde{x}_1(\psi_1(s), t)$. Formally, this can be taken into account by introducing the notation $\boldsymbol{w}_\mathrm{p}(s; t)$ for the predicted value of $\boldsymbol{w}(t+s)$, wherein t is a parameter describing the time instant for which (25) is solved. Consequently, the predictor

$$\tfrac{\mathrm{d}}{\mathrm{d}s}\boldsymbol{w}_\mathrm{p}(s; t) = \boldsymbol{f}\left(\boldsymbol{w}_\mathrm{p}(s; t),\ \tilde{x}_1(\psi_1(s), t) + \int_0^s h_1(\psi_1(s-\sigma))c(\boldsymbol{w}_\mathrm{p}(\sigma; t))\,\mathrm{d}\sigma\right) \tag{26a}$$

$$\boldsymbol{w}_\mathrm{p}(0; t) = \boldsymbol{w}(t) \tag{26b}$$

for $s \in (0, \phi_1(1)]$ (cf. (25)) is an initial value problem (IVP) with the current ODE state $\boldsymbol{w}(t)$ playing the role of the IC.

It can be seen that (26a) is a *general nonlinear Volterra integro-differential equation*, for which conditions on the existence and uniqueness of solutions are presented in [10]. Here, the existence of a unique solution $\boldsymbol{w}_\mathrm{p}(s; t)$ for (26) is guaranteed by the Lipschitz continuity of c and $\boldsymbol{f}$ (by assumption), the continuity of the kernel K (see Remark 3) and thereby the continuity of h_1 (cf. (10)).

Since $\boldsymbol{w}(t+s)$, i.e. the predicted values of $\boldsymbol{w}$, can be determined for every $t \geq 0$ by the solution of (26), the tracking controller (19) and (23) is modified by replacing $\boldsymbol{w}(t+s)$ with $\boldsymbol{w}_{\mathrm{p}}(s;t)$ to get

$$\tilde{u}(t) = \nu(t) - \int_0^{\phi_1(1)} h_1(\psi_1(\phi_1(1)-\sigma))c(\boldsymbol{w}_{\mathrm{p}}(\sigma;t))\,\mathrm{d}\sigma \tag{27a}$$

$$\nu(t) = \kappa(t+\phi_1(1), \boldsymbol{w}_{\mathrm{p}}(\phi_1(1);t)). \tag{27b}$$

Finally, the tracking controller

$$u(t) = \kappa(t+\phi_1(1), \boldsymbol{w}_{\mathrm{p}}(\phi_1(1);t)) - \int_0^{\phi_1(1)} h_1(\psi_1(\phi_1(1)-\sigma))c(\boldsymbol{w}_{\mathrm{p}}(\sigma;t))\,\mathrm{d}\sigma \\ - q_1 x_2(1,t) + \int_0^1 \boldsymbol{e}_1^{\mathrm{T}} K(1,\zeta)\boldsymbol{x}(\zeta,t)\,\mathrm{d}\zeta \tag{28}$$

for the PDE-ODE system (1) results from inserting (27) into (12), with $\boldsymbol{w}_{\mathrm{p}}(s;t)$ as the solution of the IVP (26) for $s \in [0, \phi_1(1)]$. The infinite-dimensional controller (28) involves reference trajectories of the ODE subsystem due to the explicit time-dependence of κ, and a state feedback part. The latter is a feedback of both the ODE state $\boldsymbol{w}(t)$, as apparent from the IC (26b), and the PDE state $\boldsymbol{x}(z,t)$ for $z \in [0,1]$, as $\tilde{x}_1(\psi_1(s),t) = \boldsymbol{e}_1^{\mathrm{T}}\mathcal{T}[\boldsymbol{x}(\cdot,t)](\psi_1(s))$ in (26a).

3.4 Main Result

The control result is summarized in the following theorem.

Theorem 1 *Let Assumptions 1–3 hold. Then, the tracking controller for the PDE-ODE system (1) consisting of*

$$u(t) = \kappa(t+\phi_1(1), \boldsymbol{w}_{\mathrm{p}}(\phi_1(1);t)) - \int_0^{\phi_1(1)} h_1(\psi_1(\phi_1(1)-\sigma))c(\boldsymbol{w}_{\mathrm{p}}(\sigma;t))\,\mathrm{d}\sigma \\ - q_1 x_2(1,t) + \int_0^1 \boldsymbol{e}_1^{\mathrm{T}} K(1,\zeta)\boldsymbol{x}(\zeta,t)\,\mathrm{d}\zeta \tag{29}$$

and the solution of the IVP

$$\tfrac{\mathrm{d}}{\mathrm{d}s}\boldsymbol{w}_{\mathrm{p}}(s;t) = f\left(\boldsymbol{w}_{\mathrm{p}}(s;t),\ \boldsymbol{e}_1^{\mathrm{T}}\mathcal{T}[\boldsymbol{x}(\cdot,t)](\psi_1(s)) + \int_0^s h_1(\psi_1(s-\sigma))c(\boldsymbol{w}_{\mathrm{p}}(\sigma;t))\,\mathrm{d}\sigma\right) \tag{30a}$$

$$\boldsymbol{w}_{\mathrm{p}}(0;t) = \boldsymbol{w}(t) \tag{30b}$$

for $s \in (0, \phi_1(1)]$ *guarantees (local) asymptotic stability of the origin in error coordinates* $\boldsymbol{e}_w(t) = \boldsymbol{w}(t) - \boldsymbol{w}_\mathrm{r}(t)$ *and* $\boldsymbol{e}_x(z,t) = \boldsymbol{x}(z,t) - \boldsymbol{x}_\mathrm{r}(z,t)$ *for ICs* $\boldsymbol{e}_w(0)$ *in an open subset of* $\mathbb{R}^n$, $\boldsymbol{e}_x(z,0) \in \mathbb{R}^2$ *piecewise continuous, and reference trajectories* $\boldsymbol{w}_\mathrm{r}(t)$ *and* $\boldsymbol{x}_\mathrm{r}(z,t)$ *satisfying (5a)–(5c). In addition,* $u(t) \to u_\mathrm{r}(t)$ *for* $t \to \infty$ *with the feed-forward controller* $u_\mathrm{r}(t)$ *satisfying (5d).*

Proof As discussed in Sect. 3.3, the solution $\boldsymbol{w}_\mathrm{p}(s;t)$ of the IVP (26) is unique for $s \in [0, \phi_1(1)]$ and every fixed $t \geq 0$ (see [10]). Furthermore, the integro-differential equations (25) and (26a) for $\boldsymbol{w}$ and $\boldsymbol{w}_\mathrm{p}$ are identical. Thus, the IC $\boldsymbol{w}_\mathrm{p}(0;t) = \boldsymbol{w}(t)$ in (26b) implies

$$\boldsymbol{w}_\mathrm{p}(s;t) = \boldsymbol{w}(t+s), \qquad s \in [0, \phi_1(1)] \tag{31}$$

for $t \geq 0$. By that, (27) is identical to (19) and (23), which leads to

$$\dot{\boldsymbol{w}}(t) = \boldsymbol{f}(\boldsymbol{w}(t), \kappa(t, \boldsymbol{w}(t))), \qquad t > \phi_1(1) \tag{32}$$

in light of (22). Therefore, the origin $\boldsymbol{e}_w(t) = \boldsymbol{0}$ of the ODE error subsystem is (locally) asymptotically stable by Assumption 2 (with $t_0 = \phi_1(1)$), i.e.

$$\lim_{t\to\infty} \|\boldsymbol{e}_w(t)\| = 0. \tag{33}$$

Additionally, it follows that

$$\lim_{t\to\infty} \left|e_{\tilde{x},1}(0,t)\right| = 0 \tag{34}$$

as $\boldsymbol{w}(t) \to \boldsymbol{w}_\mathrm{r}(t)$ implies $\kappa(t, \boldsymbol{w}(t)) \to \tilde{x}_{\mathrm{r},1}(0,t)$ by (20) and (23).

In order to discuss the remainder of the PDE error subsystem, introduce the reference trajectory $\tilde{\boldsymbol{x}}_\mathrm{r}(z,t) = \mathcal{T}[\boldsymbol{x}_\mathrm{r}(\cdot,t)](z)$ for $\tilde{\boldsymbol{x}}(z,t)$ by applying the backstepping transformation (8) to the reference $\boldsymbol{x}_\mathrm{r}(z,t)$. This transforms the system (5) into

$$\begin{aligned}
\dot{\boldsymbol{w}}_\mathrm{r}(t) &= \boldsymbol{f}(\boldsymbol{w}_\mathrm{r}(t), \tilde{x}_{\mathrm{r},1}(0,t)) && \text{(35a)}\\
\tilde{x}_{\mathrm{r},2}(0,t) &= q_0\tilde{x}_{\mathrm{r},1}(0,t) + c(\boldsymbol{w}_\mathrm{r}(t)) && \text{(35b)}\\
\partial_t\tilde{\boldsymbol{x}}_\mathrm{r}(z,t) &= \Lambda(z)\partial_z\tilde{\boldsymbol{x}}_\mathrm{r}(z,t) + \boldsymbol{g}(z)\tilde{x}_{\mathrm{r},1}(0,t) + \boldsymbol{h}(z)c(\boldsymbol{w}_\mathrm{r}(t)) && \text{(35c)}\\
\tilde{x}_{\mathrm{r},1}(1,t) &= \tilde{u}_\mathrm{r}(t), && \text{(35d)}
\end{aligned}$$

after using

$$u_\mathrm{r}(t) = \tilde{u}_\mathrm{r}(t) - q_1 x_{2,\mathrm{r}}(1,t) + \int_0^1 \boldsymbol{e}_1^\mathrm{T} K(1,\zeta)\boldsymbol{x}_\mathrm{r}(\zeta,t)\,\mathrm{d}\zeta \tag{36}$$

with the new input $\tilde{u}_\mathrm{r}(t) \in \mathbb{R}$. Obviously, (11) and (35) share the same structure.

Then, the PDE error system reads

$$e_{\tilde{x},2}(0,t) = q_0 e_{\tilde{x},1}(0,t) + c(\boldsymbol{w}(t)) - c(\boldsymbol{w}_\mathrm{r}(t)) \tag{37a}$$

$$\partial_t \boldsymbol{e}_{\tilde{x}}(z,t) = \Lambda(z)\partial_z \boldsymbol{e}_{\tilde{x}}(z,t) + \boldsymbol{g}(z) e_{\tilde{x},1}(0,t) + \boldsymbol{h}(z)(c(\boldsymbol{w}(t)) - c(\boldsymbol{w}_\mathrm{r}(t))) \tag{37b}$$

$$e_{\tilde{x},1}(1,t) = e_{\tilde{u}}(t) \tag{37c}$$

with transformed error coordinates $\boldsymbol{e}_{\tilde{x}}(z,t) = \mathcal{T}[\boldsymbol{e}_x(\cdot,t)](z)$ and $e_{\tilde{u}} = \tilde{u} - \tilde{u}_\mathrm{r}$.

From (37a) it is clear that

$$\left|e_{\tilde{x},2}(0,t)\right| \leq |q_0| \left|e_{\tilde{x},1}(0,t)\right| + L_c \left\|\boldsymbol{e}_w(t)\right\| \tag{38}$$

with the Lipschitz constant $L_c > 0$ of the function c on the domain of interest. As a consequence of (33) and (34), $e_{\tilde{x},2}(0,t) = 0$ is (locally) asymptotically stable, i.e.

$$\lim_{t\to\infty} \left|e_{\tilde{x},2}(0,t)\right| = 0. \tag{39}$$

The first component of (37b) describes an inhomogeneous transport equation (cf. (16)) that is excited by a weighted ODE output error $\boldsymbol{h}(z)(c(\boldsymbol{w}(t)) - c(\boldsymbol{w}_\mathrm{r}(t)))$. Its (unique) solution follows analogously to (24). Therefore, the relation

$$e_{\tilde{x},1}(\psi_1(s),t) = e_{\tilde{x},1}(0,t+s) - \int_0^s h_1(\psi_1(s-\sigma))(c(\boldsymbol{w}(t+\sigma)) - c(\boldsymbol{w}_\mathrm{r}(t+\sigma)))\,\mathrm{d}\sigma \tag{40}$$

holds for $s \in [0, \phi_1(1)]$ and $t \geq 0$. This is used to find an upper bound according to

$$\begin{aligned}
\left|e_{\tilde{x},1}(\psi_1(s),t)\right| &\leq \left|e_{\tilde{x},1}(0,t+s)\right| + L_c \int_0^s |h_1(\psi_1(s-\sigma))| \left\|\boldsymbol{e}_w(t+\sigma)\right\| \,\mathrm{d}\sigma \\
&\leq \left|e_{\tilde{x},1}(0,t+s)\right| + L_c\, h_{1,\max} \int_0^s \left\|\boldsymbol{e}_w(t+\sigma)\right\| \,\mathrm{d}\sigma \\
&\leq \left|e_{\tilde{x},1}(0,t+s)\right| + L_c\, h_{1,\max}\, \phi_1(1) \sup_{\sigma\in[0,\phi_1(1)]} \left\|\boldsymbol{e}_w(t+\sigma)\right\|
\end{aligned} \tag{41}$$

with

$$h_{1,\max} := \max_{s\in[0,\phi_1(1)]} |h_1(\psi_1(s))| \geq 0. \tag{42}$$

As $\left|e_{\tilde{x},1}(0,t)\right|$ and $\left\|\boldsymbol{e}_w(t)\right\|$ decay asymptotically for $t \geq \phi_1(1)$ (cf. (33) and (34)), it follows that

$$\lim_{t\to\infty} \sup_{z\in[0,1]} \left|e_{\tilde{x},1}(z,t)\right| = 0. \tag{43}$$

Note that the second component of (37b) describes an inhomogeneous transport equation for $e_{\tilde{x},2}$. The inhomogeneous part is bounded by the decaying errors $|e_{\tilde{x},1}(0,t)|$ and $\|\boldsymbol{e}_w(t)\|$. Therefore, and by similar arguments, $|e_{\tilde{x},2}(z,t)|$ is asymptotically stabilized for $t \geq \phi_1(1) + \phi_2(1)$, i.e.

$$\lim_{t\to\infty} \sup_{z\in[0,1]} |e_{\tilde{x},2}(z,t)| = 0. \tag{44}$$

By (43)–(44) and the invertibility of the backstepping transformation $\boldsymbol{e}_{\tilde{x}}(z,t) = \mathcal{T}[\boldsymbol{e}_x(\cdot,t)](z)$, the origin of the PDE error subsystem in original coordinates $\boldsymbol{e}_x$ is (locally) asymptotically stable. This implies in particular that $e_{x,1}(1,t) \to 0$ and $e_{x,2}(1,t) \to 0$ for $t \to \infty$, by which the property $u(t) \to u_{\mathrm{r}}(t)$ for $t \to \infty$ of the tracking controller (29) is obtained as a consequence of (1d). □

In the case that (1a) is flat, an alternative proof of Theorem 1 is presented, which does not require an inspection of the PDE error system (37) in the closed loop. For this, note that (33) implies $y(t) \to y_{\mathrm{r}}(t)$ as $t \to \infty$ for a sufficiently smooth function φ (see the flatness discussion in the context of Assumption 3). As all quantities of the PDE-ODE system (1), and in particular $\boldsymbol{x}(z,t)$, are parameterized by the flat output y, its derivatives as well as (lumped and distributed) delays and predictions thereof, it follows that $\boldsymbol{x}(z,t) \to \boldsymbol{x}_{\mathrm{r}}(z,t)$ for $t \to \infty$. Therefore, the PDE error subsystem (37) is asymptotically stabilized as well.

4 Simulation

The tracking controller proposed in Theorem 1 is demonstrated in the following for the PDE-ODE system

$$\dot{\boldsymbol{w}}(t) = \begin{pmatrix} \sin(w_1(t)) + w_2(t) - 1 \\ \cos^2(w_1(t)) - w_2(t) + x_1(0,t) \end{pmatrix} \tag{45a}$$

$$x_2(0,t) = x_1(0,t) + |w_2(t)|\sqrt{w_1^2(t) + 1} \tag{45b}$$

$$\partial_t \boldsymbol{x}(z,t) = \begin{pmatrix} \frac{1}{2} + \frac{z}{2} & 0 \\ 0 & -\left(\frac{1}{3} + \frac{z}{2}\right) \end{pmatrix} \partial_z \boldsymbol{x}(z,t) + \begin{pmatrix} 0 & \left(\frac{1}{2} + \frac{z}{2}\right) \mathrm{e}^{\frac{z}{3}} \\ (2+3z)\,\mathrm{e}^{-\frac{z}{3}} & 0 \end{pmatrix} \boldsymbol{x}(z,t) \tag{45c}$$

$$x_1(1,t) = -x_2(1,t) + u(t). \tag{45d}$$

From (2) it follows that $\phi_1(z) = 2\ln(1+z)$ and $\phi_2(z) = 2\ln(1+\frac{3}{2}z)$. The nonlinear ODE subsystem (45a) is obviously flat and has a flat output $y(t) = w_1(t)$. This is apparent from the parameterizations

$$w_1(t) = y(t) \tag{46a}$$
$$w_2(t) = \dot{y}(t) + 1 - \sin(y(t)) \tag{46b}$$
$$x_1(0,t) = \ddot{y}(t) + (1-\cos(y(t)))\dot{y}(t) - (1-\sin(y(t)))\sin(y(t)) \tag{46c}$$

of $\boldsymbol{w}(t)$ and $x_1(0,t)$ in terms of the flat output and its derivatives (cf. (6)).

In the subsequent results, the reference trajectory $y_r(t)$ is chosen as the polynomial transition (7) with $\tilde{y}_1 = 0$, $\tilde{y}_2 = \pi$, $T = 3$ and $\mu = 4$. The planned transition happens between the time $t_1 = \phi_1(1) \approx 1.39$ and the time $t_2 = t_1 + T \approx 4.39$. The reference trajectories $\boldsymbol{w}_r(t)$ and $x_{1,r}(0,t)$ are obtained by inserting the reference trajectory $y_r(t)$ and its derivatives in (46), i.e.

$$\boldsymbol{w}_r(t) = \begin{pmatrix} y_r(t) \\ \dot{y}_r(t) + 1 - \sin(y_r(t)) \end{pmatrix} \tag{47a}$$
$$x_{1,r}(0,t) = \ddot{y}_r(t) + (1-\cos(y_r(t)))\dot{y}_r(t) - (1-\sin(y_r(t)))\sin(y_r(t)). \tag{47b}$$

Furthermore, the reference trajectory $x_{2,r}(0,t)$ is found by inserting (47) into (5b). In addition, the nontrivial ICs

$$x_1(z,0) = \tfrac{2}{5}\sin^3(2\pi z), \quad x_2(z,0) = -\tfrac{2}{5}\sin^3(2\pi z), \quad w_1(0) = \tfrac{3}{4}, \quad w_2(0) = 0 \tag{48}$$

are chosen for the PDE state $\boldsymbol{x}(z,t)$ and the ODE state $\boldsymbol{w}(t)$, respectively.

In the following, simulation results for the controlled system are presented. Note that the structure of the tracking controller in (29) allows for a degree of freedom in the sense that κ, i.e. the tracking controller for the ODE subsystem, can be interchanged without any impact on the remainder of the design. This is illustrated in Sects. 4.2 and 4.3 by two scenarios, the first one using a flatness-based static state feedback and the second one using a linear dynamic state feedback with additive feed-forward control. However, before presenting these results, some aspects about the implementation are discussed.

4.1 Implementation

The simulation of the closed-loop system is implemented in MATLAB. A schematic overview of this implementation is sketched in Fig. 3. Therein, the controller is divided into an online and an offline part. The latter consists of the solution of the kernel equations (9) and the planning of a suitable reference y_r for the flat output y. The solution $K(z,\zeta)$ of the kernel equations (9) is determined for 101 equidistant

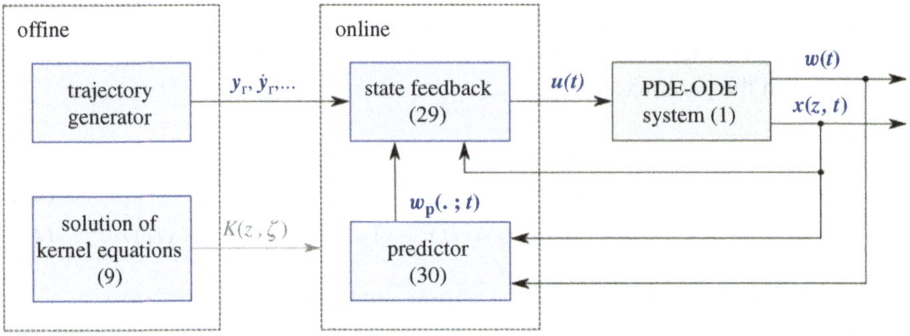

Fig. 3 A schematic overview of the implementation of the controller (29) with the online IVP (30) in the closed loop

spatial points by the method of successive approximations (cf. [6]). In this context, several functions of the *coni* MATLAB library [11] are used[2].

As previously mentioned, the tracking controller (29) uses the solution $\boldsymbol{w}_p(\cdot\,; t)$ of the IVP (30) at time t. This means that the prediction step must happen online. Moreover, determining the solution of a general nonlinear Volterra integro-differential equation in real time is tedious (see [10] for numeric methods). However, here, the numeric realization is done in a simple manner without any obvious loss in control accuracy. The integrals over the predicted state $\boldsymbol{w}_p$, appearing in (29) and (30), are approximated using the trapezoidal rule. The online predictor is implemented by numerically solving (30) via the explicit (forward) Euler method. This means that the integro-differential equation is treated as an ODE since both arguments of the nonlinear function $\boldsymbol{f}$ are evaluated at the previous time step, i.e., only previously determined values are used.

The controller κ for the ODE subsystem is an essential part of (29). It is implemented as a MATLAB function with the required reference trajectories $y_r(t)$, $\dot{y}_r(t)$ and $\ddot{y}_r(t)$ and the predicted state $\boldsymbol{w}_p(\phi_1(1); t)$ at the end of the prediction horizon as its arguments. As this MATLAB function is only set once before starting the simulation, interchanging different control strategies for the ODE subsystem is a trivial task. Two scenarios, a flatness-based nonlinear controller and a linear dynamic state feedback, are presented in the following.

Remark 4 Special emphasis is put on the simulation of the PDE subsystem (45c) due to the spatial dependence of the transport velocities. First, an equidistant time grid is defined for both the implementation of the closed loop and the trajectory generator. Based on that, the spatial grid follows from a piecewise linear approximation of the characteristic curves. More precisely, the spatial grids defined for the state components x_1 and x_2 are different, which is in accordance with the difference in their corresponding entries $\lambda_1(z)$ and $-\lambda_2(z)$ in the matrix $\Lambda(z)$ of transport velocities.

[2] The authors would like to express their special thanks to Jakob Gabriel for his valuable assistance with the numerical solution of the kernel equations.

Then, a finite difference scheme with discretization points along the characteristic curves is used for the simulation of the PDE subsystem. With that, the in-domain coupling requires an interpolation of each PDE state component on the spatial grid of the respective other. Note that the evaluation of the backstepping integral in the tracking controller (29) is based on these interpolated values as well. □

4.2 Controller 1: Flatness-Based Static State Feedback

Designing a tracking controller for the flat ODE subsystem (45a) is straightforward by using the parameterization (46c) of $x_1(0,t)$ in terms of the flat output $y(t)$ and its derivatives. From that, it can be verified that the exactly linearizing virtual controller

$$\begin{aligned} x_1(0,t) = \ddot{y}_r(t) - k_1\left(\dot{y}(t) - \dot{y}_r(t)\right) - k_0\left(y(t) - y_r(t)\right) \\ + (1-\cos(y(t)))\dot{y}(t) - (1-\sin(y(t)))\sin(y(t)) \end{aligned} \tag{49}$$

with $k_0, k_1 > 0$ leads to the (globally) exponentially stable dynamics

$$\ddot{e}_y + k_1\dot{e}_y + k_0 e_y = 0 \tag{50}$$

for the error $e_y = y - y_r$. This controller may also be written as the state feedback

$$\begin{aligned} \kappa(t, \boldsymbol{w}(t)) = \ddot{y}_r(t) - k_1\left(\sin(w_1(t)) + w_2(t) - 1 - \dot{y}_r(t)\right) - k_0\left(w_1(t) - y_r(t)\right) \\ + w_2(t) - \cos^2(w_1(t)) - \cos(w_1(t))\left(\sin(w_1(t)) + w_2(t) - 1\right), \end{aligned} \tag{51}$$

which is then used in (29).

Simulation results for the ODE controller (51) with $k_0 = 10$ and $k_1 = 6$ are depicted in Figs. 4 and 5. Note that the explicit computation of the reference trajectories $x_{1,r}(z,t)$ and $x_{2,r}(z,t)$ is not necessary for the control design. The convergence of the PDE state $\boldsymbol{x}$ to its reference $\boldsymbol{x}_r$ is inferred by the flat output $y = w_1$ converging to its reference $y_r = w_{1,r}$ (as seen in Fig. 5). However, for the sake of validation, the stationary reference profile $\lim_{t\to\infty}\boldsymbol{x}_r(z,t)$ for the PDE state $\boldsymbol{x}$ is illustrated in Fig. 4 by the red curves at $t = 8$. On a further note, the trajectory of the ODE state in Fig. 5 between both vertical dash-dotted lines depicts the behavior during the polynomial transition between $t_1 = \phi_1(1)$ and $t_2 = t_1 + T$ (see (7)). This part of the trajectory, in turn, determines the input $u(t)$ on the interval $(0, \phi_1(1) + \phi_2(1) + T)$, i.e. the input trajectory until the dash-dotted line.

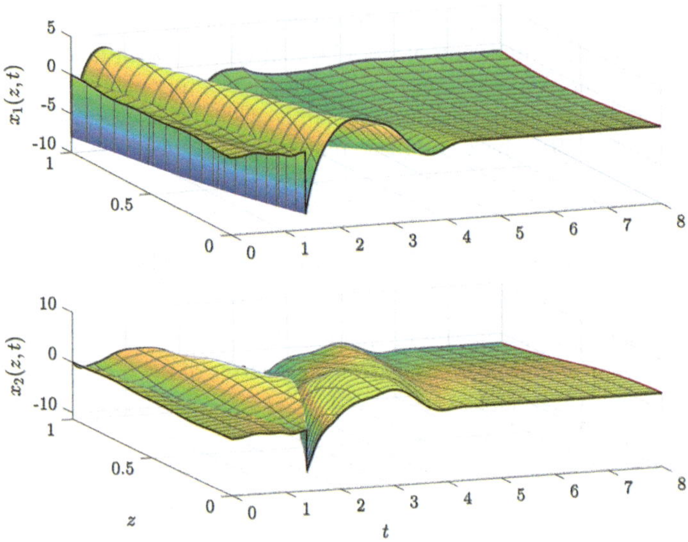

Fig. 4 Flatness-based static state feedback (51): profiles of the PDE states $x_1(z,t)$ and $x_2(z,t)$ converging to their corresponding reference (red)

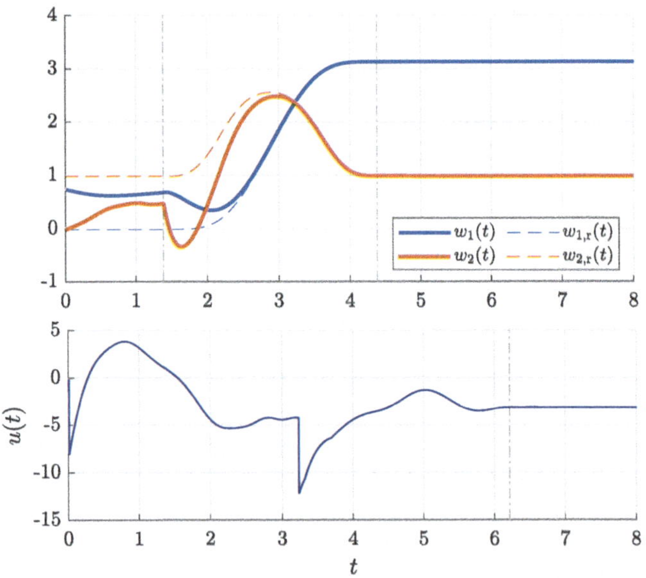

Fig. 5 Flatness-based static state feedback (51): ODE state $\boldsymbol{w}(t)$ converging to its reference trajectory $\boldsymbol{w}_{\mathrm{r}}(t)$, and control input $u(t)$

4.3 Controller 2: Linear Dynamic State Feedback with Additive Feed-Forward Control

The flatness-based static state feedback (51) can easily be extended to a dynamic one by introducing a controller state that accounts for an additional integral term. This is not pursued here since the control performance of the flatness-based controller is excellent without the dynamic extension. Hence, presenting a different control strategy for dynamic state feedback is more instructive.

The second simulated scenario uses the linear dynamic state feedback

$$\kappa(t, \boldsymbol{w}(t)) = x_{1,\mathrm{r}}(0,t) - l_2\left(w_2(t) - w_{2,\mathrm{r}}(t)\right) - l_1\left(w_1(t) - w_{1,\mathrm{r}}(t)\right) - l_0 \int_{t_0}^{\max(t_0,t)} \left(w_1(\tau) - w_{1,\mathrm{r}}(\tau)\right) \mathrm{d}\tau \qquad (52)$$

with $x_{1,\mathrm{r}}(0,t)$ and $\boldsymbol{w}_\mathrm{r}(t)$ defined in (47) and the parameters $l_0 = 8, l_1 = 15, l_2 = 6$ and $t_0 = \phi_1(1)$. A detailed discussion of the corresponding simulation results depicted in Figs. 6 and 7 is omitted for the sake of brevity. Both the PDE state $\boldsymbol{x}(z,t)$ and the ODE state $\boldsymbol{w}(t)$ converge to their corresponding reference trajectories.

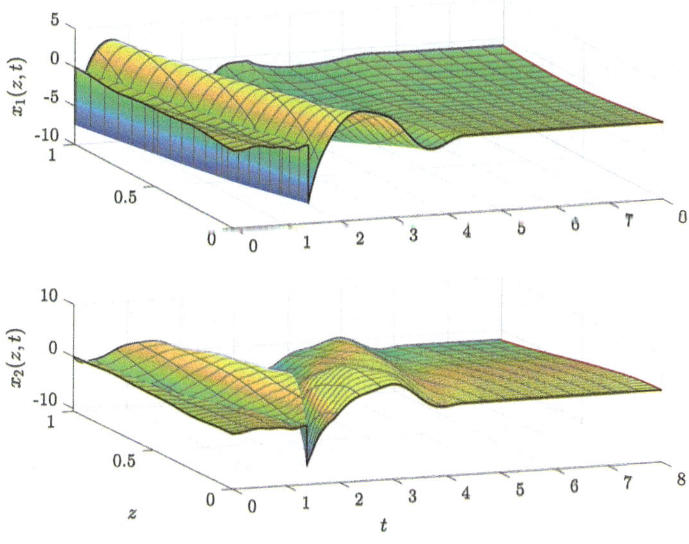

Fig. 6 Linear dynamic state feedback (52): profiles of the PDE states $x_1(z,t)$ and $x_2(z,t)$ converging to their corresponding reference (red)

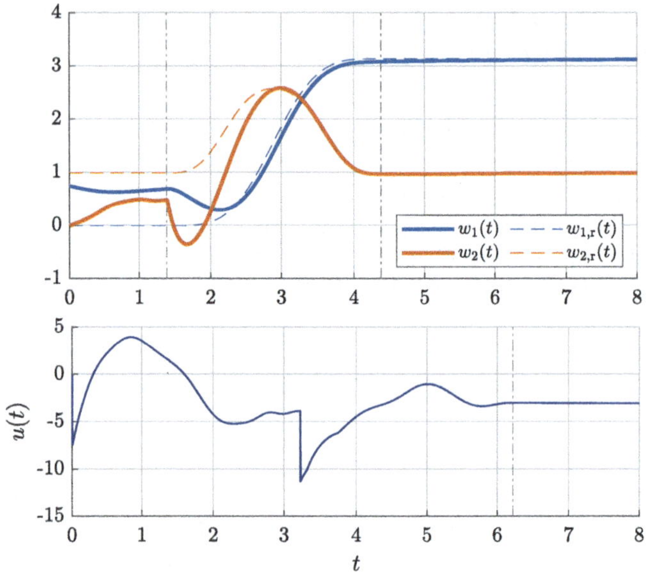

Fig. 7 Linear dynamic state feedback (52): ODE state $\boldsymbol{w}(t)$ converging to its reference trajectory $\boldsymbol{w}_r(t)$, and control input $u(t)$

5 Special Case: Linear PDE-ODE System

The case of linear PDE-ODE systems (1) with

$$f(\boldsymbol{w}(t), x_1(0,t)) = F\boldsymbol{w}(t) + \boldsymbol{b}x_1(0,t) \tag{53a}$$

$$c(\boldsymbol{w}(t)) = \boldsymbol{c}^{\mathrm{T}}\boldsymbol{w}(t), \tag{53b}$$

and the pair $(F, \boldsymbol{b})$ controllable, is special for several reasons. First, the online solution of a Volterra integro-differential equation becomes unnecessary as there exists a prediction-free, i.e. explicit, representation of its solution in terms of the ODE state $\boldsymbol{w}(t)$ and the PDE state $\boldsymbol{x}(z,t)$ only. This leads to significant simplifications of the design. Second, in the case of a linear stabilizing feedback, the introduced solution-based controller and the backstepping controller from [7] are identical. Hence, the proposed approach includes the results in [7] as a special case.

5.1 Explicit Representation of the Solution-Based Tracking Controller

The linearity of $\boldsymbol{f}$ (w. r. t. both its arguments) and the controllability of the pair $(F, \boldsymbol{b})$ imply that Assumptions 1–3 hold. In particular, the ODE controller can simply be chosen to be linear, i.e.

$$\kappa(t, \boldsymbol{w}(t)) = \boldsymbol{k}^{\mathrm{T}}(\boldsymbol{w}(t) - \boldsymbol{w}_{\mathrm{r}}(t)) + x_{1,\mathrm{r}}(0, t), \tag{54}$$

with $\boldsymbol{k}^{\mathrm{T}}$ such that $F + \boldsymbol{b}\boldsymbol{k}^{\mathrm{T}}$ is Hurwitz, and reference trajectories $\boldsymbol{w}_{\mathrm{r}}(t)$ and $x_{1,\mathrm{r}}(0, t)$ satisfying

$$\dot{\boldsymbol{w}}_{\mathrm{r}}(t) = F\boldsymbol{w}_{\mathrm{r}}(t) + \boldsymbol{b}x_{1,\mathrm{r}}(0, t). \tag{55}$$

Then, in light of (53), the tracking controller (28) for the linear PDE-ODE system (1) takes the form

$$\begin{aligned} u(t) = \boldsymbol{k}^{\mathrm{T}}\boldsymbol{w}_{\mathrm{p}}(\phi_1(1); t) &- \int_0^{\phi_1(1)} h_1(\psi_1(\phi_1(1) - \sigma))\boldsymbol{c}^{\mathrm{T}}\boldsymbol{w}_{\mathrm{p}}(\sigma; t)\,\mathrm{d}\sigma \\ &+ x_{1,\mathrm{r}}(0, t + \phi_1(1)) - \boldsymbol{k}^{\mathrm{T}}\boldsymbol{w}_{\mathrm{r}}(t + \phi_1(1)) - q_1x_2(1, t) + \int_0^1 \boldsymbol{e}_1^{\mathrm{T}}K(1, \zeta)\boldsymbol{x}(\zeta, t)\,\mathrm{d}\zeta, \end{aligned} \tag{56}$$

wherein $\boldsymbol{w}_{\mathrm{p}}(s; t)$ solves the IVP (26), i.e.

$$\tfrac{\mathrm{d}}{\mathrm{d}s}\boldsymbol{w}_{\mathrm{p}}(s; t) = F\boldsymbol{w}_{\mathrm{p}}(s; t) + \boldsymbol{b}\tilde{x}_1(\psi_1(s), t) + \int_0^s \boldsymbol{b}h_1(\psi_1(s - \sigma))\boldsymbol{c}^{\mathrm{T}}\boldsymbol{w}_{\mathrm{p}}(\sigma; t)\,\mathrm{d}\sigma \tag{57a}$$

$$\boldsymbol{w}_{\mathrm{p}}(0; t) = \boldsymbol{w}(t) \tag{57b}$$

for $s \in (0, \phi_1(1)]$. Its formal solution leads to the linear integral equation

$$\begin{aligned} \boldsymbol{w}_{\mathrm{p}}(s; t) = \mathrm{e}^{Fs}\boldsymbol{w}(t) &+ \int_0^s \mathrm{e}^{F(s-\tau)}\boldsymbol{b}\tilde{x}_1(\psi_1(\tau), t)\,\mathrm{d}\tau \\ &+ \int_0^s \mathrm{e}^{F(s-\tau)}\boldsymbol{b}\int_0^\tau h_1(\psi_1(\tau - \sigma))\boldsymbol{c}^{\mathrm{T}}\boldsymbol{w}_{\mathrm{p}}(\sigma; t)\,\mathrm{d}\sigma\,\mathrm{d}\tau \end{aligned} \tag{58}$$

for $\boldsymbol{w}_{\mathrm{p}}(s; t)$ due to the linearity of the Volterra integro-differential equation (57a). In fact, (58) is equivalent to the IVP (57) in the sense that its solution satisfies the linear Volterra integro-differential equation (57a) and the IC (57b). Thereby, (56) can be implemented by solving the integral equation (58) instead of the IVP (57).

It is even possible to eliminate $\boldsymbol{w}_\mathrm{p}(s;t)$ entirely in order to express the controller (56) as an explicit feedback of the states $\boldsymbol{w}(t)$ and $\boldsymbol{x}(z,t)$ only. For that, note that the integral equation (58) is of Volterra type and can be written as

$$\boldsymbol{w}_\mathrm{p}(s;t) = \mathrm{e}^{Fs}\boldsymbol{w}(t) + \int_0^s \mathrm{e}^{F(s-\tau)}\boldsymbol{b}\tilde{x}_1(\psi_1(\tau),t)\,\mathrm{d}\tau + \int_0^s M(s-\sigma)\boldsymbol{w}_\mathrm{p}(\sigma;t)\,\mathrm{d}\sigma \tag{59}$$

with the convolution kernel

$$M(s) = \int_0^s \mathrm{e}^{F(s-\tau)}\boldsymbol{b}h_1(\psi_1(\tau))\boldsymbol{c}^\mathrm{T}\,\mathrm{d}\tau \tag{60}$$

defined for $s \in [0, \phi_1(1)]$. This follows from (58) after changing the order of integration of the double integral and substituting the integration variable. The continuity of the kernel M, inferred by the continuity of the kernel K (cf. (10) and (60)), guarantees the existence of a unique solution [18], which takes the form

$$\boldsymbol{w}_\mathrm{p}(s;t) = \boldsymbol{\vartheta}(s;t) + \int_0^s M_\mathrm{I}(s-\sigma)\boldsymbol{\vartheta}(\sigma;t)\,\mathrm{d}\sigma \tag{61a}$$

with

$$\boldsymbol{\vartheta}(s;t) = \mathrm{e}^{Fs}\boldsymbol{w}(t) + \int_0^s \mathrm{e}^{F(s-\tau)}\boldsymbol{b}\tilde{x}_1(\psi_1(\tau),t)\,\mathrm{d}\tau \tag{61b}$$

and the inverse kernel M_I satisfying the reciprocity relation

$$M_\mathrm{I}(s) = M(s) + \int_0^s M(s-\tau)M_\mathrm{I}(\tau)\,\mathrm{d}\tau. \tag{61c}$$

While (61a) is already an explicit representation of $\boldsymbol{w}_\mathrm{p}(s;t)$ in terms of $\boldsymbol{\vartheta}(s;t)$, it can be simplified further. For that, introduce the matrix

$$P(s) = \mathrm{e}^{Fs} + \int_0^s M_\mathrm{I}(s-\sigma)\mathrm{e}^{F\sigma}\,\mathrm{d}\sigma \tag{62}$$

such that the predicted state $\boldsymbol{w}_\mathrm{p}(s;t)$ can be written as

$$\boldsymbol{w}_\mathrm{p}(s;t) = P(s)\boldsymbol{w}(t) + \int_0^s P(s-\tau)\boldsymbol{b}\tilde{x}_1(\psi_1(\tau),t)\,\mathrm{d}\tau, \tag{63}$$

which immediately follows from (61a)–(61b) by changing the order of integration and applying a substitution of variables. This means that the online prediction step is not required in the linear case, as the predicted state $\boldsymbol{w}_\mathrm{p}(s;t)$ can be determined by evaluating (63), wherein $P(s)$ can be computed offline. In the end, by taking (63) into account, the tracking controller (56) then reads

$$u(t) = \boldsymbol{n}^{\mathrm{T}}(\phi_1(1))\boldsymbol{w}(t) + \int_0^{\phi_1(1)} \boldsymbol{n}^{\mathrm{T}}(\phi_1(1) - \tau)\boldsymbol{b}\tilde{x}_1(\psi_1(\tau), t)\,\mathrm{d}\tau$$
$$+ x_{1,\mathrm{r}}(0, t + \phi_1(1)) - \boldsymbol{k}^{\mathrm{T}}\boldsymbol{w}_{\mathrm{r}}(t + \phi_1(1)) - q_1 x_2(1, t) + \int_0^1 \boldsymbol{e}_1^{\mathrm{T}} K(1, \zeta)\boldsymbol{x}(\zeta, t)\,\mathrm{d}\zeta \quad (64)$$

with the row vector $\boldsymbol{n}^{\mathrm{T}}$ defined by

$$\boldsymbol{n}^{\mathrm{T}}(s) = \boldsymbol{k}^{\mathrm{T}} P(s) - \int_0^s h_1(\psi_1(s - \sigma))\boldsymbol{c}^{\mathrm{T}} P(\sigma)\,\mathrm{d}\sigma. \quad (65)$$

5.2 Equivalence to Known Backstepping Controllers

In [7], the stabilization of linear PDE-ODE systems[3] (1) with (53) via a multi-step backstepping design method is proposed. For that, a backstepping transformation (8) maps the PDE subsystem into a form without in-domain coupling, which is equivalent[4] to the preliminary backstepping transformation in Sect. 3.1. Then, a transformation of the PDE state $\tilde{\boldsymbol{x}}$ into so-called decoupling coordinates

$$\bar{\boldsymbol{x}}(z, t) = \tilde{\boldsymbol{x}}(z, t) - \int_0^z \bar{P}(z, \zeta)\tilde{\boldsymbol{x}}(\zeta, t)\,\mathrm{d}\zeta - \bar{N}(z)\boldsymbol{w}(t) \quad (66)$$

is used, wherein $\bar{P}(z, \zeta) \in \mathbb{R}^{2\times 2}$ and $\bar{N}(z) \in \mathbb{R}^{2\times n}$ are the unique (and continuous) solutions of the so-called decoupling equations. In particular, these are described by a boundary value problem (BVP), consisting of transport equations for $\bar{P}$, coupled with an IVP, consisting of an ODE for $\bar{N}$. Whereas the BVP is explicitly solved, the IVP is traced back to a linear Volterra integral equation. Note the obvious similarities to the inhomogeneous transport equation in Sect. 3.2 and the IVP (57) for the predicted state $\boldsymbol{w}_{\mathrm{p}}$, which is traced back to the linear Volterra integral equation (58).

Using the solution of the decoupling equations, the backstepping controller in [7] reads

$$\bar{u}(t) = \bar{\boldsymbol{n}}^{\mathrm{T}}(1)\boldsymbol{w}(t) + \int_0^1 \frac{1}{\lambda_1(z)}\bar{\boldsymbol{n}}^{\mathrm{T}}(\psi_1(\phi_1(1) - \phi_1(z)))\boldsymbol{b}\tilde{x}_1(z, t)\,\mathrm{d}z$$
$$+ \int_0^1 \boldsymbol{e}_1^{\mathrm{T}} K(1, \zeta)\boldsymbol{x}(\zeta, t)\,\mathrm{d}\zeta \quad (67)$$

---

[3] Note that, in [7], the roles of the PDE state components x_1 and x_2 are interchanged and $q_1 = 0$ in (1d). In order to prevent confusion, the results therein are presented with the notation of this work.

[4] The kernel equations proposed in [7] for the kernel $K(z, \zeta)$ account for the case $q_0 = 0$. They are equivalent to the kernel equations (9) of this work in the special case that (13) is used instead of the second component of (9c) (see Remark 3).

with $\bar{\boldsymbol{n}}^{\mathrm{T}}(z) = \boldsymbol{e}_1^{\mathrm{T}} \bar{N}(z)$ satisfying the linear Volterra integral equation

$$\bar{\boldsymbol{n}}^{\mathrm{T}}(z) = \boldsymbol{k}^{\mathrm{T}} \mathrm{e}^{F\phi_1(z)} - \int_0^z \frac{h_1(\zeta)}{\lambda_1(\zeta)} \boldsymbol{c}^{\mathrm{T}} \mathrm{e}^{F(\phi_1(z)-\phi_1(\zeta))} \,\mathrm{d}\zeta + \int_0^z \bar{\boldsymbol{n}}^{\mathrm{T}}(\zeta) \bar{Q}(z,\zeta) \,\mathrm{d}\zeta, \quad \text{(68a)}$$

where[5]

$$\bar{Q}(z,\zeta) = \int_\zeta^z \frac{1}{\lambda_1(\xi)\lambda_1(\zeta)} \boldsymbol{b} h_1(\psi_1(\phi_1(\xi) - \phi_1(\zeta))) \boldsymbol{c}^{\mathrm{T}} \mathrm{e}^{F(\phi_1(z)-\phi_1(\xi))} \,\mathrm{d}\xi. \quad \text{(68b)}$$

Note that the resulting backstepping controller (67) and the solution-based controller (64) share the same structure. In fact, the backstepping controller (67) and its solution-based counterpart (64) are equivalent for $q_1 = 0$ in the special case of stabilization, i.e. $\boldsymbol{w}_{\mathrm{r}}(t) = \boldsymbol{0}$ and therefore $x_{1,\mathrm{r}}(0,t) = 0$. Next, this equivalence is shown based on the implicit representation of the solution-based controller (56).

In order to simplify the comparison, define $\tilde{\boldsymbol{n}}^{\mathrm{T}}(s) = \bar{\boldsymbol{n}}^{\mathrm{T}}(\psi_1(s))$. By substituting the integration variables, the backstepping controller (67) may be expressed as

$$\bar{u}(t) = \tilde{\boldsymbol{n}}^{\mathrm{T}}(\phi_1(1))\boldsymbol{w}(t) + \int_0^{\phi_1(1)} \tilde{\boldsymbol{n}}^{\mathrm{T}}(\phi_1(1) - \tau) \boldsymbol{b} \tilde{x}_1(\psi_1(\tau), t) \,\mathrm{d}\tau + \int_0^1 \boldsymbol{e}_1^{\mathrm{T}} K(1,\zeta) \boldsymbol{x}(\zeta, t) \,\mathrm{d}\zeta \quad \text{(69)}$$

with $\tilde{\boldsymbol{n}}^{\mathrm{T}}(s)$ being the unique solution of the much simpler looking integral equation

$$\tilde{\boldsymbol{n}}^{\mathrm{T}}(s) = \boldsymbol{k}^{\mathrm{T}} \mathrm{e}^{Fs} - \int_0^s h_1(\psi_1(\tau)) \boldsymbol{c}^{\mathrm{T}} \mathrm{e}^{F(s-\tau)} \,\mathrm{d}\tau + \int_0^s \tilde{\boldsymbol{n}}^{\mathrm{T}}(\sigma) \tilde{Q}(s-\sigma) \,\mathrm{d}\sigma \quad \text{(70a)}$$

of Volterra type with the convolution kernel

$$\tilde{Q}(s) = \int_0^s \boldsymbol{b} h_1(\psi_1(\tau)) \boldsymbol{c}^{\mathrm{T}} \mathrm{e}^{F(s-\tau)} \,\mathrm{d}\tau. \quad \text{(70b)}$$

Note that (70) is the result of using (68), the definition of $\tilde{\boldsymbol{n}}^{\mathrm{T}}(s)$ and a substitution of integral variables. Next, define $\Delta u = u - \bar{u}$. Then, it follows from (56) and (69) that

$$\Delta u(t) = \boldsymbol{k}^{\mathrm{T}} \boldsymbol{w}_{\mathrm{p}}(\phi_1(1); t) - \int_0^{\phi_1(1)} h_1(\psi_1(\phi_1(1) - \sigma)) \boldsymbol{c}^{\mathrm{T}} \boldsymbol{w}_{\mathrm{p}}(\sigma; t) \,\mathrm{d}\sigma - \tilde{\boldsymbol{n}}^{\mathrm{T}}(\phi_1(1))\boldsymbol{w}(t) - \int_0^{\phi_1(1)} \tilde{\boldsymbol{n}}^{\mathrm{T}}(\phi_1(1) - \tau) \boldsymbol{b} \tilde{x}_1(\psi_1(\tau), t) \,\mathrm{d}\tau. \quad \text{(71)}$$

[5] In [7], the negative sign in the definition of $\bar{Q}$ can be traced back to a typo.

This motivates the definition of $q(s;t) = \tilde{\boldsymbol{n}}^{\mathrm{T}}(\phi_1(1) - s)\boldsymbol{w}_{\mathrm{p}}(s;t)$ with

$$q(0;t) = \tilde{\boldsymbol{n}}^{\mathrm{T}}(\phi_1(1))\boldsymbol{w}(t), \qquad q(\phi_1(1);t) = \boldsymbol{k}^{\mathrm{T}}\boldsymbol{w}_{\mathrm{p}}(\phi_1(1);t), \tag{72}$$

as apparent from (57b) and (70a). Also, notice that

$$\begin{aligned} q(\phi_1(1);t) - q(0;t) &= \int_0^{\phi_1(1)} \frac{\mathrm{d}q}{\mathrm{d}s}(s;t)\,\mathrm{d}s \\ &= \int_0^{\phi_1(1)} \tilde{\boldsymbol{n}}^{\mathrm{T}}(\phi_1(1) - s)\left(\tfrac{\mathrm{d}}{\mathrm{d}s}\boldsymbol{w}_{\mathrm{p}}(s;t)\right)\mathrm{d}s \\ &\quad + \int_0^{\phi_1(1)} \left(\tfrac{\mathrm{d}}{\mathrm{d}s}\tilde{\boldsymbol{n}}^{\mathrm{T}}(\phi_1(1) - s)\right)\boldsymbol{w}_{\mathrm{p}}(s;t)\,\mathrm{d}s. \end{aligned} \tag{73}$$

Using (71)–(73) and the linear Volterra integro-differential equations (57a) and

$$\tfrac{\mathrm{d}}{\mathrm{d}s}\tilde{\boldsymbol{n}}^{\mathrm{T}}(s) = \tilde{\boldsymbol{n}}^{\mathrm{T}}(s)F - h_1(\psi_1(s))\boldsymbol{c}^{\mathrm{T}} + \int_0^s \tilde{\boldsymbol{n}}^{\mathrm{T}}(\sigma)\boldsymbol{b}h_1(\psi_1(s-\sigma))\boldsymbol{c}^{\mathrm{T}}\,\mathrm{d}\sigma, \tag{74}$$

which is the derivative of (70a) w.r.t. s, yields

$$\begin{aligned} \Delta u(t) = &\int_0^{\phi_1(1)}\int_0^s \tilde{\boldsymbol{n}}^{\mathrm{T}}(\phi_1(1) - s)\boldsymbol{b}h_1(\psi_1(s-\sigma))\boldsymbol{c}^{\mathrm{T}}\boldsymbol{w}_{\mathrm{p}}(\sigma;t)\,\mathrm{d}\sigma\,\mathrm{d}s \\ - &\int_0^{\phi_1(1)}\int_0^{\phi_1(1)-s} \tilde{\boldsymbol{n}}^{\mathrm{T}}(\sigma)\boldsymbol{b}h_1(\psi_1(\phi_1(1) - s - \sigma))\boldsymbol{c}^{\mathrm{T}}\boldsymbol{w}_{\mathrm{p}}(s;t)\,\mathrm{d}\sigma\,\mathrm{d}s. \end{aligned} \tag{75}$$

Changing the order of integration and substituting the integration variable immediately shows that $\Delta u(t) = 0$ for all $t > 0$. By that, the equivalence of the stabilizing backstepping controller $\bar{u}$ in [7] and its solution-based counterpart is implied. As an immediate consequence, $\boldsymbol{n}^{\mathrm{T}}(s)$, defined in (65), is the unique solution of the linear Volterra integral equation (70) in the sense that $\boldsymbol{n}^{\mathrm{T}}(s) = \tilde{\boldsymbol{n}}^{\mathrm{T}}(s)$ for $s \in [0, \phi_1(1)]$.

6 Concluding Remarks

The proposed control strategy can easily be adjusted to account for a nonlinear dependence on $x_1(0,t)$ in (1b), i.e. $c(\boldsymbol{w}(t), x_1(0,t))$ instead of $c(\boldsymbol{w}(t))$. The case of a $(1+p)\times(1+p)$ PDE subsystem, in the sense that $p > 1$ state components are propagated in the positive z-direction, only necessitates formal modifications. Both extensions are left out in this work for the sake of brevity and simplicity. The more challenging extension to heterodirectional hyperbolic PDE subsystems of arbitrary order is the subject of current research. First results on a matching solution-based design strategy for the collocated state observation are given in [14].

References

1. Angeli, D., Sontag, E.: Forward completeness, unboundedness observability, and their Lyapunov characterizations. Syst. Control Lett. **38**(4–5), 209–217 (1999)
2. Auriol, J., Bribiesca Argomedo, F., Bou Saba, D., Di Loreto, M., Di Meglio, F.: Delay-robust stabilization of a hyperbolic PDE-ODE system. Automatica **95**, 494–502 (2018)
3. Bekiaris-Liberis, N., Krstic, M.: Compensation of wave actuator dynamics for nonlinear systems. IEEE Trans. Autom. Control **59**(6), 1555–1570 (2014)
4. Bekiaris-Liberis, N., Krstic, M.: Compensation of transport actuator dynamics with input-dependent moving controlled boundary. IEEE Trans. Autom. Control **63**(11), 3889–3896 (2018)
5. Cai, X., Wan, L., Yang, Q., Wu, J., Liu, L.: Backstepping control of ODE/wave PDE cascaded systems with dynamic boundary conditions. Int. J. Contr., 1–9 (2021)
6. Coron, J.M., Vazquez, R., Krstic, M., Bastin, G.: Local exponential H^2 stabilization of a 2×2 quasilinear hyperbolic system using backstepping. SIAM J. Control Optim. **51**(3), 2005–2035 (2013)
7. Deutscher, J., Gehring, N., Kern, R.: Backstepping control of linear 2×2 hyperbolic systems with dynamic boundary conditions. IFAC-PapersOnLine **50**(1), 4522–4527 (2017)
8. Deutscher, J., Gehring, N., Kern, R.: Output feedback control of general linear heterodirectional hyperbolic PDE-ODE systems with spatially-varying coefficients. Int. J. Control **92**(10), 2274–2290 (2019)
9. Di Meglio, F., Bribiesca Argomedo, F., Hu, L., Krstic, M.: Stabilization of coupled linear heterodirectional hyperbolic PDE-ODE systems. Automatica **87**, 281–289 (2018)
10. Feldstein, A., Sopka, J.: Numerical methods for nonlinear Volterra integro-differential equations. SIAM J. Numer. Anal. **11**(4), 826–846 (1974)
11. Fischer, F., Gabriel, J., Kerschbaum, S.: coni - a Matlab toolbox facilitating the solution of control problems (2021)
12. Gehring, N., Woittennek, F.: Flatness-based output feedback tracking control of a hyperbolic distributed-parameter system. IEEE Control Syst. Lett., 992–997 (2022)
13. Hu, L., Di Meglio, F., Vazquez, R., Krstic, M.: Control of homodirectional and general heterodirectional linear coupled hyperbolic PDEs. IEEE Trans. Autom. Control **61**(11), 3301–3314 (2016)
14. Irscheid, A., Gehring, N., Deutscher, J., Rudolph, J.: Observer design for 2×2 linear hyperbolic PDEs that are bidirectionally coupled with nonlinear ODEs. In: 2021 European Control Conference (ECC), 2506–2511 (2021)
15. Irscheid, A., Gehring, N., Rudolph, J.: Trajectory tracking control for a class of 2×2 hyperbolic PDE-ODE systems. IFAC-PapersOnLine **54**(9), 416–421 (2021)
16. Knüppel, T., Woittennek, F.: Control design for quasi-linear hyperbolic systems with an application to the heavy rope. IEEE Trans. Autom. Control **60**, 5–18 (2015)
17. Krstic, M., Smyshlyaev, A.: Boundary Control of PDEs: A Course on Backstepping Designs, vol. 16. SIAM (2008)
18. Linz, P.: Analytical and Numerical Methods for Volterra Equations. SIAM (1985)
19. Rudolph, J.: Flatness Based Control of Distributed Parameter Systems. Shaker Verlag (2003)
20. Rudolph, J.: Flatness-Based Control: An Introduction. Shaker Verlag (2021)
21. Strecker, T., Aamo, O.: Output feedback boundary control of 2×2 semilinear hyperbolic systems. Automatica **83**, 290–302 (2017)
22. Vazquez, R., Krstic, M.: Marcum Q-functions and explicit kernels for stabilization of 2×2 linear hyperbolic systems with constant coefficients. Syst. Control Lett. **68**, 33–42 (2014)
23. Vazquez, R., Krstic, M., Coron, J.: Backstepping boundary stabilization and state estimation of a 2×2 linear hyperbolic system. In: 2011 50th IEEE Conference on Decision and Control and European Control Conference, pp. 4937–4942 (2011)
24. Woittennek, F.: Flatness based feedback design for hyperbolic distributed parameter systems with spatially varying coefficients. IFAC Proc. Vol. **46**(26), 37–42 (2013)

Practical Output Regulation and Tracking for Linear ODE-hyperbolic PDE-ODE Systems

Jeanne Redaud, Federico Bribiesca-Argomedo, and Jean Auriol

Abstract In this chapter, we consider the problem of practical output regulation and output tracking for a linear 2×2 hyperbolic Partial Differential Equation (PDE) system with actuation and load dynamics. Indeed, it is actuated via an Ordinary Differential Equation (ODE) at one boundary and the output to be controlled is the output of an ODE at the other boundary. The main focus is on load tracking. Here, we propose to extend existing results on approximate output regulation to a class of systems similar to that considered in [8] and to extend filtering techniques to a dynamically augmented system with finite-dimensional exosystems considering possible trajectory and disturbance inputs. Issues with respect to small delays in the state reconstruction and feedback loop are considered. Due to the nature of the disturbances, the state estimation and disturbance reconstruction problems are also considered. This scenario finds applications in many systems of engineering interest, such as drilling systems [3], pneumatic systems [17], or electric transmission lines [22].

1 Introduction

In this chapter, we consider a linear 2×2 hyperbolic Partial Differential Equation (PDE) coupled at one end with an actuator and on the other end with a load, both modeled by potentially unstable Ordinary Differential Equations (ODE). The main focus is on load tracking (i.e. tracking an output depending on the states of the

J. Redaud (✉) · J. Auriol
Université Paris-Saclay, CNRS, CentraleSupélec, Inria Saclay, Laboratoire des Signaux et Systèmes, 91190 Gif-sur-Yvette, France
e-mail: jeanne.redaud@centralesupelec.fr

J. Auriol
e-mail: jean.auriol@centralesupelec.fr

F. Bribiesca-Argomedo
Université de Lyon, INSA Lyon, Laboratoire Ampère (CNRS UMR5005), 69621 Villeurbanne, France
e-mail: federico.bibriesca-argomedo@insa-lyon.fr

J. Auriol et al. (eds.), *Advances in Distributed Parameter Systems*, Advances in Delays and Dynamics 14, https://doi.org/10.1007/978-3-030-94766-8_7

unactuated ODE, while only controlling the opposite ODE system) and disturbance rejection. More precisely, we design a dynamic full-state feedback control law acting only on the first ODE system (the actuator) and ensuring that an output, depending on the states of the opposite ODE system (the unactuated load), follows a prescribed trajectory generated by a finite-dimensional exosystem. This tracking property holds in the presence of exogenous disturbances (modeled by a known finite-dimensional system) acting on the load.

Since transport phenomena and delays are common in industrial applications, many systems can be modeled by hyperbolic PDEs, such as oil drilling pipes [2, 25], pneumatic systems [17], networks of hydraulic distribution lines or electrical lines [6, 22]. Such systems can be coupled with ODE dynamics that represent loads or actuator dynamics. In different applications, the distal ODE state can model the dynamics of a suspended object or payload for Unmanned Aerial Vehicles (UAV) [15] or the Bottom Hole Assembly at the end of a drilling pipe [5], to cite some examples. More specifically, the ODE-PDE-ODE structure naturally arises when considering linear systems of balance laws with finite-dimensional actuator and load dynamics, as for the UAV-cable-payload structure [25].

Therefore, coupled hyperbolic PDEs and ODEs have been an active research field in the previous years. The stabilization of each independent subsystem is a well-known task. Indeed, PID or Smith Predictor controllers have been extensively developed for linear ODEs [23]. Moreover, the well-known backstepping technique has been adapted to provide constructive designs of control laws for hyperbolic systems [18]. Finally, output regulation problems have already been solved for hyperbolic systems without actuator and load dynamics [1].

However, new problems have to be overcome when ODEs and PDEs are interconnected. The backstepping method has then been adapted to such systems. First, only PDE-ODE interconnections were considered [12]. Next, some results have been extended to ODE-PDE-ODE interconnections, and full-state feedback controllers have been designed for such coupled systems. In most cases, the constructive design is based on several invertible backstepping transforms. They are used to cancel the reflection term on the actuated boundary such that the PDE has no longer an impact onto the state of the controlled ODE [7, 10].

However, suppression of the reflection term may lead to non strictly proper control laws. This can cause robustness issues, leading in particular to unstable behaviors in the presence of arbitrary small delays [4]. This is, therefore, a significant concern for further application to real systems. Consequently, the focus has been made on robustly stabilizing such systems. Instead of canceling all the reflection terms at the actuated boundary, [13] assigned it the dynamics of a low-pass filter. In [8], the authors proposed a strictly proper control law guaranteeing a non-zero delay margin in closed-loop and stabilizing the coupled ODE-PDE-ODE structure, using filtering techniques. However, the stabilizing control law from [8] requires the knowledge of all the states. In this chapter, we extend the proposed approach to an output feedback control law and design an observer for the coupled system. This result is closely related to that in [25]. However, the specific observer design proposed here does not require multiple transforms. Unlike [25], all the error injection terms used in the proposed observer are strictly proper, which avoids having arbitrarily big gains at high frequencies (typically amplifying measurement noise).

Furthermore, instead of focusing on stabilizing the interconnected system (bringing it back to an equilibrium state), we solve the problem of practical output regulation and output tracking. Indeed, the ODE system opposite to the control input can be subject to perturbation. We extend the results in [8] to a dynamically augmented system with a finite-dimensional exosystem corresponding to a possible trajectory reference or/and to a disturbance input. The problem of disturbance reconstruction and the state estimation are also addressed in the proposed observer.

More specifically, the contribution of this chapter is to propose a constructive approach for the design of a dynamic, strictly proper, output-feedback control law for practical output regulation and output tracking, guaranteeing a non-zero delay margin for a large class of interconnected ODE-hyperbolic PDEs-ODE systems. This approach can be easily implemented on real systems. Unlike [10, 25], it does not require to apply multiple successive state transformations, which simplifies the design and implementation procedure for field engineers who are not familiar with multi-step approaches. Moreover, it is based on assumptions that can be easily tested in practice.

The proposed approach can be summarized as follows. Based on some structural assumptions, and inspired by the approach proposed in [8], we first design a state-feedback controller to stabilize an output depending on the states of the unactuated ODE, solving thereby the practical output tracking and regulation problem. We follow the backstepping methodology and use a general invertible integral transform to map our initial system (dynamically augmented with finite-dimensional exosystems) to a target system. We use a frequency analysis to design a feedback controller. Using structural assumptions similar to those in [25], we then propose a state observer design for the system and the disturbances. Unlike [25], we use filtering techniques to guarantee that all the dynamic error injection gains are strictly proper. Finally, the two designs are coupled in order to obtain a dynamic output feedback controller.

The layout of this chapter is the following: in Sect. 2, we present the structure of the system and give some structural assumptions. Next, we design a robust state feedback controller for practical output regulation in Sect. 3. We design an observer for the state and the perturbation in Sect. 4. Finally, we propose some illustrative examples with test-cases simulations in Sect. 5.

Notation

In the following, we denote $\mathcal{T}^+$ the upper-triangular domain defined in $\mathbb{R}^2$ by $\mathcal{T}^+ \doteq \{(x, y) \in [0, 1] \times [0, 1],\ y \geq x\}$, and $\mathcal{T}^-$ the lower-triangular domain defined by $\mathcal{T}^- \doteq \{(x, y) \in [0, 1] \times [0, 1],\ y \leq x\}$. Given a set $[a, b] \in \mathbb{R}$, we define the characteristic function $\mathbb{1}_{[a,b]}(x) \doteq 1$, if $x \in [a, b]$, 0 otherwise. The euclidean norm of a vector $\varphi \in \mathbb{R}^r$, $r \in \mathbb{N}\backslash\{0\}$ is denoted by $\|\varphi\|_{\mathbb{R}^r} \doteq (\varphi^\top \varphi)^{1/2}$. The L_2-norm of a function $\phi \in L^2([0, 1]; \mathbb{R})$ is taken in the usual sense $\|\phi\|_{L^2} = \left(\int_0^1 \phi^2(s)\mathrm{d}s\right)^{1/2}$. We denote $\mathcal{X} \doteq \mathbb{R}^n \times L^2([0, 1]; \mathbb{R})^2 \times \mathbb{R}^{m+p}$ the space of the states of the system

$(X(t), u(t, \cdot), v(t, \cdot), Y(t))$. We define the $\mathcal{X}$−norm of the state (X, u, v, Y) as the sum of these norms: $||(X, u, v, Y)\|_{\mathcal{X}} \doteq \|X\|_{\mathbb{R}^n} + \|Y\|_{\mathbb{R}^{m+p}} + \|u\|_{L^2} + \|v\|_{L^2}$.
The notation I_r represents the $r \times r$ identity matrix. For any matrix A, we denote its transpose A^T and its conjugate transpose (or Hermitian transpose) A^*. We use $\mathscr{L}(\phi)(s)$ to denote the Laplace transform of $\phi(t)$. For any proper and stable transfer matrix $G(s)$, $\bar{\sigma}(G(j\omega))$ stands for the largest singular value of $G(j\omega)$ at frequency ω, and the H_∞-norm of G is $\|G\|_\infty = \text{ess sup}_{\omega\in\mathbb{R}}\bar{\sigma}(G(j\omega))$.

2 Problem Statement

In this section, we present the system considered herein in more details, and we outline the control strategy. We also give some structural assumptions used in the following sections.

2.1 System Presentation

Let us consider a linear 2 × 2 hetero-directional hyperbolic PDE system interconnected at both ends with an ODE system. The first ODE system ($X(t) \in \mathbb{R}^{n\times 1}$), is actuated by a control input $U(t) \in \mathbb{R}^{c\times 1}$, and the second ODE system ($Y_1(t) \in \mathbb{R}^{m\times 1}$) is dynamically augmented by an exosystem ($Y_2(t) \in \mathbb{R}^{p\times 1}$), such that $Y(t) = \begin{bmatrix} Y_1(t) \\ Y_2(t) \end{bmatrix} \in \mathbb{R}^{(m+p)\times 1}$. The exogenous input encompasses a disturbance Y_{pert} and/or a known reference trajectory Y_{ref}, as illustrated in Sect. 5. Following the notations given in Fig. 1, we can model the system dynamics by the following set of equations

$$\dot{X}(t) = A_0X(t) + E_0v(t, 0) + B_0U(t), \tag{1}$$
$$u_t(t, x) + \lambda u_x(t, x) = \sigma^+(x)v(t, x), \tag{2}$$
$$v_t(t, x) - \mu v_x(t, x) = \sigma^-(x)u(t, x), \tag{3}$$
$$\dot{Y}(t) = \begin{bmatrix} A_{11} & A_{12} \\ 0_{p\times m} & A_{22} \end{bmatrix} Y(t) + \begin{bmatrix} E_1 \\ 0_{p\times 1} \end{bmatrix} u(t, 1), \tag{4}$$

with $(t, x) \in [0, +\infty) \times [0, 1]$, and the boundary conditions

$$v(t, 1) = \rho u(t, 1) + \begin{bmatrix} C_{11} & C_{12} \end{bmatrix} Y(t), \tag{5}$$
$$u(t, 0) = qv(t, 0) + C_0X(t), \tag{6}$$

with non scalar ODE dynamics of dimension $n, m, p, c \in \mathbb{N}\backslash\{0\}$, and $A_0 \in \mathbb{R}^{n\times n}$, $E_0 \in \mathbb{R}^{n\times 1}$, $B_0 \in \mathbb{R}^{n\times c}$, $A_{11} \in \mathbb{R}^{m\times m}$, $A_{12} \in \mathbb{R}^{m\times p}$, $A_{22} \in \mathbb{R}^{p\times p}$, $E_1 \in \mathbb{R}^{m\times 1}$, $C_0 \in \mathbb{R}^{1\times n}$, $C_{11} \in \mathbb{R}^{1\times m}$ and $C_{12} \in \mathbb{R}^{1\times p}$. We suppose that the transport velocities $\lambda, \mu >$

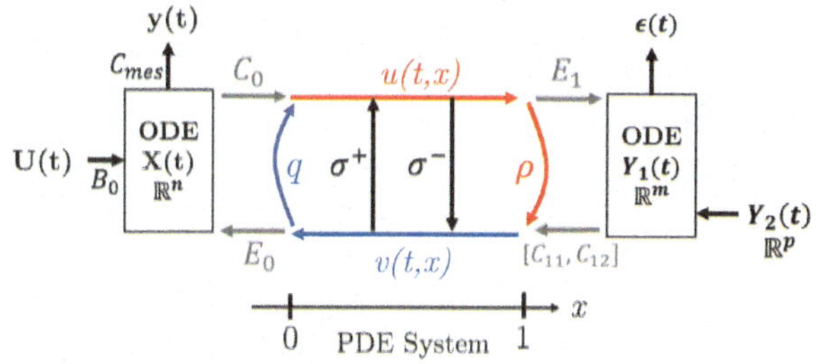

Fig. 1 Schematic presentation of the system

0 are constant for sake of simplicity. Similarly, the boundary couplings $q, \rho \in \mathbb{R}$ are assumed constant, but the in-domain couplings between the PDEs $\sigma^+, \sigma^- \in C([0,1]; \mathbb{R})$ are space-dependent functions. Let us consider any initial conditions $(X_0, u_0, v_0, Y_0) \in \mathcal{X}$. The system (1)–(6) is well-posed in a weak sense [6, Appendix A], i.e. it admits a solution in $\mathcal{X}$. We assume that we measure a part of the first ODE state, such that $y(t) = C_{\text{mes}} X(t) \in \mathbb{R}^{n' \times 1}$. The structure presented herein is motivated by drilling applications. Indeed, the ODEs can model the non negligeable dynamics of the top-drive at the surface, and the Bottom Hole Assembly (BHA) at the end of the pipe. The hyperbolic PDEs can represent torsional or axial wave-like propagation. Measurement and actuation are then only available at the surface $(x = 0)$. However, due to the symmetry of the problem, we could have considered measurement in $x = 1$. In this paper, we assume that the rock interaction has an impact on the BHA and the boundary of the PDE, such that the friction along the pipe is neglected. It allows a simpler regulation design, but distributed disturbance was considered for output regulation in [11].

2.2 Control Strategy

The objective of this chapter is to design a control law $U(t)$ guaranteeing the approximate regulation to zero of a virtual output $\epsilon(t)$ defined by

$$\epsilon(t) \doteq C_e Y(t), \quad \text{with } C_e = \begin{bmatrix} C_{e1} & C_{e2} \end{bmatrix}. \tag{7}$$

In most cases, $\epsilon(t) \in \mathbb{R}$ is a scalar function ($C_{e1} \in \mathbb{R}^{1 \times m}$, $C_{e2} \in \mathbb{R}^{1 \times p}$). Therefore, we can only stabilize a linear combination of components of the extended state.

The stabilization of ϵ fulfills the trajectory tracking and perturbation rejection objectives, depending on the value of C_e. Generally speaking, we want the unactuated ODE state to robustly converge towards a known reference trajectory even in the presence of a disturbance.

The design of a robust output-feedback controller is schematically represented on Fig. 2. First we follow the backstepping methodology to map our initial system to a simpler target system, using an invertible integral transform. Then, we use frequency

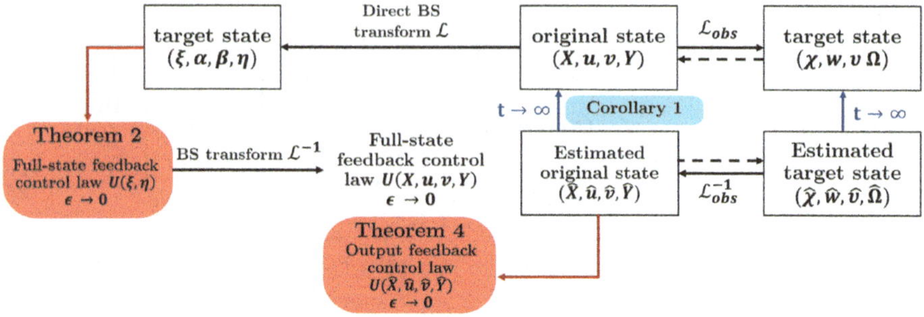

Fig. 2 Schematic presentation of the output-feedback control design

analysis to show that this constructive approach leads to a control law solving the practical output regulation/output tracking problem. Inspired by [8], we apply filtering techniques to guarantee the delay-robustness of the proposed controller. Next, we solve the problem of state estimation and disturbance reconstruction.

2.3 *Structural Assumptions*

Throughout this chapter, we use several assumptions on the system.

Assumption 1 The coefficients ρ, q satisfy $|\rho q| < 1$ and $\rho q \neq 0$.

This assumption is sufficient and can be easily tested, compared to numerical computation of eigenvalues. The fact that $|\rho q| < 1$ is nearly necessary to guarantee a delay-robustness margin [4] (in the case $|\rho q| = 1$, it could be robust under specific conditions). Else, we could have an infinite number of poles in the closed right-half plane, in which case any (potentially dynamic) linear control law would lead to a zero delay margin [20], meaning that it could not be implementable and relevant for industrial applications.

Assumption 2 The pairs (A_0, B_0) and (A_{11}, E_1) are stabilizable (i.e. there exist $F_0 \in \mathbb{R}^{p \times n}$, $F_1 \in \mathbb{R}^{1 \times m}$ such that $\bar{A}_0 \doteq A_0 + B_0 F_0$ and $\bar{A}_{11} \doteq A_{11} + E_1 F_1$ are Hurwitz).

The two conditions in Assumption 2 are quite natural and can be easily numerically tested when implemented. The stabilizability of (A_{11}, E_1) is necessary and implies that the second ODE system can be stabilized in absence of perturbation independently of the PDE or interconnection structure. The condition on (A_0, B_0) allows for a simpler design of the control law, as it will appear in Sect. 3.1.

Assumption 3 The matrices $(\bar{A}_0, B_0, C_0)$ satisfy

$$\operatorname{rank}\left(\begin{bmatrix} sI - \bar{A}_0 & B_0 \\ C_0 & 0_{1 \times c} \end{bmatrix}\right) = n + 1, \quad \forall s \in \mathbb{C}, \operatorname{Re}(s) \geq 0.$$

As before, this assumption allows a simpler controller design, and serves several purposes. First, it implies that C_0 is not identically zero, and in particular, admits a right inverse since it is necessarily full row rank. Indeed, if it was not the case, it would have obstructed the stabilization of the PDE and Y_1 subsystems through X in the absence of perturbation. Then, it is also equivalent (under Assumption 2) with the transfer matrix $P_0(s) \doteq C_0(sI - \bar{A}_0)^{-1}B_0$ having no zeros in the complex right-half plane that are common to all its components. Consequently, $P_0(s)$ admits a stable right-inverse (which, in general, is not proper). We denote $P_0^+(s)$ any such right inverse.

Assumption 4 The matrix A_{22} is marginally stable. Also, there exist matrices $T_a \in \mathbb{R}^{m\times p}$, $F_a \in \mathbb{R}^{1\times p}$ solutions to the *regulator equations*:

$$\begin{cases} -A_{11}T_a + T_aA_{22} + A_{12} = -E_1F_a, \\ -C_{e1}T_a + C_{e2} = 0. \end{cases} \tag{8}$$

This assumption gives a sufficient structural condition for the existence of a solution for the output regulation problem [14]. This is a condition on the invariant zeros of the plant and the spectrum of the exosystem. More precisely, A_{11} and A_{22} have disjoint spectra, and the number of outputs we regulate (one in the case ϵ scalar) is coherent with the number of inputs (c). The matrices T_a, F_a can be computed using a Schur triangulation.

Since observation is a dual problem to stabilization, sufficient assumptions corresponding to Assumptions 2, 3 are formulated for the observer design.

Assumption 5 The pairs (A_0, C_{mes}) and $\left(\begin{bmatrix} A_{11} & A_{12} \\ 0_{p\times m} & A_{22} \end{bmatrix}, \begin{bmatrix} C_{11} & C_{12} \end{bmatrix}\right)$ are detectable (i.e. there exist $L_X \in \mathbb{R}^{p\times n'}$, $L_1 \in \mathbb{R}^{m\times 1}$, $L_2 \in \mathbb{R}^{p\times 1}$ such that $A_0^{\text{obs}} \doteq A_0 + L_XC_{\text{mes}}$ and $A_1^{\text{obs}} = \begin{bmatrix} A_{11} & A_{12} \\ 0_{p\times m} & A_{22} \end{bmatrix} + \begin{bmatrix} L_1 \\ L_2 \end{bmatrix}\begin{bmatrix} C_{11} & C_{12} \end{bmatrix}$ are Hurwitz).

Assumption 6 The matrices $(A_0^{\text{obs}}, E_0, C_{\text{mes}})$ satisfy

$$\text{rank}\left(\begin{bmatrix} sI - A_0^{\text{obs}} & E_0 \\ C_{\text{mes}} & 0_{n'\times 1} \end{bmatrix}\right) = n + 1, \quad \forall s \in \mathbb{C}, \text{Re}(s) \geq 0.$$

This assumption implies that the column vector E_0 is not identically zero, and in particular admits a left inverse. Moreover, it implies that the transfer matrix $P_{\text{mes}}(s) \doteq C_{\text{mes}}(sI - \bar{A}_0)^{-1}E_0$ has no zeros in the complex right-half plane and admits a stable left-inverse (which is not proper).

Note that all assumptions can be easily tested before implementation, in order to simplify the use of this methodology for field engineers.

3 State-Feedback Controller Design

In this section, we extend the results from [8] to a dynamically augmented system with a finite-dimensional exosystem whose state is denoted $Y_2(t)$. First, we follow the backstepping methodology to map our initial system to a simpler target system. We show that the integral transform we use is invertible. Next, a frequency analysis of the target system is made. Filtering techniques are used to guarantee the robustness of the full-state feedback control law, designed to solve the initial practical output tracking/output regulation problem. Our objective is to stabilize the virtual output $\epsilon(t)$, which is a combination of the unactuated ODE state $Y_1(t)$ and the exogenous state $Y_2(t)$.

3.1 Backstepping Transform

In order to map the initial system (1)–(6) onto a simpler target system, we use the following integral transform

$$X(t) = \xi(t) + \int_0^1 M^{12}(y)\alpha(t,y) + M^{13}(y)\beta(t,y)\mathrm{d}y + \begin{bmatrix} M^{14} & M^{15}\end{bmatrix}\eta(t), \tag{9}$$

$$u(t,x) = \alpha(t,x) + \int_x^1 M^{22}(x,y)\alpha(y) + M^{23}(x,y)\beta(y)\mathrm{d}y + \begin{bmatrix} M^{24}(x) & M^{25}(x)\end{bmatrix}\eta(t),$$

$$v(t,x) = \beta(t,x) + \int_x^1 M^{32}(x,y)\alpha(y) + M^{33}(x,y)\beta(y)\mathrm{d}y + \begin{bmatrix} M^{34}(x) & M^{35}(x)\end{bmatrix}\eta(t),$$

$$Y(t) = \eta(t), \tag{10}$$

with kernel gains and kernel functions M^{12}, $M^{13} \in C([0,1]; \mathbb{R}^{n\times 1})$, $M^{14} \in \mathbb{R}^{n\times m}$, M^{24}, $M^{34} \in C([0,1]; \mathbb{R}^{1\times m})$ (resp. $M^{15} \in \mathbb{R}^{n\times p}$, M^{25}, $M^{35} \in C([0,1]; \mathbb{R}^{1\times p})$), and M^{22}, M^{23}, M^{32}, $M^{33} \in C(\mathcal{T}^+; \mathbb{R})$. Note that this change of variables is well-defined, since the transform is composed of identity functions, integral operators with regular kernels and matrices, and is therefore bounded. Due to its inner upper triangular structure, with identities and invertible Volterra integral operators on the diagonal, it is invertible [24, 26]. The inverse transform $\mathcal{L}$ (using a direct formulation) has the same structure, with coefficients K^{ij} expressed in terms of M^{ij} (and reciprocally) [19].

3.2 Target System

Let us consider the target state $(\xi, \alpha, \beta, \eta) \in \mathcal{X}$, where $\eta(t) = \begin{bmatrix} \eta_1(t) \\ \eta_2(t)\end{bmatrix}$ is decomposed into two parts. These new variables satisfy the following set of equations

$$\dot{\xi}(t) = \bar{A}_0\xi(t) - \lambda M^{12}(0)C_0\xi(t) + \bar{E}_1\alpha(t,1) + \bar{E}_0\beta(t,0)$$
$$+ \begin{bmatrix} M_1 & M_p \end{bmatrix}\eta(t) + \int_0^1 M^{\alpha}(y)\alpha(t,y) + M^{\beta}(y)\beta(t,y)\mathrm{d}y + B_0\tilde{U}(t), \quad (11)$$
$$\alpha_t(t,x) + \lambda\alpha_x(t,x) = 0, \quad \beta_t(t,x) - \mu\beta_x(t,x) = 0, \quad (12)$$
$$\dot{\eta}(t) = \begin{bmatrix} \bar{A}_{11} & \bar{A}_{12} \\ 0 & A_{22} \end{bmatrix}\eta(t) + \begin{bmatrix} E_1 \\ 0_{p\times 1} \end{bmatrix}\alpha(t,1), \quad (13)$$

with the boundary conditions

$$\alpha(t,0) = q\beta(t,0) + C_0\xi(t), \quad \beta(t,1) = \rho\alpha(t,1), \quad (14)$$

with $\bar{A}_0$, $\bar{A}_{11}$ are defined by Assumption 2, and the other coefficients by

$$\bar{A}_{12} = A_{12} + E_1(F_a + F_1T_a), \quad \bar{E}_0 = E_0 - q\lambda M^{12}(0) + \mu M^{13}(0), \quad (15)$$
$$\bar{E}_1 = \lambda M^{12}(1) - M^{14}E_1 - \rho\mu M^{13}(1), \quad (16)$$
$$M_1 = -M^{14}\bar{A}_{11} + A_0M^{14} + E_0M^{34}(0), \quad (17)$$
$$M_p = -M^{15}A_{22} - M^{14}\bar{A}_{12} + A_0M^{15} + E_0M^{35}(0), \quad (18)$$
$$M^{\alpha}(y) = -\lambda\frac{\mathrm{d}}{\mathrm{d}y}M^{12}(y) + A_0M^{12}(y) + E_0M^{32}(0,y), \quad (19)$$
$$M^{\beta}(y) = \mu\frac{\mathrm{d}}{\mathrm{d}y}M^{13}(y) + A_0M^{13}(y) + E_0M^{33}(0,y). \quad (20)$$

The initial conditions $(\xi_0, \alpha_0, \beta_0, \eta_0) \in \mathcal{X}$ are defined from the initial conditions of the system (1)–(6) using the integral transform. The different coefficients naturally derive from the backstepping procedure. We also define a new actuation $\tilde{U}(t) \doteq U(t) - F_0\xi(t)$ in (11), with F_0 defined in Assumption 2. Note that in absence of disturbance, namely $\eta_2 = 0$, this target system was shown to be stabilizable in [8].

3.3 Kernel Equations

Following the backstepping procedure, we derive Eqs. (9)–(10) and integrate by parts to find the equations satisfied by the different kernels defined on $\mathcal{T}^+$. We obtain

$$\lambda M_x^{22}(x,y) + \lambda M_y^{22}(x,y) = \sigma^+(x)M^{32}(x,y), \quad (21)$$
$$\lambda M_x^{23}(x,y) - \mu M_y^{23}(x,y) = \sigma^+(x)M^{33}(x,y), \quad (22)$$
$$\mu M_x^{32}(x,y) - \lambda M_y^{32}(x,y) = -\sigma^-(x)M^{22}(x,y), \quad (23)$$
$$\mu M_x^{33}(x,y) + \mu M_y^{33}(x,y) = -\sigma^-(x)M^{23}(x,y), \quad (24)$$

with the boundary conditions

$$M^{23}(x,x) = -\frac{\sigma^+(x)}{\lambda+\mu}, \quad M^{32}(x,x) = \frac{\sigma^-(x)}{\lambda+\mu}, \tag{25}$$

$$M^{22}(x,1) = \frac{1}{\lambda}\left(M^{24}(x)E_1 + \rho\mu M^{23}(x,1)\right), \tag{26}$$

$$M^{33}(x,1) = \frac{1}{\mu\rho}\left(\lambda M^{32}(x,1) - M^{34}(x)E_1\right). \tag{27}$$

The kernels associated to the state $\eta(t)$ are defined on [0, 1] and must satisfy the set of ODEs

$$\lambda\frac{\mathrm{d}}{\mathrm{dx}}M^{24}(x) + M^{24}(x)\bar{A}_{11} = \sigma^+(x)M^{34}(x), \tag{28}$$

$$\lambda\frac{\mathrm{d}}{\mathrm{dx}}M^{25}(x) + M^{25}(x)A_{22} = \sigma^+(x)M^{35}(x) - M^{24}(x)\bar{A}_{12}, \tag{29}$$

$$\mu\frac{\mathrm{d}}{\mathrm{dx}}M^{34}(x) - M^{34}(x)\bar{A}_{11} = -\sigma^-(x)M^{24}(x), \tag{30}$$

$$\mu\frac{\mathrm{d}}{\mathrm{dx}}M^{35}(x) - M^{35}(x)A_{22} = -\sigma^-(x)M^{25}(x) + M^{34}(x)\bar{A}_{12}, \tag{31}$$

with the boundary conditions

$$M^{24}(1) = F_1, \quad M^{25}(1) = F_a + F_1T_a, \tag{32}$$

$$M^{34}(1) = C_{11} + \rho F_1, \quad M^{35}(1) = C_{12} + \rho M^{25}(1), \tag{33}$$

where F_1 is defined by Assumption 2 and (F_a, T_a) is defined by Assumption 4. Finally, the following set of algebraic relations is fulfilled

$$C_0M^{12}(y) = M^{22}(0,y) - qM^{32}(0,y), \tag{34}$$

$$C_0M^{13}(y) = M^{23}(0,y) - qM^{33}(0,y), \tag{35}$$

$$C_0M^{14} = M^{24}(0) - qM^{34}(0), \tag{36}$$

$$C_0M^{15} = M^{25}(0) - qM^{35}(0). \tag{37}$$

The kernels are well-defined on their definition domain.

3.4 Frequency Analysis of the Target System

Let us denote $\tau = \frac{1}{\mu} + \frac{1}{\lambda}$. We use the method of characteristics to rewrite the target system as a *time-delay system*. Then, we analyze the properties of such system in the frequency domain. We express the Laplace transform of $\alpha(\cdot, 0)$, $\beta(\cdot, 1)$ and η_1 in function of the ones of η_2, ξ. After some technical computations, we finally obtain for any $s \in \mathbb{C}$ with $\mathrm{Re}(s) \geq 0$

$$(sI - \bar{A}_0)\mathscr{L}(\xi)(s) = G(s)C_0\mathscr{L}(\xi)(s) + H(s)\mathscr{L}(\eta_2)(s) + B_0\mathscr{L}(\tilde{U})(s), \quad (38)$$

with

$$G(s) = -\lambda M^{12}(0) \quad (39)$$
$$+(1-\rho q e^{-\tau s})^{-1}\left[\left(\bar{E}_1 + M_1(sI - \bar{A}_{11})^{-1}E_1\right)e^{-\frac{s}{\lambda}} + \rho e^{-\tau s}\bar{E}_0 + \int_0^{\tau} M^{\xi}(\theta)e^{-s\theta}\mathrm{d}\theta\right],$$
$$M^{\xi}(\theta) = \lambda M^{\alpha}(\lambda\theta)\mathbb{1}_{[0,\frac{1}{\lambda}]}(\theta) + \rho\mu M^{\beta}\left(1 - \mu\theta + \frac{\mu}{\lambda}\right)\mathbb{1}_{(\frac{1}{\lambda},\tau]}(\theta), \quad (40)$$
$$H(s) = M_1(sI - \bar{A}_{11})^{-1}\bar{A}_{12} + M_p. \quad (41)$$

This will be further used in the design of the control law stabilizing $\epsilon(t)$. In the next section, we use the transfer functions we defined to determine an adequate input $\tilde{U}(t)$, corresponding to $\mathscr{L}(\tilde{U})(s)$ in the Laplace domain.

3.5 *Full-State Feedback Controller Design*

We now design a control law for the practical output regulation of the target system $(\xi, \alpha, \beta, \eta)$, and more precisely its Laplace transform $\mathscr{L}(\tilde{U})(s)$. It is decomposed into two parts: $\mathscr{L}(\tilde{U})(s) = U_{\xi}(s) + U_{\eta}(s)$. First, we design a stabilizing state feedback control law U_{ξ} for the system in absence of the disturbance or reference signal (i.e. $\eta_2 = Y_2 \equiv 0$), using the results from [8]. Then, we design a robust bounded control law U_{η} depending on the dynamics of the perturbation, to cancel its effects on the dynamics of ξ. The practical output tracking/output regulation problems are therefore solved.

3.5.1 Target System Stabilization in Absence of Disturbance

Following the procedure in [8] for the extended system, we design a state feedback control law of form $U_{\xi}(s) = F_{\xi}(s)C_0\mathscr{L}(\xi)(s)$ stabilizing the output $C_0\xi(.)$. It implies the stabilization of the target system.

When $\eta_2 \equiv 0$, (38) rewrites on the complex right half plane

$$C_0\mathscr{L}(\xi)(s) = C_0(sI - \bar{A}_0)^{-1}G(s)C_0\mathscr{L}(\xi)(s) + P_0\mathscr{L}(\tilde{U})(s), \quad (42)$$

with P_0 defined in Assumption 3. As previously mentioned, $P_0(s)$ admits a stable right inverse $P_0^+(s)$, for instance the Moore-Penrose right inverse $P_0^+(s) \doteq P_0^T(s)(P_0(s)P_0^T(s))^{-1}$.

We now define the transfer function $F_{\xi}(s) = -P_0^+(s)C_0(sI - \bar{A}_0)^{-1}G(s)$. Since it is not necessarily proper, we can use a filtering technique as in [8]. Let us decompose $G(s)$ in (39) into $G(s) = \mathrm{w}(s)G(s) + (1 - \mathrm{w}(s))G(s)$, with $\mathrm{w}(s)$ a (SISO) stable low-pass filter of sufficient order. A candidate (proper) controller can therefore be defined as

$$\tilde{F}_\xi(s) = -P_0^+(s)\left[C_0(sI - \bar{A}_0)^{-1}\mathrm{w}(s)G(s)\right]. \tag{43}$$

We have the following theorem [8]:

Lemma 1 *Let* $\mathrm{w}(s)$ *be any low-pass filter, with sufficiently high relative degree, and* $0 < \delta < 1$ *sufficiently small, such that*

$$\forall x \in \mathbb{R},\ |1 - \mathrm{w}(jx)| \leq \frac{1-\delta}{\|G\|_\infty \bar{\sigma}(C_0(jxI - \bar{A}_0)^{-1})}. \tag{44}$$

Then the dynamic output feedback $\tilde{U}_\xi(s) = \tilde{F}_\xi(s)C_0\mathscr{L}(\xi)(s)$ *with* $\tilde{F}_\xi(s)$ *given in* (43) *exponentially stabilizes* $C_0\xi(.)$.

Proof The proof is omitted here due to space restriction. More details are given in [8]. □

Then, let us show that the target system is exponentially stabilized by the state feedback (43).

Theorem 1 *Assume that* $C_0\xi(\cdot)$ *is exponentially stabilized by a dynamic output feedback of the form* $U_\xi(s) = F_\xi(s)C_0\mathscr{L}(\xi)(s)$ *where* $F_\xi(s)$ *is a stable strictly-proper transfer matrix. Then, under Assumptions 1–3, the target system* (11)–(14) *in absence of perturbation is exponentially stable in the* $\mathcal{X}$*-norm.*

Proof The proof is omitted here due to space restriction. □

Using Lemma 1 and Theorem 1, we have designed a stabilizing control input $\tilde{U}$ for the target system. We can then deduce the corresponding control input for the original system (1)–(5) as $U(t) = \tilde{U}(t) + F_0\xi(t)$.

3.5.2 Practical Output Regulation

Following the same procedure, we can now design a control law based on the supposedly known perturbation to cancel the effects of the disturbance η_2 on the dynamics of the output $C_0\xi(.)$.
In presence of perturbation, (38) rewrites for any $s \in \mathbb{C}$ with $\mathrm{Re}(s) \geq 0$

$$\begin{aligned} C_0\mathscr{L}(\xi)(s) = {} & C_0(sI - \bar{A}_0)^{-1}G(s)C_0\mathscr{L}(\xi)(s) + P_0(s)U_\xi(s) \\ & + C_0(sI - \bar{A}_0)^{-1}H(s)\mathscr{L}(\eta_2)(s) + P_0(s)U_\eta(s). \end{aligned} \tag{45}$$

First, let us define the transfer function $F_\eta(s) \doteq -P_0^+(s)C_0(sI - \bar{A}_0)^{-1}H(s)$, s.t. knowing the perturbation, the control law $U_\eta(s) = F_\eta(s)\mathscr{L}(\eta_2)(s)$ cancels the effect of the perturbation on the target system.

As above, since the obtained transfer function is not proper in general, we can regularize it to design a strictly proper transfer function

$$\tilde{F}_\eta(s) = -P_0^+(s)C_0(sI - \bar{A}_0)^{-1}H(s)w(s), \tag{46}$$

using an adequately chosen low-pass filter $w(s)$. We now show the following theorem:

Theorem 2 *Assume that in the absence of the disturbance, the function $C_0\mathscr{L}(\xi)(s)$ is exponentially stabilized by a dynamic full-state feedback of the form $\tilde{U}_\xi(s) = \tilde{F}_\xi(s)C_0\mathscr{L}(\xi)(s)$. Consider the extended control law*

$$\mathscr{L}(U)(s) = (\tilde{F}_\xi(s)C_0 + F_0)\mathscr{L}(\xi)(s) + \tilde{F}_\eta(s)\mathscr{L}(\eta_2)(s), \tag{47}$$

where $\tilde{F}_\xi(s)$, $\tilde{F}_\eta(s)$ are two stable (proper) transfer matrices. Then, under Assumptions 1, 2, 3, 4, the virtual output $\epsilon(t)$ is practically regulated to zero. This is, given the norm of the initial condition of the exosystem $\eta_2(0)$, and given a desired asymptotic bound for the virtual output $\epsilon(t)$ ($\equiv$ tracking error), we can choose an adequate low-pass filter $w(s)$ such that the practical tracking objective is attained. Furthermore, under these assumptions, the control action and the trajectories of ξ and η remain bounded.

Proof In the absence of disturbances or tracking signals, the state of the system converges exponentially to zero, due to the exponential stability of $C_0\mathscr{L}(\xi)(s)$. Since the disturbance signal remains bounded (Assumption 4), so does the state in presence of the disturbance (or reference trajectory) term. More precisely, with the control law (47), we have

$$\begin{aligned} &C_0\mathscr{L}(\xi)(s) - C_0(sI - \bar{A}_0)^{-1}(1 - w(s))(G(s)C_0\mathscr{L}(\xi)(s) + H(s)\mathscr{L}(\eta_2)(s)) \\ &\implies (1 - \Phi(s))C_0\mathscr{L}(\xi)(s) = C_0(sI - \bar{A}_0)^{-1}(1 - w(s))H(s)\mathscr{L}(\eta_2)(s)). \end{aligned}$$

The gain between the disturbance or trajectory term and the residual term $C_0\mathscr{L}(\xi)(s)$ can be chosen to be arbitrarily small at the frequencies contained in the disturbance or trajectories, in order to guarantee that the virtual output $\epsilon(t)$ converges practically to zero.

The stability of $\alpha(\cdot, 0)$, $\beta(\cdot, 1)$ is deduced from the Laplace transform of (14), which implies a bounded gain between the exo-system state (reference/reference trajectory) and α and β in the L^2-norm.

Let us now study the behaviour of $C_{e1}\eta_1 + C_{e2}\eta_2$ in the absence of disturbances or reference trajectories. The dynamics of η_1 rewrites

$$\begin{aligned}\dot{\eta}_1(t) &= (A_{11} + E_1F_1)\eta_1(t) + A_{12}\eta_2(t) + E_1(F_a + F_1T_a)\eta_2(t) + E_1\alpha(t,1),\\ &= (A_{11} + E_1F_1)\eta_1(t) + (-E_1F_a + A_{11}T_a - T_aA_{22})\eta_2(t)\\ &\quad + E_1(F_a + F_1T_a)\eta_2(t) + E_1\alpha(t,1) \quad \text{by Assumption 10}\\ &\Longrightarrow \overbrace{(\eta_1 + T_a\eta_2)}^{\cdot}(t) = \bar{A}_{11}(\eta_1 + T_a\eta_2) + \underbrace{E_1\alpha(t,1)}_{\longrightarrow 0}. \end{aligned} \tag{48}$$

Therefore, the dynamics of $\eta_1 + T_a\eta_2$ is stable. It implies that $C_{e1}(\eta_1 + T_a\eta_2)(t) = C_{e1}(Y_1 + T_aY_2)(t) = C_{e1}Y_1(t) + C_{e2}Y_2(t) = \epsilon(t)$ has a bounded gain w.r.t. $\alpha(t,1)$. If, $\alpha(t,1)$ converges to an arbitrarily small value (selected by adequately choosing the low-pass filter $w(s)$), then, $C_{e1}\eta_1 + C_{e2}\eta_2$ can be made to converge to a desired neighborhood of zero. □

Theorem 2 proves that there exists a dynamic feedback in the form (47), the parameters of which can be chosen depending only on the norm of the exo-system state and the desired precision for the tracking, such that the virtual output $\epsilon(t)$ converges to any desired neighborhood of zero. Given the control input $\tilde{U}$ for the target system, the corresponding control input for the original system can be obtained $U(t) = \tilde{U}(t) + F_0\xi(t)$.

Remark that the controller has been chosen as strictly proper which means that it is robust to small delays in the input, which is not the case in some designs that include derivative terms. The proof follows the same ideas as those in [4].

In the case where the exogenous state $Y_2(t) = \eta_2(t)$ is a reference trajectory supposedly known, we can make the output $\epsilon(t)$ converge arbitrarily close to zero knowing the evolution of $\xi(t)$ using the robust full-state feedback control law $U(t) = U_\xi(t) + U_\eta(t) + F_0\xi(t)$. However, this full-state feedback controller requires the knowledge of all the states (X, u, v, Y) at any time. For further applications on real systems, it is then necessary to *estimate* the state (X, u, v, Y) using the measure $y(t)$.

In the case where the exogenous state Y_2 is an unknown disturbance, the control law we propose is not fully satisfactory. We need to reconstruct the perturbation since the knowledge of η_2 is required to design the control law $U(t)$.

4 Output Dynamic Feedback Controller Design

In this section, we first use the backstepping approach to map the original system (1)–(6) to a simpler target system, for which we design an observer. Following [13], we use an invertible transform to map the error system to a stable target system and design the corresponding observer gains. Next, we integrate the observers in the previously designed control law to obtain an output-feedback controller solving the practical output tracking/output regulation problem.

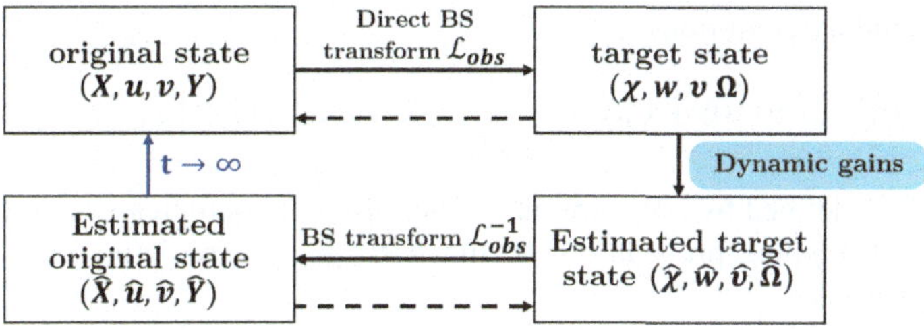

Fig. 3 Observer design strategy for system (1)–(6)

4.1 *Observer Design*

We now design an adequate invertible transform to map system (1)–(6) to a simpler target system, as schematized on Fig. 3.

4.1.1 Invertible Transform and Target System

In order to map the initial system (1)–(6) to an simpler target system, we use the following backstepping transform

$$X(t) = \chi(t), \tag{49}$$

$$u(t,x) = w(t,x) + \int_0^x L^{22}(x,y)w(y) + L^{23}(x,y)\upsilon(y)dy, \tag{50}$$

$$v(t,x) = \upsilon(t,x) + \int_0^x L^{32}(x,y)w(y) + L^{33}(x,y)\upsilon(y)dy, \tag{51}$$

$$Y(t) = \Omega(t) + \int_0^1 \begin{bmatrix} L^{42} \\ L^{52} \end{bmatrix}(y)w(y) + \begin{bmatrix} L^{43} \\ L^{53} \end{bmatrix}(y)\upsilon(y)dy, \tag{52}$$

where the kernel functions $L^{i2}, L^{i3} \in C([0,1];\mathbb{R}^{*\times 1})$ ($* = m$ if $i = 4$, $* = p$ if $i = 5$) and $L^{22}, L^{23}, L^{32}, L^{33} \in C(\mathcal{T}^-;\mathbb{R})$ have yet to be defined. For the same reasons given in Sect. 3.1, this transform is invertible. The inverse transform $\mathcal{L}_{\text{obs}}$ (with direct formulation) has the same structure that the transform (49)–(52) with kernels N^{ij} being expressed as functions of L^{ij}.

Let us define a target state denoted $(\chi, w, \upsilon, \Omega) \in \mathcal{X}$, where $\Omega(t) = \begin{bmatrix}\Omega_1(t) & \Omega_p(t)\end{bmatrix}^T$ is decomposed into two parts. This new set of variables satisfies

$$\dot{\chi}(t) = A_0\chi(t) + E_0\upsilon(t,0) + B_0U(t), \tag{53}$$

$$w_t(t,x) + \lambda w_x(t,x) = g_w(x)\upsilon(t,0) + h_w(x)\chi(t), \tag{54}$$

$$\upsilon_t(t,x) - \mu\upsilon_x(t,x) = g_\upsilon(x)\upsilon(t,0) + h_\upsilon(x)\chi(t), \tag{55}$$

$$\dot{\Omega}(t) = A_1^{\text{obs}}\Omega(t) + K_\Omega^{\upsilon,0}\upsilon(t,0) + K_\Omega^{\chi}\chi(t), \tag{56}$$

with the boundary conditions

$$w(t,0) = qv(t,0) + C_0\chi(t), \quad v(t,1) = \rho w(t,1) + \begin{bmatrix} C_{11} & C_{12} \end{bmatrix} \Omega(t), \quad (57)$$

where A_1^{obs} is defined by Assumption 5. The different coefficients naturally derive from the backstepping procedure. The functions g_w, g_v are defined by the set of integral equations[1]

$$g_w(x) + \int_0^x L^{22}(x,y)g_w(y) + L^{23}(x,y)g_v(y)\mathrm{d}y = \mu L^{23}(x,0) - \lambda q L^{22}(x,0),$$
$$g_v(x) + \int_0^x L^{32}(x,y)g_w(y) + L^{33}(x,y)g_v(y)\mathrm{d}y = \mu L^{33}(x,0) - \lambda q L^{32}(x,0),$$

and the functions h_w, h_v by

$$h_w(x) + \int_0^x L^{22}(x,y)h_w(y) + L^{23}(x,y)h_v(y)\mathrm{d}y = -\lambda L^{22}(x,0)C_0,$$
$$h_v(x) + \int_0^x L^{32}(x,y)h_w(y) + L^{33}(x,y)h_v(y)\mathrm{d}y = -\lambda L^{32}(x,0)C_0.$$

These coupled integral equations admit a unique solution [26]. The terms $K_\Omega^{v,0}$, K_Ω^χ are defined by

$$K_\Omega^{v,0} = \mu \begin{bmatrix} L^{43}(0) \\ L^{53}(0) \end{bmatrix} - \lambda q \begin{bmatrix} L^{42}(0) \\ L^{52}(0) \end{bmatrix} - \int_0^1 \begin{bmatrix} L^{42}(y) \\ L^{52}(y) \end{bmatrix} g_w(y) + \begin{bmatrix} L^{43}(y) \\ L^{53}(y) \end{bmatrix} g_v(y)\mathrm{d}y,$$
$$K_\Omega^\chi = -\lambda \begin{bmatrix} L^{42}(0) \\ L^{52}(0) \end{bmatrix} C_0 - \int_0^1 \begin{bmatrix} L^{42}(y) \\ L^{52}(y) \end{bmatrix} h_w(y) + \begin{bmatrix} L^{43}(y) \\ L^{53}(y) \end{bmatrix} h_v(y)\mathrm{d}y.$$

4.1.2 Kernel Equations

We derive Eqs. (53)–(56) and integrate by parts to find the equations satisfied by the kernels L^{ij} on their definition domain. The equations in w, v imply that kernels $L^{22}, L^{23}, L^{32}, L^{33}$ are defined by the following set of PDEs

$$\lambda L_x^{22}(x,y) + \lambda L_y^{22}(x,y) = \sigma^+(x)L^{32}(x,y), \quad (58)$$
$$\lambda L_x^{23}(x,y) - \mu L_y^{23}(x,y) = \sigma^+(x)L^{33}(x,y), \quad (59)$$
$$\mu L_x^{32}(x,y) - \lambda L_y^{32}(x,y) = -\sigma^-(x)L^{22}(x,y), \quad (60)$$
$$\mu L_x^{33}(x,y) + \mu L_y^{33}(x,y) = -\sigma^-(x)L^{23}(x,y), \quad (61)$$

[1] Note that an explicit expression of g_w, g_v could have been obtained using the transform $\mathcal{L}_{\text{obs}}$ with direct formulation. However, it would have led to more complex terms in the boundary conditions (57).

with the boundary conditions

$$L^{23}(x,x) = \frac{\sigma^+(x)}{\lambda+\mu}, \quad L^{32}(x,x) = -\frac{\sigma^-(x)}{\lambda+\mu}, \tag{62}$$

$$L^{22}(1,y) = \frac{1}{\rho}(L^{32}(1,y) - \begin{bmatrix} C_{11} & C_{12} \end{bmatrix} \begin{bmatrix} L^{42}(y) \\ L^{52}(y) \end{bmatrix}), \tag{63}$$

$$L^{33}(1,y) = \rho L^{23}(1,y) + \begin{bmatrix} C_{11} & C_{12} \end{bmatrix} \begin{bmatrix} L^{43}(y) \\ L^{53}(y) \end{bmatrix}. \tag{64}$$

The equations in Ω imply that L^{42}, L^{43}, L^{52}, L^{53} satisfy the following set of ODEs

$$\lambda \frac{\mathrm{d}}{\mathrm{d}y} \begin{bmatrix} L^{42} \\ L^{52} \end{bmatrix}(y) = \begin{bmatrix} A_{11} & A_{12} \\ 0_{p\times m} & A_{22} \end{bmatrix} \begin{bmatrix} L^{42} \\ L^{52} \end{bmatrix}(y) + \begin{bmatrix} E_1 \\ 0_{p\times 1} \end{bmatrix} L^{22}(1,y), \tag{65}$$

$$-\mu \frac{\mathrm{d}}{\mathrm{d}y} \begin{bmatrix} L^{43} \\ L^{53} \end{bmatrix}(y) = \begin{bmatrix} A_{11} & A_{12} \\ 0_{p\times m} & A_{22} \end{bmatrix} \begin{bmatrix} L^{43} \\ L^{53} \end{bmatrix}(y) + \begin{bmatrix} E_1 \\ 0_{p\times 1} \end{bmatrix} L^{23}(1,y), \tag{66}$$

with the boundary conditions

$$\begin{bmatrix} L^{42} \\ L^{52} \end{bmatrix}(1) = -\frac{1}{\lambda}(\rho \begin{bmatrix} L_1 \\ L_2 \end{bmatrix} + \begin{bmatrix} E_1 \\ 0_{p\times 1} \end{bmatrix}), \quad \begin{bmatrix} L^{43} \\ L^{53} \end{bmatrix}(1) = -\frac{1}{\mu} \begin{bmatrix} L_1 \\ L_2 \end{bmatrix}, \tag{67}$$

with $\begin{bmatrix} L_1 \\ L_2 \end{bmatrix}$ defined by Assumption 5.

The kernels are well-defined on their domains of definition. More precisely, we simultaneously solve the set of coupled PDEs and ODEs $L^{22}, L^{23}, L^{32}, L^{33}$ and $L^{42}, L^{43}, L^{52}, L^{53}$. Indeed, the kernels $L^{42}, L^{43}, L^{52}, L^{53}$ can be *embedded* onto the triangular domain $\mathcal{T}^+$, by defining $\tilde{L}^{ij}(x,y) \doteq \mathbb{1}_{x=1}(x,y) L^{ij}(x)$, $i \in \{2,3\}$, $j \in \{4,5\}$. Following the proof of the main result from [12], stating the existence of a solution of a general class of kernel equations, we conclude that (58)–(67) admits a unique L^∞ solution. Higher regularity can be obtained by following the same procedure and requiring higher regularity in the coefficients.

4.1.3 Observer and Error State

We recall that we have access to the measurement $y(t) = C_{\text{mes}} X(t) = C_{\text{mes}} \chi(t) \in \mathbb{R}^{n'}$. We introduce the observer state system as a copy of the target dynamics, with dynamic output feedback gains $\mathcal{P}_\cdot$, the structure of which will be precised later this section.

$$\dot{\hat{\chi}}(t) = A_0\hat{\chi}(t) + E_0\hat{\upsilon}(t,0) + B_0U(t) - P_\chi\left(y(t) - C_{\text{mes}}\hat{\chi}(t)\right), \quad (68)$$
$$\hat{w}_t(t,x) + \lambda\hat{w}_x(t,x) = g_w(x)\hat{\upsilon}(t,0) + h_w(x)\hat{\chi}(t) - \mathcal{P}_w(t,x), \quad (69)$$
$$\hat{\upsilon}_t(t,x) - \mu\hat{\upsilon}_x(t,x) = g_\upsilon(x)\hat{\upsilon}(t,0) + h_\upsilon(x)\hat{\chi}(t) - \mathcal{P}_\upsilon(t,x), \quad (70)$$
$$\dot{\hat{\Omega}}(t) = A_1^{\text{obs}}\hat{\Omega}(t) + K_\Omega^{\upsilon,0}\hat{\upsilon}(t,0) + K_\Omega^\chi\hat{\chi}(t) - \mathcal{P}_\Omega(t), \quad (71)$$

with boundary conditions

$$\hat{w}(t,0) = q\hat{\upsilon}(t,0) + C_0\hat{\chi}(t) - \mathcal{P}_w^0(t), \;\; \hat{\upsilon}(t,1) = \rho\hat{w}(t,1) + \begin{bmatrix}C_{11} & C_{12}\end{bmatrix}\hat{\Omega}(t), \quad (72)$$

with any initial conditions in $\mathcal{X}$. The observer gain P_χ and injected signals $\mathcal{P}_w, \mathcal{P}_\upsilon, \mathcal{P}_\Omega, \mathcal{P}_w^0$ will be chosen to stabilize the error system, whose state is defined by $(\tilde{\chi}, \tilde{w}, \tilde{\upsilon}, \tilde{\Omega}) \doteq (\chi, w, \upsilon, \Omega) - (\hat{\chi}, \hat{w}, \hat{\upsilon}, \hat{\Omega})$.
The error state satisfies the set of equations

$$\dot{\tilde{\chi}}(t) = A_0\tilde{\chi}(t) + E_0\tilde{\upsilon}(t,0) + P_\chi C_{\text{mes}}\tilde{\chi}(t), \quad (73)$$
$$\tilde{w}_t(t,x) + \lambda\tilde{w}_x(t,x) = g_w(x)\tilde{\upsilon}(t,0) + h_w(x)\tilde{\chi}(t) + \mathcal{P}_w(t,x), \quad (74)$$
$$\tilde{\upsilon}_t(t,x) - \mu\tilde{\upsilon}_x(t,x) = g_\upsilon(x)\tilde{\upsilon}(t,0) + h_\upsilon(x)\tilde{\chi}(t) + \mathcal{P}_\upsilon(t,x), \quad (75)$$
$$\dot{\tilde{\Omega}}(t) = A_1^{\text{obs}}\tilde{\Omega}(t) + K_\Omega^{\upsilon,0}\tilde{\upsilon}(t,0) + K_\Omega^\chi\tilde{\chi}(t) + \mathcal{P}_\Omega(t), \quad (76)$$

with the boundary conditions

$$\tilde{w}(t,0) = q\tilde{\upsilon}(t,0) + C_0\tilde{\chi}(t) + \mathcal{P}_w^0(t), \quad (77)$$
$$\tilde{\upsilon}(t,1) = \rho\tilde{w}(t,1) + \begin{bmatrix}C_{11} & C_{12}\end{bmatrix}\tilde{\Omega}(t). \quad (78)$$

First, we can define $P_\chi \doteq L_X$, with L_X given in Assumption 5, such that (73) rewrites $\dot{\tilde{\chi}}(t) = A_0^{\text{obs}}\tilde{\chi}(t) + E_0\tilde{\upsilon}(t,0)$.

4.1.4 Frequency Analysis of the Error System

In this section, we determine the functions $\mathcal{P}_w, \mathcal{P}_\upsilon, \mathcal{P}_\Omega, \mathcal{P}_w^0$ to ensure the stability of the error system (73)–(78) in the sense of the $\mathcal{X}-$norm. The Laplace transform of (73) gives

$$(sI - A_0^{\text{obs}})\mathscr{L}(\tilde{\chi})(s) = E_0\mathscr{L}(\tilde{\upsilon}(.,0))(s) \quad (79)$$
$$\implies \mathscr{L}(\tilde{\chi})(s) = (sI - A_0^{\text{obs}})^{-1}E_0\mathscr{L}(\tilde{\upsilon}(.,0))(s) \quad \forall s \in \mathbb{C}^+,$$

since $sI - A_0^{\text{obs}}$ is non-singular on the complex right-half plane by Assumption 5. Then, let us determine the signals $\mathcal{P}_w(t,x), \mathcal{P}_\upsilon(t,x)$. Applying the variation of constants formula in (74)–(75), we have

$$\tilde{w}(t,1) = \tilde{w}(t-\frac{1}{\lambda},0) + \int_0^{\frac{1}{\lambda}} g_w(1-\lambda\theta)\tilde{v}(t-\theta,0) + h_w(1-\lambda\theta)\tilde{\chi}(t-\theta) + \mathcal{P}_w(t-\theta, 1-\lambda\theta)\mathrm{d}\theta, \quad (80)$$

$$\tilde{v}(t,0) = \tilde{v}(t-\frac{1}{\mu},0) + \int_0^{\frac{1}{\mu}} g_v(\mu\theta)\tilde{v}(t-\theta,0) + h_v(\mu\theta)\tilde{\chi}(t-\theta) + \mathcal{P}_v(t-\theta, \mu\theta)\mathrm{d}\theta. \quad (81)$$

Taking the Laplace transform of (80), and incorporating therein the Laplace transform of $\tilde{\chi}$ deduced from (79), we have

$$\mathscr{L}(\tilde{w}(.,1))(s) = e^{-\frac{s}{\lambda}}\mathscr{L}(\tilde{w}(.,0))(s) + \int_0^{\frac{1}{\lambda}} (g_w(1-\lambda\theta) + h_w(1-\lambda\theta)(sI - A_0^{\text{obs}})^{-1}E_0)\mathscr{L}(\tilde{v}(.,0))(s) + \mathcal{P}_w(s, 1-\lambda\theta))e^{-s\theta}\mathrm{d}\theta. \quad (82)$$

Assume we have a gain $\mathcal{P}_w$ of the form $\mathcal{P}_w(s,x) = P_w(s,x)C_{\text{mes}}\mathscr{L}(\tilde{\chi})(s)$, with $x = 1-\lambda\theta \in [0,1]$. By (79), this implies $\mathcal{P}_w(s,x) = P_w(s,x)P_{\text{mes}}(s)\mathscr{L}(\tilde{v}(.,0))(s)$ for all $s \in \mathbb{C}^+$, with $P_{\text{mes}}(s)$ defined after Assumption 6. This transfer function admits a stable left-inverse denoted $P_{\text{mes}}^-(s)$ (which is not necessarily proper). In our examples, we will use the Moore-Penrose left inverse

$$P_{\text{mes}}^-(s) \doteq (P_{\text{mes}}^T(s)P_{\text{mes}}(s))^{-1}P_{\text{mes}}^T(s)$$

which is verified to be stable. In order to rewrite (74) as a transport equation, that is to say, in the Laplace domain, $\mathscr{L}(\tilde{w}(.,1))(s) = e^{-\frac{s}{\lambda}}\mathscr{L}(\tilde{w}(.,0))(s)$, we thus define the transfer function

$$P_w(s,x) \doteq -(g_w(x) + h_w(x)(sI - A_0^{\text{obs}})^{-1}E_0)P_{\text{mes}}^-(s). \quad (83)$$

Similarly, taking the Laplace transform of (81), and incorporating therein the Laplace expression (79), we have

$$\mathscr{L}(\tilde{v}(.,0))(s) = e^{-\frac{s}{\mu}}\mathscr{L}(\tilde{v}(.,1))(s) + \int_0^{\frac{1}{\mu}} (g_v(\mu\theta) + h_v(\mu\theta)(sI - A_0^{\text{obs}})^{-1}E_0)\mathscr{L}(\tilde{v}(.,0))(s) + \mathcal{P}_v(s,\mu\theta))e^{-s\theta}\mathrm{d}\theta.$$

In order to rewrite (75) as a transport equation, that is to say, in the Laplace domain, $\mathscr{L}(\tilde{v}(.,0))(s) = e^{-\frac{s}{\mu}}\mathscr{L}(\tilde{v}(.,1))(s)$, we can choose the observer gain of form $\mathcal{P}_v(s,x) = P_v(s,x)C_{\text{mes}}\mathscr{L}(\tilde{\chi})(s)$. To cancel the terms in the integral, we thus define

$$P_v(s,x) \doteq -(g_v(x) + h_v(x)(sI - A_0^{\text{obs}})^{-1}E_0)P_{\text{mes}}^-(s). \quad (84)$$

In order to guarantee that for all $x \in [0, 1]$, $P_w(s, x)$, $P_\upsilon(s, x)$ are strictly proper transfer functions, it is still possible to use an adequate low-pass filter $\omega(s)$ of sufficient order. Taking the Laplace transform of (77) and incorporating therein (79), we have $\mathscr{L}(\tilde{w}(.,0))(s)(q + C_0(sI - A_0^{\text{obs}})^{-1}E_0)\ \mathscr{L}(\tilde{\upsilon}(.,0))(s) + \mathcal{P}_w^0(s)$. Therefore, choosing $\mathcal{P}_w^0(s)$ of the form $\mathcal{P}_w^0(s) \doteq P_w^0(s)C_{\text{mes}}\mathscr{L}(\tilde{\chi})(s)$, we define the transfer function

$$P_w^0(s) \doteq -(q + C_0(sI - A_0^{\text{obs}})^{-1}E_0)P_{\text{mes}}^-(s), \tag{85}$$

such that the reflection terms at the boundary are cancelled.[2] Finally, taking the Laplace transform of (76) and incorporating therein (79), we have

$$(sI - A_1^{\text{obs}})\mathscr{L}(\tilde{\Omega})(s) = (K_\Omega^{\upsilon,0} + K_\Omega^\chi(sI - A_0^{\text{obs}})^{-1}E_0)\mathscr{L}(\tilde{\upsilon}(.,0))(s) + \mathcal{P}_\Omega(s),$$

with $sI - A_1^{\text{obs}}$ non-singular on the right half plane due to Assumption 5. Once again, we assume we have a signal of the form $\mathcal{P}_\Omega(s) = P_\Omega(s)\ C_{\text{mes}}\mathscr{L}(\tilde{\chi})(s)$. In order to guarantee the convergence of $\tilde{\Omega}$ to zero, we consequently define

$$P_\Omega(s) = -(K_\Omega^{\upsilon,0} + K_\Omega^\chi(sI - A_0^{\text{obs}})^{-1}E_0)P_{\text{mes}}^-(s). \tag{86}$$

Again, since $P_{\text{mes}}^-(s)$ is not necessarily proper, we can use the low-pass filter $\omega(s)$ to filter all the transfer functions. In the following, we describe the construction of a stabilizing output feedback of form

$$\mathcal{P}_w(s, \theta) = \omega(s)P_w(s, \theta)C_{\text{mes}}\mathscr{L}(\tilde{\chi})(s), \tag{87}$$

$$\mathcal{P}_\upsilon(s, \theta) = \omega(s)P_\upsilon(s, \theta)C_{\text{mes}}\mathscr{L}(\tilde{\chi})(s), \tag{88}$$

$$\mathcal{P}_\Omega(s) = \omega(s)P_\Omega(s)C_{\text{mes}}\mathscr{L}(\tilde{\chi})(s), \tag{89}$$

$$\mathcal{P}_w^0(s) = \omega(s)P_w^0(s)C_{\text{mes}}\mathscr{L}(\tilde{\chi})(s). \tag{90}$$

We have the following theorem:

Theorem 3 *Let $\omega(s)$ be any low pass filter with sufficiently high relative degree, and $0 < \tilde{\delta} < 1$ sufficiently small, such that*

$$\forall x \in \mathbb{R},\ |1 - \omega(jx)| < \frac{1 - \tilde{\delta}}{||G_{obs}||_\infty \bar{\sigma}(P_{mes}(jx))}, \tag{91}$$

[2] Due to the use of an adequate low-pass filter, this will however not prevent the output feedback control law to be robust with regard to small delays.

$$\text{with } G_{obs}(s) = \int_0^{\frac{1}{\mu}} P_v(s, \mu\theta)e^{-s\theta}\mathrm{d}\theta + e^{-\frac{s}{\mu}}[[C_{11}\ C_{12}](sI - A_1^{obs})^{-1}P_\Omega(s)$$
$$+\rho(e^{-\frac{s}{\lambda}}P_w^0(s) + \int_0^{\frac{1}{\lambda}} P_w(s, 1-\lambda\theta)e^{-s\theta}\mathrm{d}\theta)]. \quad (92)$$

Consider the dynamic output feedback gains (87)–(90) *with* $P_w(s, \theta)$, $P_v(s, \theta)$, $P_\Omega(s)$, $P_w^0(s)$ *defined by* (83)–(86). *Then, under Assumptions 5 and 6, the error system* (73)–(78) *with any initial conditions in* $\mathcal{X}$ *is exponentially stable in the sense of the* $\mathcal{X}$*−norm.*

Proof The proof is omitted here due to space restrictions. □

Thanks to Assumption 6, we are thus able to design dynamically observer gains stabilizing the target error system. Let us now define the original observer state

$$(\hat{X}, \hat{u}, \hat{v}, \hat{Y}) = \mathcal{L}_{\text{obs}}^{-1}(\hat{\chi}, \hat{w}, \hat{v}, \hat{\Omega}), \quad (93)$$

with $\mathcal{L}_{\text{obs}}^{-1}$ defined by (49)–(52). We have the following corollary

Corollary 1 *Let* $\omega(s)$ *be any low pass filter with sufficiently high relative degree, satisfying* (91) *and the dynamic output feedback of form* (87)–(90) *with* $P_w(s, \theta)$, $P_v(s, \theta)$, $P_\Omega(s)$, $P_w^0(s)$ *defined by* (83)–(86). *Then, under Assumptions 5, 6, the observer state* (93) *converges towards the original state* (X, u, v, Y) *at an exponential rate.*

Proof Under the corollary assumptions, the target error state converges to zero at an exponential rate by Theorem 4. Consequently, the target observer state converges towards the target state. We therefore have access to an estimation of the state (χ, w, v, Ω) with the observer state. Using the invertible backstepping transform $\mathcal{L}_{\text{obs}}^{-1}$ defined by (49)–(52), we then obtain an observer of the original state (X, u, v, Y). Indeed, we can define the original error state as $(\tilde{X}, \tilde{u}, \tilde{v}, \tilde{Y}) = (X, u, v, Y) - (\hat{X}, \hat{u}, \hat{v}, \hat{Y})$. Since the backstepping transform is invertible, the original error system shares same stability properties with the target error system, and is thus exponentially stable. Since the original error state converges to zero, the original observer is a correct estimation of the original state. □

4.2 Output-Feedback Controller

We can now state the main result of this chapter

Theorem 4 *Consider the system* (1)–(6) *with the observer* (93), (68), (72) *and the control law*

$$U(s) = (\tilde{F}_\xi(s)C_0 + F_0)[\hat{X}(s) - \int_0^1 K^{12}(y)\hat{u}(s,y) + K^{13}(y)\hat{v}(s,y)\mathrm{d}y$$
$$+[-K^{14}\hat{Y}_1(s)]\tilde{F}_\eta(s) - (\tilde{F}_\xi(t)C_0 + F_0)K^{15}]\hat{Y}_2(s), \tag{94}$$

with $\tilde{F}_\xi(t)$ *defined by* (43) *and* $\tilde{F}_\eta(s)$ *defined by* (46). *Then, with any arbitrary initial conditions* $(X_0, u_0, v_0, Y_0) \in \mathcal{X}$, *the virtual output* $\epsilon(t)$ *is practically stabilized, i.e.* $|C_{e1}Y_1 + C_{e2}Y_2| \longrightarrow 0$, *and the norm of the state is bounded.*

Proof Using Theorem 2, we just need to show that the dynamics of $C_0\xi$ are stabilized by the output feedback law (94) in the absence of disturbance. By Corollary 1, the error state $(\tilde{X}, \tilde{u}, \tilde{v}, \tilde{Y})$ exponentially converges to zero. Due to the invertibility of the backstepping transform $\mathcal{L}$, the target error state $(\tilde{\xi}, \tilde{\alpha}, \tilde{\beta}, \tilde{\eta}) = \mathcal{L}(\tilde{X}, \tilde{u}, \tilde{v}, \tilde{Y})$ exponentially converges to zero. In the Laplace domain, (45) with the proposed control law rewrites

$$C_0\mathscr{L}(\xi)(s) = C_0(sI - \bar{A}_0)^{-1}(1 - w(s))G(s)C_0\mathscr{L}(\xi)(s)$$
$$-B_0\tilde{F}_\xi(s)\mathscr{L}(\tilde{\xi})(s) - B_0\tilde{F}_\eta(s)\mathscr{L}(\tilde{\eta}_p)(s). \tag{95}$$

Since the error terms converge to zero due to Theorem 3, we use the variations of constants [16] to conclude that $C_0\xi$ converges to zero in the absence of disturbance. □

5 Numerical Simulation

In this section, we illustrate the performances of the state feedback controller on a test case. Due to lack of time, the observer has not been tested in simulation. The proposed approach was implemented using Matlab and Simulink. The transport PDE systems' evolution was simulated using an explicit in time, first-order, upwind finite difference method with 101 spatial discretization points. The ODE states were simulated using a stiff solver based on the second order Rosenbrock method (`ode23s`). The transfer functions in the control law were transformed to a state-space representation for implementation (`tf2ss`). The evolution of the systems were computed on a 50s time scale, with a CFL number equal to 0.9.

For the same initial system, we consider the tracking of a sinusoide reference trajectory in presence of a constant perturbation.

5.1 Presentation of the System

Let us first introduce the system described in Sect. 2.1, with the following numerical values: $\lambda = 2$, $\mu = 0.7$, $\sigma^+ = 1$, $\sigma^- = 0.5$, $\rho = 0.5$, $q = 1.2$. The ODE dynamics are in dimension $n = 4$, $m = 3$, $c = 2$,[3] and defined by the matrices

$$A_0 = \begin{bmatrix} 0 & 0.14 & 0 & 0.1 \\ 0 & 0 & 0.14 & 0 \\ 0.29 & -0.43 & 0.57 & 0.2 \\ 0 & 0 & 0 & -1.1 \end{bmatrix}, B_0 = \begin{bmatrix} 0 & 0 \\ 0 & -1 \\ 1 & -1 \\ 0 & 0 \end{bmatrix}, C_0 = \begin{bmatrix} 1 & 0 & 0 & -0.5 \end{bmatrix}, E_0 = \begin{bmatrix} 2 \\ -1 \\ 0.1 \\ 0 \end{bmatrix},$$

$$A_{11} = \begin{bmatrix} 0.29 & 0.14 & 0 \\ 0.14 & 0 & 0.1 \\ 0 & 0 & -0.9 \end{bmatrix}, E_1 = \begin{bmatrix} -1 \\ 1 \\ 0 \end{bmatrix}, C_{11} = \begin{bmatrix} 0 & 1 & 0.5 \end{bmatrix}.$$

With the numerical values chosen for the system, each ODE system and the PDE subsystem are independently unstable (and remain so when interconnected). Note that the system satisfies Assumption 1 since $\rho q = 0.6$. We can find matrices from Assumption 2 using pole placement. For the simulations, we used

$$F_0 = \begin{bmatrix} 41.71 & 5.43 & -1.93 & 0 \\ 42 & 5 & 0.14 & 0 \end{bmatrix}, F_1 = \begin{bmatrix} 12 & 8.71 & 0 \end{bmatrix}.$$

The kernels M^{ij}, K^{ij} are previously computed using a fixed-point algorithm (successive approximation technique). Their domains of definition are discretized using the same space mesh than (u, v)-state. The integral terms are numerically approximated by the trapezoidal method. Moreover, an input delay of 2.5ms was introduced in the control action to show the robustness of the design to small delays in the loop. The chosen w(s) is a simple 4th order low-pass filter with a bandwidth of approximately 100 rad/s.

5.2 Test Case: Trajectory Tracking in Presence of a Disturbance

Let us consider a tracking problem, in which the second ODE state is subject to a constant disturbance $Y_{\text{pert}} = 0.1$. We want to follow a sinusoidal trajectory $Y_{\text{ref}}(t) = 0.1\sin(t)$. We have an exogenous state with dynamics

$$\dot{Y}_2(t) = A_{22}Y_2(t) = \begin{bmatrix} 0 & 0 & 0 \\ 0 & 0 & 1 \\ 0 & -1 & 0 \end{bmatrix} \begin{bmatrix} Y_{\text{pert}} \\ Y_{\text{ref}} \\ \dot{Y}_{\text{ref}} \end{bmatrix}.$$

As previously, the constant perturbation acts on the first component of $Y_1(t)$ and on the boundaries of the PDE system

[3] In this section, note that since the first component of $X(t)$ is a flat output, a scalar control should be sufficient.

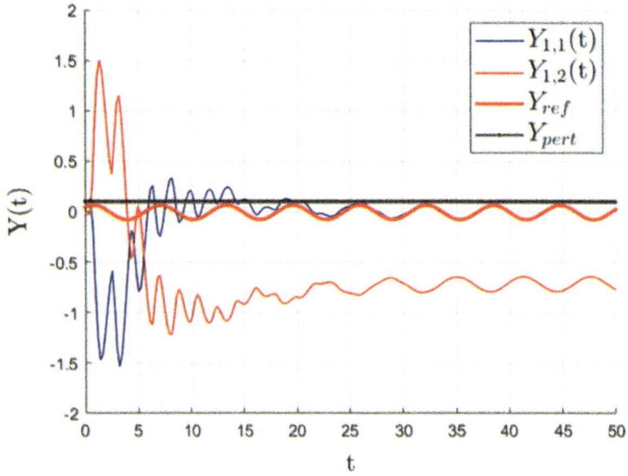

Fig. 4 Evolution of the state $Y(t)$

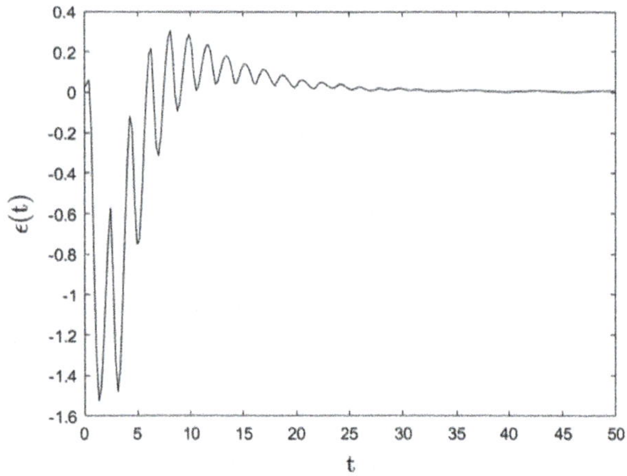

Fig. 5 Evolution of the virtual output $\epsilon(t)$

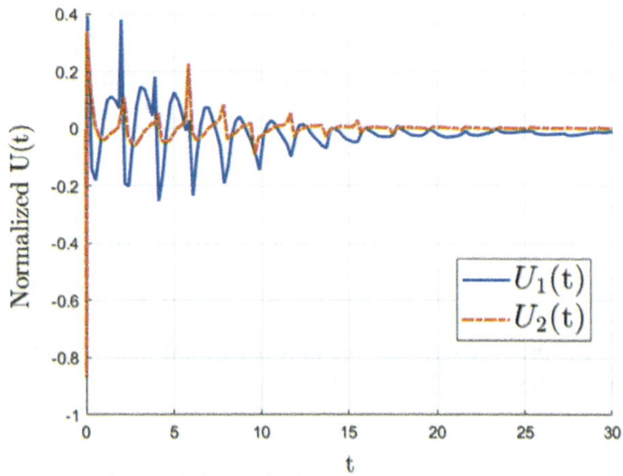

Fig. 6 Evolution of the normalized state-feedback effort $U(t)$

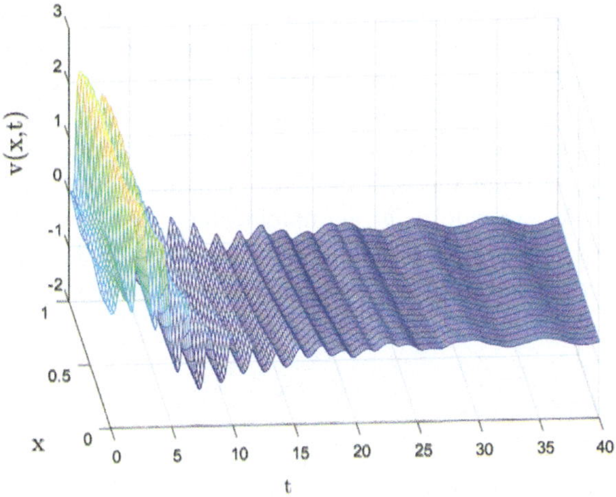

Fig. 7 Evolution of the leftward convecting PDE state $v(t, x)$

$$A_{12} = \begin{bmatrix} 1\,0\,0 \\ 0\,0\,0 \\ 0\,0\,0 \end{bmatrix}, \quad C_{12} = \begin{bmatrix} 0.1\;0\;0 \end{bmatrix}.$$

Notice that $A_{12} \notin Im(E_1)$ (unmatched disturbance), $E_1 C_1 = 0$, $C_0 B_0 = 0$, $E_0 \notin Im(B_0)$ and (A_0, B_0) is not controllable but is stabilizable. The numerical values were chosen such that the disturbance cannot be algebraically cancelled by the control law, which must take into account the dynamics of the system. The initial conditions are given by $u_0(x) = 0.05\sin(10x)$, $v_0(x) = 0.1\cos(13x)$ and $X(0) = \begin{bmatrix} -0.2 & -1 & -2 & -2 \end{bmatrix}^T$, $Y(0) = 0.5 \times \begin{bmatrix} 0.15 & -0.05 & 0.1 & 0.2 & 0.1 & 0.1 \end{bmatrix}^T$. We have $C_e = \begin{bmatrix} 1 & 0 & 0 & 0 & -1 & 0 \end{bmatrix}$, such that the first component of the output follows the sinusoidal trajectory by stabilizing $\epsilon(t) = Y_{1,1}(t) - Y_{\text{ref}}(t)$ (Figs. 4 and 5). The control effort is represented on Fig. 6. Note that the input ODE system and the PDE systems do not converge towards a constant value, but follow a sinusoidal dynamics as expected (Fig. 7). Due to the high level of instability of the original system, the initial effort presents high values. The effect of saturation is out of the scope of this chapter, but should be investigated for further applications on real systems. The control effort remains continuous for any L^2 initial conditions.

6 Notes, Comments and Concluding Remarks

In this chapter, a strictly proper dynamic output-feedback controller was designed for the practical output regulation and output tracking of a class of 2×2 linear hyperbolic ODE-PDE-ODE systems. The load dynamics (the unactuated end of the PDE) were dynamically augmented with a finite-dimensional exosystem modeling possible trajectory and disturbance inputs. The proposed approach was based on a backstepping transform that allowed us to reformulate the regulation problem in terms of a time-delay system with pointwise and distributed delays. We also followed the backstepping methodology to design an observer for the state and disturbance reconstruction. We used frequency analysis to design a feedback controller robust to small delays by extending the filtering techniques used in [8].

The omitted proofs and more computational details can be found in [21]. In future works, we will focus on implementation aspects (saturation of the effort, numerical observer design) to apply the proposed methodology to industrial applications. The performances of the feedback controller will be compared to PI controllers.

References

1. Aamo, O.M.: Disturbance rejection in 2×2 linear hyperbolic systems. IEEE Trans. Autom. Control **58**(5), 1095–1106 (2013)
2. Aarsnes, U.J., Di Meglio, F., Shor, R.J.: Avoiding stick slip vibrations in drilling through startup trajectory design. J. Process Control **70**, 24–35 (2018)
3. Aarsnes, U.J.F., Di Meglio, F., Evje, S., Aamo, O.-M.: Control-oriented drift-flux modeling of single and two-phase flow for drilling. In: ASME 2014 Dynamic Systems and Control Conference (2014)
4. Auriol, J., Aarsnes, U.J.F., Martin, P., Di Meglio, F.: Delay-robust control design for heterodirectional linear coupled hyperbolic PDEs. IEEE Trans. Autom. Control (2018)
5. Auriol, J., Kazemi, N., Innanen, K., Shor, R.: Combining formation seismic velocities while drilling and a PDE-ODE observer to improve the drill-string dynamics estimation. In: IEEE American Control Conference (ACC) (2020)
6. Bastin, G., Coron, J.-M.: Stability and Boundary Stabilization of 1-D Hyperbolic Systems. Springer (2016)
7. Bou Saba, D., Bribiesca-Argomedo, F., Di Loreto, M., Eberard, D.: Backstepping stabilization of 2 × 2 linear hyperbolic PDEs coupled with potentially unstable actuator and load dynamics. In: IEEE 56th Annual Conference on Decision and Control (CDC), pp. 2498–2503 (2017)
8. Saba, D.B., Bribiesca-Argomedo, F., Di Loreto, M., Eberard, D.: Strictly proper control design for the stabilization of 2 × 2 linear hyperbolic ODE-PDE-ODE systems. In: IEEE 58th Conference on Decision and Control (CDC) (2019)
9. Deutscher, J., Gehring, N., Kern, R.: Output feedback control of general linear heterodirectional hyperbolic ODE-PDE-ODE systems. Elsevier Autom. **95**, 472–480 (2018)
10. Deutscher, J., Gehring, N., Kern, R.: Output feedback control of general linear heterodirectional hyperbolic ODE-PDE-ODE systems. Automatica **95**, 472–480 (2018)
11. Deutscher, J., Gabriel, J.: A backstepping approach to output regulation for coupled linear wave-ODE systems. Elsevier Autom. **123**, 109338 (2021)
12. Di Meglio, F., Argomedo, F.B., Hu, L., Krstic, M.: Stabilization of coupled linear heterodirectional hyperbolic PDE–ODE systems. Elsevier Autom. **87**, 281–289 (2018)
13. Di Meglio, F., Lamare, P.-O., Aarsnes, U.J.F.: Robust output feedback stabilization of an ODE-PDE-ODE interconnection. Elsevier Autom. **119**, 109059 (2020)
14. Francis, B.A., Wonham, W.M.: The internal model principle for linear multivariable regulators. Appl. Math. Optim. **2**(2), 170–194 (1975)
15. Guerrero, M.E., Mercado, D.A., Lozano, R., García, C.D.: Passivity based control for a quadrotor UAV transporting a cable-suspended payload with minimum swing. In: 54th IEEE Conference on Decision and Control (CDC), pp. 6718–6723 (2015)
16. Hale, J.K., Lunel, S.M.V.: Introduction to Functional Differential Equations. Springer (1993)

17. Kern, R., Gehring, N., Deutscher, J., Meissner, M.: Design and experimental validation of an output feedback controller for a pneumatic system with distributed parameters. In: S18th International Conference on Control, Automation and Systems (ICCAS), pp. 1391–1396 (2018)
18. Krstic, M., Smyshlyaev, A.: Backstepping boundary control for first-order hyperbolic PDEs and application to systems with actuator and sensor delays. Syst. Control Lett. **57**(9), 750–758 (2008)
19. Krstic, M., Smyshlyaev, A.: Boundary control of PDEs: a course on backstepping designs. SIAM 16 (2008)
20. Logemann, H., Rebarber, R., Weiss, G.: Conditions for robustness and nonrobustness of the stability of feedback systems with respect to small delays in the feedback loop. SIAM J. Control Optim. **34**(2), 572–600 (1996)
21. Redaud, J., Argomedo, F.B., Auriol, J.: Output regulation and tracking for linear ODE-hyperbolic PDE-ODE systems. Elsevier Autom. (to be submitted) (2022)
22. Schmuck, C., Woittennek, F., Gensior, A., Rudolph, J.: Flatness-based feed-forward control of an HVDC power transmission network. In: IEEE 33rd International Telecommunications Energy Conference (INTELEC), pp. 1–6 (2011)
23. Smith, O.: A controller to overcome dead time. ISA J. **6**, 28–33 (1959)
24. Vazquez, R., Krstic, M., Coron, J.-M.: Backstepping boundary stabilization and state estimation of a 2×2 linear hyperbolic system. In: 50th IEEE Conference on Decision and Control (CDC), pp. 4937–4942 (2011)
25. Wang, J., Krstic, M.: Delay-compensated control of sandwiched ODE-PDE-ODE hyperbolic systems for oil drilling and disaster relief. Elsevier Autom. **120**, 109131 (2020)
26. Yoshida, K.: Lectures on differential and integral equations. Intersci. Publ. 10 (1960)

A Systematic Backstepping Design of Tracking Controllers for ODE-PDE-ODE Systems with Nonlinear Actuator Dynamics

Nicole Gehring

Abstract A backstepping design of tracking controllers is addressed for hyperbolic and parabolic bidirectionally coupled ODE-PDE-ODE systems with nonlinear ODEs at the actuated boundary. The stabilization of a reference trajectory is achieved by successively stabilizing the different ODE and PDE subsystems by means of virtual feedback laws. Appropriate state transformations allow to map the overall system into error coordinates, from which stability is easily inferred based on the cascaded structure and the stability of each of the subsystems. In the end, the presented design solely relies on the strict feedback form of the ODE-PDE-ODE system. Using classical concepts such as integrator backstepping, output zeroing and differential flatness, all of which are well known in the ODE world, a multi-step modular and systematic design algorithm is shown to be essentially identical for the tracking control of both hyperbolic and parabolic ODE-PDE-ODE systems.

1 Introduction

This contribution presents a systematic and modular backstepping-based design of trajectory tracking controllers for so-called ODE-PDE-ODE systems, with partial differential equations (PDEs) of either hyperbolic or parabolic type. This system class is motivated by infinite-dimensional models in one spatial variable with dynamic boundary conditions (BCs), where the spatially-distributed process is described by PDEs. In turn, the BCs result in ordinary differential equations (ODEs) at both boundaries of the PDE. For example, such ODEs may arise from finite-dimensional actuator or sensor dynamics. Importantly, if the coupling between the PDE and each of the respective ODEs is in both directions, it is called bidirectional. One has a cascade of subsystems if the coupling is in one direction only. Here, both hyperbolic and parabolic ODE-PDE-ODE systems with bidirectional coupling are considered.

N. Gehring (✉)
Johannes Kepler University Linz, Institute of Automatic Control and Control Systems Technology, Altenberger Straße 69, 4040 Linz, Austria
e-mail: nicole.gehring@jku.at

J. Auriol et al. (eds.), *Advances in Distributed Parameter Systems*, Advances in Delays and Dynamics 14, https://doi.org/10.1007/978-3-030-94766-8_8

Although the PDE and the ODE at the distal boundary are assumed to be linear, the ODE at the actuated boundary is nonlinear. Note that ODE-PDE-ODE systems also include the special cases of PDE-ODE and ODE-PDE systems, where the control input acts on the boundary of the PDE in the former scenario and on the ODE in the latter. Overall, technological examples are manifold and include, for example, coupled open channels and transmission lines (e.g. [5]), as well as chemical and biochemical processes (e.g. [20]).

1.1 Literature Overview

Approaches for the stabilization of linear hyperbolic and parabolic ODE-PDE-ODE, PDE-ODE and ODE-PDE systems include a design based on Fuzzy logic (e.g. [28]) or the one in [25] using optimality principles. Furthermore, both the flatness-based design for a hyperbolic PDE-ODE system in [29] and the solution-based approach for the same type of system in [18] allow one to even track a reference trajectory.

Still, in most cases, backstepping controllers are used for the stabilization of linear hyperbolic and parabolic ODE-PDE-ODE, PDE-ODE and ODE-PDE systems. They strongly vary w.r.t. the system class, i.e. the order of the respective ODEs and PDEs, their coupling structure, as well as any assumptions imposed for the design. In the hyperbolic case, first results consider PDEs that are bidirectionally coupled with an ODE system, i.e. PDE-ODE systems (e.g. [6, 8, 11]), with generalizations to (simple) ODE-PDE-ODE structures soon after (e.g. [1]). Backstepping controllers for coupled parabolic systems are more scarce. In [26] and related works, PDE-ODE systems with a single heat equation are stabilized. An additional ODE at the other boundary is taken into account in [27].

Backstepping controller for very general ODE-PDE-ODE systems are derived in [9] for the hyperbolic case and in [7] for the parabolic one. In both contributions, the PDE subsystems are of arbitrary order with very minimal assumptions imposed towards the system structure. In contrast to the aforementioned references, [7, 9] suggest a multi-step backstepping design to stabilize and decouple the different subsystems. For that, it makes use of the solution of the kernel equations derived in [16] and [10], respectively. The multi-step design in [7, 9] is made even more systematic in [15] by exploiting the naturally arising strict feedback form of the ODE-PDE-ODE systems. By that, it answers the question about the proper choice and order of successive transformations in the multi-step design and allows for a generalization to significantly broader classes of systems. Similar to [7, 9], the stability of the closed-loop system in [15] is inferred from its cascaded structure and the stability of each of the subsystems.

1.2 Motivation

It is well known, that backstepping for PDEs (see, e.g., [24]) originated from extending the integrator backstepping design for ODEs (see, e.g., [22]) to systems with distributed parameters. This is beautifully motivated in [4], where a backstepping controller is designed for a scalar reaction-diffusion equation

$$\partial_t x(z,t) = \lambda \partial_z^2 x(z,t) + a x(z,t) \tag{1a}$$

$$x(0,t) = 0 \tag{1b}$$

$$x(1,t) = u(t), \tag{1c}$$

with $x(z,t), u(t) \in \mathbb{R}$, $\lambda > 0$ and $a \in \mathbb{R}$. Using a semi-discretization of the normalized spatial variable $z \in [0,1]$ by means of the central difference quotient, (1) is approximated by the finite-dimensional model

$$x_0(t) = 0 \tag{2a}$$

$$\dot{x}_i(t) = \lambda \frac{x_{i+1}(t) - 2x_i(t) + x_{i-1}(t)}{\Delta z} + a x_i(t), \qquad i = 1, \ldots, n \tag{2b}$$

$$x_{n+1}(t) = u(t) \tag{2c}$$

of order n, where $x_i(t) = x(i\Delta z, t)$, $i = 0, \ldots, n+1$ and $\Delta z = \frac{1}{n+1}$. Notably, the ODEs in (2) (or the dynamic matrix of a corresponding state-space representation) exhibit a band structure resulting in a special kind of strict feedback form.[1] A strict feedback form is also required for the integrator backstepping design (see, e.g., [22]), where the idea is to successively stabilize subordinate parts of the system by considering $x_{i+1}(t)$ as the virtual control input in the i-th step of the design. By choosing a (virtual) state feedback

$$x_{i+1}(t) = \alpha_i(x_1(t), \ldots, x_i(t)), \qquad i = 1, \ldots, n, \tag{3}$$

the subsystem with the state $[x_1(t), \ldots, x_i(t)]^T$ is stabilized (in some appropriate sense). As this feedback cannot be implemented, an error variable

$$\tilde{x}_{i+1}(t) = x_{i+1}(t) - \alpha_i(x_1(t), \ldots, x_i(t)), \qquad i = 0, \ldots, n-1 \tag{4}$$

is defined in each step. Once it goes to zero, which is ensured by the design of the subsequent virtual feedback, it guarantees (3).

---

[1] In general, in a strict feedback form, the i-th derivative $\dot{x}_i(t)$, $i = 1, \ldots, n$ is a function of the state components $x_1(t), \ldots, x_{i+1}(t)$ with $x_{n+1}(t) = u(t)$.

Applying integrator backstepping to (2) and letting $n \to \infty$, [4] shows based on (4) that the Volterra integral transformation

$$\bar{x}(z,t) = x(z,t) - \int_0^z k(z,\zeta)x(\zeta,t)\,\mathrm{d}\zeta \tag{5}$$

may directly be applied to the PDE (1). Note the correlation between the triangular structure of (4) and the triangular domain $0 < \zeta < z < 1$ the kernel $k(z,\zeta) \in \mathbb{R}$ in (5) is defined on. Whereas one would basically choose the virtual feedbacks (3) such that a potentially destabilizing reaction term $ax_i(t)$ is removed, appropriate choice of $k(z,\zeta)$ (see [4]) does the same for the PDE (1), i.e., (5) maps (1) into:

$$\partial_t \bar{x}(z,t) = \lambda \partial_z^2 \bar{x}(z,t) \tag{6a}$$

$$\bar{x}(0,t) = 0 \tag{6b}$$

$$\bar{x}(1,t) = u(t) - \int_0^1 k(1,z)x(z,t)\,\mathrm{d}z. \tag{6c}$$

Based on that, it is straightforward to choose $u(t) = \int_0^1 k(1,z)x(z,t)\,\mathrm{d}z$ (which corresponds to (3) for $i = n$) in order to stabilize (6) and thus equivalently (1). Notably, in the framework of integrator backstepping and in light of (3), the integral transformation (5) can be interpreted as introducing the error state $\bar{x}(z,t)$ based on the boundary value $x(1,t)$ serving as the virtual control input.

All of that is of course well known. However, this motivation is meant to highlight several aspects. First, the underlying ideas of (integrator) backstepping should be applicable not only to ODEs in strict feedback form or to PDEs but also to (potentially bidirectionally) coupled structures of ODEs and PDEs. This basically only requires that an ODE-PDE-ODE system or another combination of finite- and infinite-dimensional subsystems is in an appropriate strict feedback form. This will be the case for the systems considered in this contribution. Note that most backstepping results for ODE-PDE-ODE and PDE-ODE systems make use of only one combined transformation of both ODE and PDE states (e.g. [6, 12]). Also, a semi-discretization of such transformations does not produce a triangular structure[2] as in (4) and it is oftentimes very complicated to show existence and uniqueness of the corresponding transformation equations. The second emphasis should be on how a feedback (3) is designed. Whereas integrator backstepping is usually closely related to the construction of Lyapunov functions, one could use different design tools in each step i, including high gain or sliding mode control as well as passivity or differential flatness, to name but a few. This too will be exploited in the following, i.e., the control method used for the stabilization of one subsystem is independent of the approach used for another subsystem.

[2] One could even question whether such a design should still be called backstepping or whether referring to it simply as a transformation-based design would be more appropriate.

1.3 Contribution and Content

The modular design philosophy mentioned at the end of the previous section was first suggested in [14, 15] in order to take backstepping for PDEs back to its ODE origins. This allows one to (easily) stabilize a larger class of systems, in comparison to other known backstepping results. It also offers a multitude of possible generalizations, without the need of having to solve new transformation equations for each new system class. Most importantly, in this contribution, the results in [15] are extended to the case of a nonlinear actuator dynamics, and trajectory tracking is considered instead of a more simple stabilization problem. Moreover, the additional space here is used to thoroughly motivate and discuss the design strategy.

In particular, ODE-PDE-ODE systems with a linear heterodirectional hyperbolic PDE subsystem with spatially varying coefficients, a linear ODE subsystem at the distal boundary and a nonlinear one at the actuated boundary are considered in Sect. 2. Under minimal assumptions, a tracking controller is designed in three consecutive steps, that explicitly exploits the strict feedback form of the ODE-PDE-ODE system by successively designing (virtual) controllers for each of the subsystems. The stability of the closed loop directly follows from the stability of each of the subsystems and their cascaded structure. The modular design is again highlighted in Sect. 3, where a linear parabolic PDE of order n replaces the hyperbolic one in Sect. 2. The results in [15] are additionally generalized, if only slightly, by considering Robin-type boundary conditions (instead of Neumann conditions as in [15]). Due to the shared structure between both types of ODE-PDE-ODE systems, the design steps in Sect. 3 are basically the same as in Sect. 2.

Note that the emphasis here is not on a specific control method used for the stabilization of a subsystem but on the modular and systematic design. As such, some technicalities are omitted, no final theorems on the closed-loop stability are given, and no simulation results are presented. Instead, Sect. 4 gives some insight into possible extensions of the presented control philosophy.

2 Hyperbolic ODE-PDE-ODE Systems

The hyperbolic case is based on the ODE-PDE-ODE system in [9, 15], with a nonlinear ODE at the actuated boundary instead of a linear one. In order to stabilize this system along a reference trajectory, three successive design steps are taken. First, the ODE at the unactuated boundary is stabilized using a (virtual) linear state feedback. The design parameters are chosen, for example, by eigenvalue assignment or optimality principles. The stabilization of the PDE subsystem follows the classical PDE backstepping methodology. However, any other (virtual) boundary controller could be used instead. Finally, the nonlinear actuator dynamics have to be controlled. Here, by rewriting the ODE in Byrnes-Isidori normal form and solving an output zeroing problem, concepts well-known for nonlinear systems are employed.

2.1 Problem Statement

The central part of the ODE-PDE-ODE system are the hyperbolic PDEs in

$$\partial_t x(z,t) = \Lambda(z)\partial_z x(z,t) + A(z)x(z,t). \tag{7}$$

They are heterodirectional due to the diagonal matrix $\Lambda(z)$ defined by $\Lambda(z) = \operatorname{diag}(\lambda_1(z), \dots, \lambda_n(z))$, where the transport velocities $\lambda_i \in C^1[0,1]$, $i = 1, \dots, n$ are ordered according to

$$\lambda_1(z) > \cdots > \lambda_p(z) > 0 > \lambda_{p+1}(z) > \cdots > \lambda_n(z). \tag{8}$$

Consequently, introducing

$$E_1 = \begin{bmatrix} I_p \\ 0 \end{bmatrix} \in \mathbb{R}^{n\times p}, \qquad E_2 = \begin{bmatrix} 0 \\ I_m \end{bmatrix} \in \mathbb{R}^{n\times m} \tag{9}$$

based on unit matrices $I_k \in \mathbb{R}^{k\times k}$, $k \in \mathbb{N}$, the transport in the negative z-direction is described by the part $x_1(z,t) = E_1^T x(z,t) \in \mathbb{R}^p$ of the PDE state $x(z,t) \in \mathbb{R}^n$, whereas $x_2(z,t) = E_2^T x(z,t) \in \mathbb{R}^m$ models the transport in the positive z-direction. The elements of the in-domain coupling matrix A are $C^1[0,1]$-functions, with $a_{ii}(z) = 0$ for $i = 1, \dots, n$. Note that the structural assumptions towards $\Lambda(z)$ and $A(z)$ hold no restriction (see [16]).

At the unactuated boundary at $z = 0$, the PDEs in (7) are bidirectionally coupled with the linear ODE system

$$\dot{w}_0(t) = F_0 w_0(t) + B_0 x_1(0,t) \tag{10}$$

via the linear boundary condition

$$x_2(0,t) = Q_0 x_1(0,t) + C_0 w_0(t). \tag{11}$$

Specifically, the ODE state $w_0(t) \in \mathbb{R}^{n_0}$ acts on the PDE through C_0, the PDE boundary term $x_1(0,t)$ on the w_0-dynamics through B_0. For the actuated boundary at $z = 1$, the nonlinear ODE system

$$\dot{w}_1(t) = f_1(w_1(t), x_2(1,t)) + B(w_1(t), x_2(1,t))u(t) \tag{12}$$

with the corresponding nonlinear BC

$$x_1(1,t) = c_1(w_1(t), x_2(1,t)) + D(w_1(t), x_2(1,t))u(t) \tag{13}$$

is considered, where the bidirectional coupling between ODE and PDE is due to the nonlinear vector- and matrix-valued functions f_1, B, c_1 and D. By the input

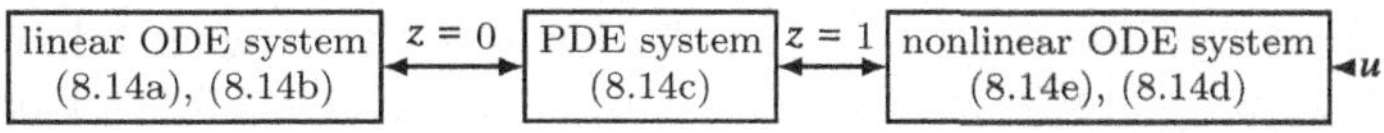

Fig. 1 Structure of the ODE-PDE-ODE system (14)

$u(t) \in \mathbb{R}^p$, the boundary at $z = 1$ is fully actuated. The dimension of the functions and matrices in (10)–(13) follows from those of the ODE and PDE states as well as from the input.

Overall, the hyperbolic ODE-PDE-ODE system with a nonlinear actuating ODE is described by

$$\dot{w}_0(t) = F_0 w_0(t) + B_0 x_1(0, t) \tag{14a}$$
$$x_2(0, t) = Q_0 x_1(0, t) + C_0 w_0(t) \tag{14b}$$
$$\partial_t x(z, t) = \Lambda(z)\partial_z x(z, t) + A(z)x(z, t) \tag{14c}$$
$$x_1(1, t) = c_1(w_1(t), x_2(1, t)) + D(w_1(t), x_2(1, t))u(t) \tag{14d}$$
$$\dot{w}_1(t) = f_1(w_1(t), x_2(1, t)) + B(w_1(t), x_2(1, t))u(t) \tag{14e}$$

(see (7) and (10)–(13)), with initial conditions (ICs) $x(z, 0) = x_0(z) \in \mathbb{R}^n, z \in [0, 1]$, $w_0(0) = w_{0,0} \in \mathbb{R}^{n_0}$ and $w_1(0) = w_{1,0} \in \mathbb{R}^{n_1}$. Inspection of (14) reveals, that the system is in strict feedback form (see also Fig. 1): the w_0-ODE (14a) is only excited by the PDE state, (specifically the boundary value $x_1(0, t)$), the PDE dynamics (14b)–(14d) by $w_1(t)$ acting on the boundary[3] at $z = 1$, and the w_1-ODE (14e) by the control input $u(t)$. In the following, this structure is exploited for the design of a tracking controller. For that, the following assumptions are imposed.

- **Assumption A1:** A reference trajectory

$$t \mapsto \big(w_{0,r}(t), x_r(\cdot, t), w_{1,r}(t), u_r(t)\big) \tag{15}$$

is known that satisfies

$$\dot{w}_{0,r}(t) = F_0 w_{0,r}(t) + B_0 x_{1,r}(0, t) \tag{16a}$$
$$x_{2,r}(0, t) = Q_0 x_{1,r}(0, t) + C_0 w_{0,r}(t) \tag{16b}$$
$$\partial_t x_r(z, t) = \Lambda(z)\partial_z x_r(z, t) + A(z)x_r(z, t) \tag{16c}$$
$$x_{1,r}(1, t) = c_1(w_{1,r}(t), x_{2,r}(1, t)) + D(w_{1,r}(t), x_{2,r}(1, t))u_r(t) \tag{16d}$$
$$\dot{w}_{1,r}(t) = f_1(w_{1,r}(t), x_{2,r}(1, t)) + B(w_{1,r}(t), x_{2,r}(1, t))u_r(t). \tag{16e}$$

- **Assumption A2:** The pair (F_0, B_0) is stabilizable.

[3] The strict feedback form seems to be broken by $u(t)$ also impacting (14d) through D. However, as will become more clear in the context of Sect. 2.5, one actually still has feedback forms for each of the components of $x_1(1, t)$, which is exploited for the control design.

- **Assumption A3:** The w_1-dynamics (14e) has a constant vector relative degree $\{r_1, \ldots, r_p\}$ w.r.t. (14d), with $\sum_{i=1}^{p} r_i \leq n_1$. The zero dynamics of order $n_1 - \sum_{i=1}^{p} r_i$ is asymptotically stable in a neighborhood of the reference (15).

The necessity of Assumption A2 is apparent from the w_0-ODE only being affected by the boundary value $x_1(0, t)$. By Assumption A1, a solution of (14) is assumed to be known. For example, such a reference trajectory (15) could be obtained for those ODE-PDE-ODE systems that are flat in the sense discussed in [23].

Remark 1 In [23], the concept of (differential) flatness, well-known for nonlinear ODEs (see, e.g., [13]), is used to design feed-forward controllers for quasi-linear hyperbolic PDEs with dynamic BCs based on a differential parametrization of the system quantities. Basically, it requires the boundary systems to be flat in the classical sense. Specifically, here, tightening Assumption A2 from stabilizability to controllability ensures flatness of (14a). Hence, a flat output $Y(t) = \varpi(w_0(t))$ exists such that

$$w_0(t) = \psi_1(Y(t), \dot{Y}(t), \ldots, Y^{(n_0-1)}(t)) \tag{17a}$$

$$x_1(0, t) = \psi_2(Y(t), \dot{Y}(t), \ldots, Y^{(n_0)}(t)), \tag{17b}$$

which in turn provides the parametrization

$$x_2(0, t) = Q_0\psi_2(Y(t), \dot{Y}(t), \ldots, Y^{(n_0)}(t)) + C_0\psi_1(Y(t), \dot{Y}(t), \ldots, Y^{(n_0-1)}(t)) \tag{17c}$$

by means of (14b). With the boundary values at $z = 0$ known as functions of Y and its derivatives, the solution of (14c) yields the parametrization of the entire distributed state $x(z, t)$ and, in particular, of the boundary value $x(1, t)$. Thus, a sufficient condition for the w_1-ODE to satisfy the conditions in [23] would be that (14d) is a flat output of (14e), which implies $D = 0$ for $n_1 \neq 0$. Then, the parametrization of $x(1, t)$ allows to express $c_1(w_1, \cdot)$ in terms of Y and, consequently, w_1 and u using (14e). In the end, the sought reference (15) directly follows from inserting a reference trajectory $Y_r(t)$ for a flat output into the differential parametrizations.

Finally, Assumption A3 is imposed on the nonlinear dynamics. This rather restrictive assumption is sufficient for the stabilization of (14) along a reference trajectory (15). It is necessary only due to the output-zeroing strategy employed in Sect. 2.5. As the use of a specific control strategy for any (and in particular the nonlinear) subsystem is not the focus here, the reader is referred to literature on that topic, e.g. [19].

Remark 2 Due to $D(w_1(t), x_2(1, t))$ in (14d), the relative degree of

$$\rho = \operatorname{rank} D(w_1(t), x_2(1, t)) \leq p \tag{18}$$

components will be zero. Also, if $D = 0$ and $c_1(w_1, \cdot)$ is a flat output of (14e) (see also Remark 1), one has $\sum_{i=1}^{p} r_i = n_1$ and no zero dynamics occur. Moreover, in the linear case (see also [15]), i.e.

$$\dot{w}_1(t) = F_1 w_1(t) + Bu(t) \tag{19a}$$

$$x_1(1, t) = C_1 w_1(t) + Du(t) \tag{19b}$$

instead of (14d) and (14e), the conditions of Assumption A3 coincide with (F_1, B, C_1, D) having only zeros with negative real parts. In [9], this assumption is shown not to be necessary for the stabilization of a linear ODE-PDE-ODE system.

In what follows, based on Assumptions A1–A3, a tracking controller is designed that stabilizes (in some appropriate sense) the ODE-PDE-ODE system (14) with the nonlinear w_1-dynamics along a reference trajectory (15).

2.2 Basic Idea

As mentioned in the introduction, one of the main ideas of the presented control design is to explicitly exploit the strict feedback form of (14) and to combine design methods known from both finite- and infinite-dimensional settings. For that, virtual control inputs are used to stabilize subordinate systems successively. By a state transformation, subsequently, the overall system is mapped into error coordinates and the next subsystem is stabilized. For that, it it mostly irrelevant what specific method is employed for the design of each virtual feedback.

To better highlight the basic principle, first, consider only the w_0-ODE (14a). Therein, the boundary value $x_1(0, t)$ takes the role of a control input, i.e.

$$x_1(0, t) = u_{\text{virtual}}(t). \tag{20}$$

Thus, to stabilize the origin $w_0(t) = 0$, one would choose the linear virtual feedback

$$u_{\text{virtual}}(t) = K_0 w_0(t) \tag{21}$$

such that $F_0 + B_0 K_0$ is a Hurwitz matrix, which is always possible due to Assumption A2. Similarly,

$$u_{\text{virtual}}(t) = K_0\big(w_0(t) - w_{0,r}(t)\big) + x_{1,r}(0, t) \tag{22}$$

is the virtual input to stabilize a reference trajectory $w_{0,r}(t)$. Consequently, the error $e_0(t) = w_0(t) - w_{0,r}(t)$ would satisfy

$$\dot{e}_0(t) = (F_0 + B_0 K_0) e_0(t) \tag{23}$$

in light of (16a). However, neither (21) nor (22) can be implemented as $x_1(0, t)$ is not a control input.

As in integrator backstepping, this motivates a change of coordinates

$$\bar{x}(z, t) = x(z, t) - N(z)w_0(t) \tag{24}$$

based on (21) or

$$\bar{x}(z, t) = x(z, t) - N(z)\big(w_0(t) - w_{0,r}(t)\big) - x_r(z, t) \tag{25}$$

based on (22). In both cases, $\bar{x}(z, t)$ is an error coordinate defined by the difference between the PDE state and the corresponding desired value inferred from the virtual feedback for $x_1(0, t)$. The matrix $N(z)$ has to satisfy (at least) $E_1^T N(0) = K_0$, leading to the ODE dynamics

$$\dot{e}_0(t) = (F_0 + B_0 K_0)e_0(t) + B_0 \bar{x}_1(0, t) \tag{26}$$

based on (25) (instead of (23)).

Note that $N(z)$ in (24) and (25) could have been chosen as a constant. However, the spatial dependence will be shown to be beneficial in the next section. Moreover, it comes as no surprise that the formal representation is simplified if instead of the tracking of the reference trajectory (15) the stabilization of the corresponding error dynamics is considered (cf., e.g., (25) or (26)). For that, introduce errors

$$e_0(t) = w_0(t) - w_{0,r}(t), \qquad e_x(z, t) = x(z, t) - x_r(z, t), \qquad e_1(t) = w_1(t) - w_{1,r}(t) \tag{27}$$

for the ODE and PDE states w.r.t. their reference values. Then, based on (14) and (16), the ODE-PDE-ODE (error) system

$$\begin{aligned}
\dot{e}_0(t) &= F_0 e_0(t) + B_0 e_{x,1}(0, t) && \text{(28a)}\\
e_{x,2}(0, t) &= Q_0 e_{x,1}(0, t) + C_0 e_0(t) && \text{(28b)}\\
\partial_t e_x(z, t) &= \Lambda(z)\partial_z e_x(z, t) + A(z)e_x(z, t) && \text{(28c)}\\
e_{x,1}(1, t) &= \bar{c}_1(t, e_1(t), e_{x,2}(1, t)) + \bar{D}(t, e_1(t), e_{x,2}(1, t))u(t) && \text{(28d)}\\
\dot{e}_1(t) &= \bar{f}_1(t, e_1(t), e_{x,2}(1, t)) + \bar{B}(t, e_1(t), e_{x,2}(1, t))u(t) && \text{(28e)}
\end{aligned}$$

admits the same structure as (14) in Fig. 1, with nonlinear functions

$$\begin{aligned}
\bar{c}_1(t, e_1(t), e_{x,2}(1, t)) &= c_1(e_1(t) + w_{1,r}(t), e_{x,2}(1, t) + x_{2,r}(1, t)) - x_{1,r}(1, t) && \text{(29a)}\\
\bar{D}(t, e_1(t), e_{x,2}(1, t)) &= D(e_1(t) + w_{1,r}(t), e_{x,2}(1, t) + x_{2,r}(1, t)) && \text{(29b)}\\
\bar{f}_1(t, e_1(t), e_{x,2}(1, t)) &= f_1(e_1(t) + w_{1,r}(t), e_{x,2}(1, t) + x_{2,r}(1, t)) - \dot{w}_r(t) && \text{(29c)}\\
\bar{B}(t, e_1(t), e_{x,2}(1, t)) &= B(e_1(t) + w_{1,r}(t), e_{x,2}(1, t) + x_{2,r}(1, t)) && \text{(29d)}
\end{aligned}$$

that represent their dependence on the reference trajectories by the explicit time argument t.

2.3 Stabilization of the ODE Subsystem at $z = 0$

In order to find $u(t)$ such that the origin of the error system (28) is stabilized, first, as in the previous section, the virtual feedback $e_{x,1}(0,t) = K_0 e_0(t)$ (see (25)) is chosen such that $F_0 + B_0 K_0$ is a Hurwitz matrix. The virtual input motivates the (invertible) transformation

$$\bar{e}_x(z,t) = e_x(z,t) - N(z)e_0(t), \qquad N(z) \in \mathbb{R}^{n \times n_0} \tag{30}$$

of the PDE state as in (25). Using (30) in (28) and omitting the boundary system at $z = 1$ for the moment, if $E_1^T N(0) = K_0$, one has (26) for $e_0(t)$ as well as the PDE dynamics

$$\bar{e}_{x,2}(0,t) = Q_0 \bar{e}_{x,1}(0,t) + \tilde{C}_0 e_0(t) \tag{31a}$$
$$\partial_t \bar{e}_x(z,t) = \Lambda(z)\partial_z \bar{e}_x(z,t) + A(z)\bar{e}_x(z,t) - N(z)B_0 \bar{e}_{x,1}(0,t) + \tilde{C}(z)e_0(t) \tag{31b}$$

where

$$\tilde{C}_0 = C_0 + Q_0 K_0 - E_2^T N(0) \tag{32a}$$
$$\tilde{C}(z) = -N(z)(F_0 + B_0 K_0) + A(z)N(z) + \Lambda(z)N'(z). \tag{32b}$$

Whereas (26) is already exponentially stable if $\bar{e}_{x,1}(0,t) = 0$, the transformation matrix $N(z)$ is not uniquely defined yet. The remaining degrees of freedom can be used to eliminate the action of the ODE state $e_0(t)$ on the PDE in (31) such that $\tilde{C}_0 = 0$ and $\tilde{C}(z) = 0$. By that, the only coupling from the e_0-dynamics to the PDE (and the e_1-dynamics) remains at the actuated boundary, where it will be compensated later by choice of the control input $u(t)$. Thus, by the transformation (30) satisfying the initial value problem (IVP)

$$\Lambda(z)N'(z) + A(z)N(z) - N(z)(F_0 + B_0 K_0) = 0 \tag{33a}$$
$$E_1^T N(0) = K_0 \tag{33b}$$
$$E_2^T N(0) = C_0 + Q_0 K_0, \tag{33c}$$

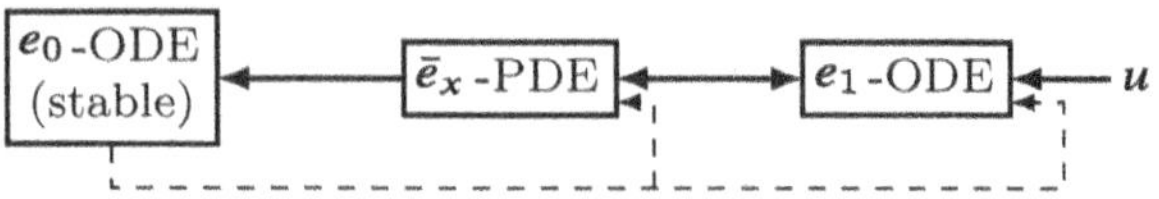

Fig. 2 Structure of the ODE-PDE-ODE system (34) following the first design step

the overall ODE-PDE-ODE system (28) takes the form

$$\dot{e}_0(t) = (F_0 + B_0 K_0)e_0(t) + B_0\bar{e}_{x,1}(0,t) \tag{34a}$$
$$\bar{e}_{x,2}(0,t) = Q_0\bar{e}_{x,1}(0,t) \tag{34b}$$
$$\partial_t \bar{e}_x(z,t) = \Lambda(z)\partial_z\bar{e}_x(z,t) + A(z)\bar{e}_x(z,t) - N(z)B_0\bar{e}_{x,1}(0,t) \tag{34c}$$
$$\begin{aligned}\bar{e}_{x,1}(1,t) &= \bar{c}_1(t, e_1(t), \bar{e}_{x,2}(1,t) + E_2^T N(1)e_0(t)) - E_1^T N(1)e_0(t)\\ &\quad + \bar{D}(t, e_1(t), \bar{e}_{x,2}(1,t) + E_2^T N(1)e_0(t))u(t)\end{aligned} \tag{34d}$$
$$\begin{aligned}\dot{e}_1(t) &= \bar{f}_1(t, e_1(t), \bar{e}_{x,2}(1,t) + E_2^T N(1)e_0(t))\\ &\quad + \bar{B}(t, e_1(t), \bar{e}_{x,2}(1,t) + E_2^T N(1)e_0(t))u(t)\end{aligned} \tag{34e}$$

in light of (30). In these transformed coordinates, one has essentially a cascade of an exponentially stable e_0-ODE (if $\bar{e}_{x,1}(0,t) = 0$) and the remaining dynamics. This is illustrated in Fig. 2.

The solvability of the matrix-valued IVP (33) is clarified in the following lemma. The proof in Appendix 1 makes use of an eigenvalue problem for $F_0 + B_0 K_0$.

Lemma 1 *The IVP (33) admits a unique $C^1[0,1]$-solution $N(z)$.*

2.4 Stabilization of the PDE Subsystem

Following the stabilization of the ODE at $z = 0$, next, the PDE is to be stabilized such that $\lim_{t\to\infty} \bar{e}_x(z,t) = 0$. For that, as motivated in the introduction, the boundary value $\bar{e}_{x,1}(1,t)$ takes the role of a virtual input, i.e.

$$\bar{e}_{x,1}(1,t) = u_{\text{virtual}}(t) \tag{35}$$

instead of (34d). Then, with regards to (34b), (34c) and (35), one has a heterodirectional PDE system as it is often considered in the backstepping context (apart from the additional coupling term due to $N(z)B_0$). Hence, the second design step simply uses the backstepping transformation

$$\tilde{e}_x(z,t) = \bar{e}_x(z,t) - \int_0^z K(z,\zeta)\bar{e}_x(\zeta,t)\,\mathrm{d}\zeta = \mathcal{T}[\bar{e}_x(t)](z), \tag{36}$$

which is a classical Volterra integral transformation with the kernel $K(z,\zeta) \in \mathbb{R}^{n\times n}$. The kernel $K_I(z,\zeta) \in \mathbb{R}^{n\times n}$ of the corresponding inverse transformation

$$\bar{e}_x(z,t) = \tilde{e}_x(z,t) + \int_0^z K_I(z,\zeta)\tilde{e}_x(\zeta,t)\,\mathrm{d}\zeta = \mathcal{T}^{-1}[\tilde{e}_x(t)](z) \tag{37}$$

follows from the reciprocity relation

$$K_I(z,\zeta) = K(z,\zeta) + \int_\zeta^z K(z,\bar{\zeta})K_I(\bar{\zeta},\zeta)\,\mathrm{d}\bar{\zeta}. \tag{38}$$

If $K(z,\zeta)$ satisfies the kernel equations

$$\Lambda(z)\partial_z K(z,\zeta) + \partial_\zeta\big(K(z,\zeta)\Lambda(\zeta)\big) = K(z,\zeta)A(\zeta) \tag{39a}$$

$$K(z,0)\Lambda(0)[E_1 + E_2Q_0] = A_0(z) + \mathcal{T}[NB_0](z) \tag{39b}$$

$$K(z,z)\Lambda(z) - \Lambda(z)K(z,z) = A(z), \tag{39c}$$

with (39a) on $0 < \zeta < z < 1$, (36) maps (34b), (34c) and (35) into

$$\tilde{e}_{x,2}(0,t) = Q_0\tilde{e}_{x,1}(0,t) \tag{40a}$$

$$\partial_t\tilde{e}_x(z,t) = \Lambda(z)\partial_z\tilde{e}_x(z,t) + A_0(z)\tilde{e}_{x,1}(0,t) \tag{40b}$$

$$\tilde{e}_{x,1}(1,t) = u_{\text{virtual}}(t) - \int_0^1 E_1^T K_I(1,z)\tilde{e}_x(z,t)\,\mathrm{d}z. \tag{40c}$$

Due to the structure of the matrix

$$A_0(z) = \begin{bmatrix} A_1(z) \\ A_2(z) \end{bmatrix} \in \mathbb{R}^{n\times p} \quad \text{with} \quad A_1(z) = \begin{bmatrix} 0 & \cdots & \cdots & 0 \\ a_{2,1}(z) & \ddots & & \vdots \\ \vdots & \ddots & \ddots & \vdots \\ a_{p,1}(z) & \cdots & a_{p,p-1}(z) & 0 \end{bmatrix} \in \mathbb{R}^{p\times p}, \tag{41}$$

choosing the virtual feedback

$$u_{\text{virtual}}(t) = \int_0^1 E_1^T K_I(1,z)\tilde{e}_x(z,t)\,\mathrm{d}z \tag{42}$$

and thus $\tilde{e}_{x,1}(1,t) = 0$ results in a cascade of p transport equations for the components of $\tilde{e}_{x,1}(z,t)$. They, in turn are cascaded with the transport equations for $\tilde{e}_{x,2}(z,t)$. By that, one has $\tilde{e}_x(z,t) = 0$ (and consequently $\bar{e}_x(z,t) = 0$) after the finite time $\sum_{i=1}^{p+1} |\int_0^z (\mathrm{d}\zeta/\lambda_i(\zeta))|$.

In contrast to the well-known kernel equations solved in [17], (39) is more general in that it involves an integral BC arising from the term $\mathcal{T}[NB_0](z)$. It is shown in [15] that (39) admits a unique piecewise $C^1[0,1]$-solution.

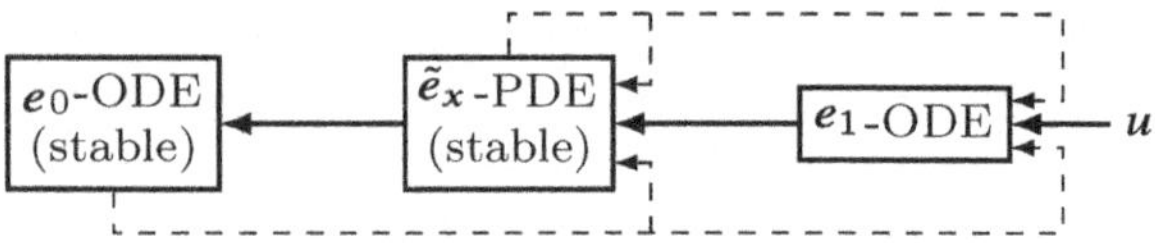

Fig. 3 Structure of the ODE-PDE-ODE system (43) following the second design step

Remark 3 The non-zero elements of the matrix $A_0(z)$ in (41) are defined by the corresponding part of (39b). Consequently, only those equations in (39b) with zero entries in $A_0(z)$ are BCs for the kernel $K(z, \zeta)$. In order to guarantee a unique solution of (39), additional BCs have to be imposed in terms of arbitrary functions (see [17] for details).

Again, the virtual feedback (42) cannot be implemented. Therefore, applying the transformation (36) to (34) yields the representation

$$\dot{e}_0(t) = (F_0 + B_0 K_0)e_0(t) + B_0\tilde{e}_{x,1}(0, t) \tag{43a}$$

$$\tilde{e}_{x,2}(0, t) = Q_0\tilde{e}_{x,1}(0, t) \tag{43b}$$

$$\partial_t\tilde{e}_x(z, t) = \Lambda(z)\partial_z\tilde{e}_x(z, t) + A_0(z)\tilde{e}_{x,1}(0, t) \tag{43c}$$

$$\begin{aligned}\tilde{e}_{x,1}(1, t) = \bar{c}_1(t, e_1(t), \mathcal{T}^{-1}[\tilde{e}_{x,2}(t)](1) + E_2^T N(1)e_0(t)) \\ - E_1^T N(1)e_0(t) - \int_0^1 E_1^T K_I(1, z)\tilde{e}_x(z, t)\,\mathrm{d}z \\ + \bar{D}(t, e_1(t), \mathcal{T}^{-1}[\tilde{e}_{x,2}(t)](1) + E_2^T N(1)e_0(t))u(t)\end{aligned} \tag{43d}$$

$$\begin{aligned}\dot{e}_1(t) = \bar{f}_1(t, e_1(t), \mathcal{T}^{-1}[\tilde{e}_{x,2}(t)](1) + E_2^T N(1)e_0(t)) \\ + \bar{B}(t, e_1(t), \mathcal{T}^{-1}[\tilde{e}_{x,2}(t)](1) + E_2^T N(1)e_0(t))u(t)\end{aligned} \tag{43e}$$

of the ODE-PDE-ODE system. Therein, as illustrated in Fig. 3, one has a cascade of a PDE and an ODE at $z = 0$, with the former finite-time stable if $\tilde{e}_{x,1}(1, t) = 0$. In the last step, the input $u(t)$ has to be chosen such that the (remaining) e_1-dynamics is stabilized in its origin, thus also compensating the potentially destabilizing terms arising from $e_0(t)$ and $\tilde{e}_x(z, t)$ (see dashed lines in Fig. 3), that have been moved to the boundary at $z = 1$ by the transformations (30) and (36).

Remark 4 The (virtual) backstepping-based feedback controller (42) was chosen to obtain $\tilde{e}_{x,1}(1, t) = 0$ and by that a PDE (target) subsystem, that is a cascade of multiple transport equations, from which finite-time stability is easily inferred. However, such a controller is not delay-robust (e.g. [3, 12]), i.e., even small delays in the actuation may result in an unstable closed loop. Though that is not the focus here, the modular design in this contribution allows one to simply replace the PDE target system that follows from using (42) in (40) by any other that is delay-robust. See [3] and other publications by the same authors for potential (delay-robust) target dynamics and the corresponding proof of stability, that is more involved if one does not have a cascade of transport equations.

2.5 Stabilization of the ODE Subsystem at $z = 1$

In the boundary system at $z = 1$, (43d) takes the role of an output equation:

$$\tilde{e}_{x,1}(1, t) = y(t). \tag{44}$$

Thus (43d)–(43e) can be written as the time-varying, affine-input nonlinear state representation

$$\dot{e}_1(t) = \tilde{f}_1(t, e_1(t)) + \tilde{B}(t, e_1(t))u(t) \tag{45a}$$

$$y(t) = \tilde{c}_1(t, e_1(t)) + \tilde{D}(t, e_1(t))u(t). \tag{45b}$$

The definition of the functions $\tilde{f}_1$, $\tilde{B}$, $\tilde{c}_1$ and $\tilde{D}$ directly follows from a comparison between (45) and (43d)–(43e). Importantly, the time argument t of the functions models not only their dependence on the reference trajectory (15) (see also (29)) but also on the states $e_0(t)$ and $\tilde{e}_x(\cdot, t)$ of those subsystems,[4] that are already stabilized (if $\tilde{e}_{x,1}(1, t) = 0$). The motivation for the representation (45) is simple: Provided $u(t)$ is found such that $y(t) = \tilde{e}_{x,1}(1, t) \to 0$ as $t \to \infty$ (which was assumed for the stabilization of the PDE in the previous section), $e_0(t)$ and $\tilde{e}_x(z, t)$ can be considered as a disturbance input w.r.t. the e_1-dynamics. This translates to an output zeroing problem[5] (see, e.g., [19] for time-invariant systems).

Remark 5 Note that the focus of this contribution is not on the design of a particular controller but on the systematic and modular design approach. As such, the (virtual) backstepping controller (42) could easily be replaced by any other feedback, stabilizing (40) (see Remark 4). The same holds for the control of the nonlinear ODE (45). Therefore, the design of a feedback achieving exact linearization, such that $\lim_{t\to\infty} y(t) = 0$, is only sketched in the following. Neither a final theorem on the overall control result nor an exhaustive, mathematically rigorous discussion on the stability of the closed-loop system is presented.

By the assumption of a constant relative degree $\{r_1, \ldots, r_p\}$ of (14e) w.r.t. (14d) in Assumption A3, the same holds for (45a) w.r.t. (45b). Thus, ρ components of $u(t)$ can directly be chosen to guarantee

$$y_i(t) = 0, \qquad i \in \{1, \ldots, p | r_i = 0\} \tag{46}$$

---

[4] See [15] for an alternative notation, that explicitly highlights the dependence of $\tilde{f}_1$, $\tilde{B}$, $\tilde{c}_1$ and $\tilde{D}$ on $e_0(t)$ and $\tilde{e}_x(\cdot, t)$ in the linear case.

[5] In Remark 4, the virtual input (35) $\tilde{e}_{x,1}(1, t)$ would be a function of $\tilde{e}_x(\cdot, t)$ (and not zero) in such a way, that the controlled PDE subsystem is delay-robust stable (see again [3]). Therefore, in this third design step, one would want $y(t)$ to converge not to zero but to this specific function.

for those $p - \rho$ components $y_i(t)$ of $y(t)$ with relative degree $r_i = 0$ (cf. (18)). Similar to [19], in general, or to [15] in the context of the stabilization of linear ODE-PDE-ODE systems, new coordinates

$$\xi_i^1(t) = y_i(t) = \phi_i(t, e_1(t)), \qquad i \in \{1, \ldots, p | r_i > 0\} \tag{47}$$

are introduced for the remaining ρ components. Then, (45a) can be written in Byrnes-Isidori normal form

$$\dot{\xi}_i^j(t) = \xi_i^{j+1}(t), \qquad j = 1, \ldots, r_i - 1 \tag{48a}$$
$$\dot{\xi}_i^{r_i}(t) = \alpha_i(t, \xi(t), \eta(t)) + \beta_i(t, \xi(t), \eta(t))u(t) \tag{48b}$$
$$\dot{\eta}(t) = \pi(t, \xi(t), \eta(t), u(t)) \tag{48c}$$

for $i \in \{1, \ldots, p | r_i > 0\}$, with[6]

$$\xi(t) = \begin{bmatrix} \xi_1(t) \\ \vdots \\ \xi_n(t) \end{bmatrix} \in \mathbb{R}^{\gamma}, \qquad \xi_i(t) = \begin{bmatrix} \xi_i^1(t) \\ \vdots \\ \xi_i^{r_i}(t) \end{bmatrix} \in \mathbb{R}^{r_i}, \tag{49}$$

$\gamma = \sum_{i=1}^{p} r_i$, and where $\eta(t) \in \mathbb{R}^{n_1 - \gamma}$ is defined such that $\Psi : w_1 \mapsto (\xi, \eta)$ is a diffeomorphism (see [19]). Based on (48), a nonlinear feedback is found to stabilize the origin $\xi(t) = 0$, thus defining $p - \rho$ components of $u(t)$. Using exact linearization by feedback, together with the input components chosen in the context of (46), the controller can be written as

$$u(t) = k(t, \xi(t), \eta(t)). \tag{50}$$

In fact, by its first argument t, the function k (just as α_i, β_i and π in (48)) depends not only on the reference trajectory (15) of the ODE-PDE-ODE system (14) but also on $e_0(t)$ and $\tilde{e}_x(\cdot, t)$ (cf. (43d)–(43e) and (45)). Moreover, it involves time derivatives of boundary values[7] $\tilde{e}_{x,1}(0, t)$ and $\tilde{e}_{x,2}(1, t)$ due to (48). Thus, (50) takes the form

$$u(t) = \bar{k}(t, \xi(t), \eta(t)), e_0(t), \tilde{e}_x(\cdot, t), \dot{\tilde{e}}_{x,1}(0, t), \dot{\tilde{e}}_{x,2}(1, t), \ldots) \tag{51}$$

where the time argument is preserved to highlight the dependence on (15).

[6] With a slight abuse of notation, $\xi(t)$ is meant to include only those $\xi_i(t)$ with $r_i > 0$.
[7] Such derivatives of the boundary values were also present in the controllers in [7, 9].

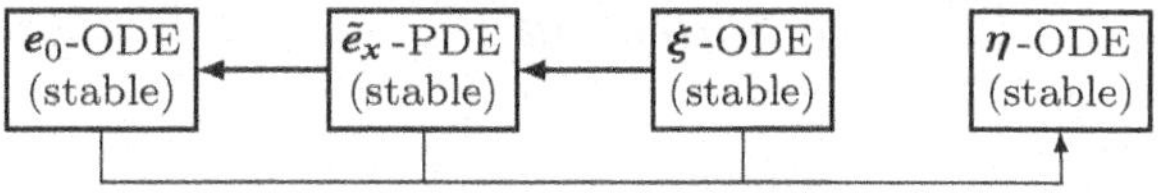

Fig. 4 Structure of the closed-loop ODE-PDE-ODE system (52)

Using (50) (or equivalently (51)) in (48), the closed-loop system

$$\dot{e}_0(t) = (F_0 + B_0 K_0)e_0(t) + B_0\tilde{e}_{x,1}(0,t) \tag{52a}$$

$$\tilde{e}_{x,2}(0,t) = Q_0\tilde{e}_{x,1}(0,t) \tag{52b}$$

$$\partial_t\tilde{e}_x(z,t) = \Lambda(z)\partial_z\tilde{e}_x(z,t) + A_0(z)\tilde{e}_{x,1}(0,t) \tag{52c}$$

$$\tilde{e}_{x,1}(1,t) = \begin{cases} 0, & i \in \{1,\ldots,p|r_i = 0\} \\ \xi_i^1(t), & i \in \{1,\ldots,p|r_i > 0\} \end{cases} \tag{52d}$$

$$\dot{\xi}(t) = P\xi(t) \tag{52e}$$

$$\dot{\eta}(t) = \pi(t,\xi(t),\eta(t),k(t,\xi(t),\eta(t))) \tag{52f}$$

follows from (43), (48) as well as (46) and (47) in view of (44). In (52e), the matrix P is Hurwitz by design of (50). Hence, the ξ-dynamics is exponentially stable. As $\lim_{t\to\infty}\tilde{e}_{x,1}(1,t) = 0$, exponential stability of the PDE subsystem (52b)–(52d) is ensured (see Sect. 2.4), which in turn guarantees exponential decay of $e_0(t)$. From that, $\eta(t) \to 0$ for $t \to \infty$ follows by Assumption A3. Thus, stability of the closed loop (52) is inferred from its structure (see Fig. 4) and the stability properties of the respective subsystems. By the invertibility of the transformation (30) and (36) as well as of Ψ, consequently, one has asymptotic converge of $w_0(t)$, $x(z,t)$ and $w_1(t)$ towards their respective reference trajectories (see (27)).

3 Parabolic ODE-PDE-ODE Systems

The main difference between Sect. 2 and the subsequently design of a tracking controller for a bidirectionally coupled ODE-PDE-ODE system is the parabolic type of PDE under consideration. It will be shown, that the same essential three steps as in Sects. 2.3–2.5 allows one to stabilize a reference trajectory. The results in [15] are extended not only by replacing the linear actuated boundary by a nonlinear one, but also by (slightly) generalizating the boundary operators. Whereas the backstepping results in [10] allow for the inclusion of mixed Dirichlet and Robin BCs, in the following, only Robin BCs are considered (with Neumann conditions in [15]), to avoid unnecessary technicalities. Overall, the focus is on extensions w.r.t. existing results and differences to the hyperbolic case in Sect. 2, with additional remarks on the modular design philosophy that, though also valid for hyperbolic ODE-PDE-ODE systems, might have been confusing if stated earlier.

3.1 Problem Statement

The parabolic PDEs in

$$\partial_t x(z,t) = \Lambda(z)\partial_z^2 x(z,t) + A(z)x(z,t) \tag{53}$$

only take diffusion and reaction into account, as a convection term could be removed by a Hopf-Cole type transformation of the PDE state $x(z,t) \in \mathbb{R}^n$ (see, e.g., [10]). The diagonal elements $\lambda_i \in C^2[0,1]$, $i = 1,\dots,n$ of the diffusion matrix $\Lambda(z) = \mathrm{diag}(\lambda_1(z),\dots,\lambda_n(z))$ satisfy $\lambda_1(z) > \cdots > \lambda_n(z)$. No special structure is assumed for the coupling matrix $A(z)$, whose entries $a_{ij} \in C^1[0,1]$ model the reaction.

The linear system

$$\dot{w}_0(t) = F_0 w_0(t) + B_0 x(0,t) \tag{54a}$$
$$\partial_z x(0,t) = Q_0 x(0,t) + C_0 w_0(t) \tag{54b}$$

at the unactuated boundary at $z = 0$ consists of an ODE system for the state $w_0(t) \in \mathbb{R}^{n_0}$ and a Robin BC. In the latter, the matrix $Q_0 = \mathrm{diag}(q_1,\dots,q_n)$ is diagonal, an assumption that is frequently imposed in this context and is often met in a chemical or biochemical setup (see, e.g., [10] and references therein). The w_0-ODE and the PDE (via the boundary at $z = 0$) are bidirectionally coupled through the matrices B_0 and C_0. The Robin-type BC

$$\partial_z x(1,t) = c_1(w_1(t), x(1,t)) + D(w_1(t), x(1,t))u(t) \tag{55}$$

is considered at the actuated boundary, where the ODE state $w_1(t) \in \mathbb{R}^{n_1}$ satisfies the nonlinear dynamics

$$\dot{w}_1(t) = f_1(w_1(t), x(1,t)) + B(w_1(t), x(1,t))u(t). \tag{56}$$

In order to have full boundary actuation at $z = 1$, the dimension of the control input $u(t) \in \mathbb{R}^n$ equals the dimension of the PDE state. By that, the dimensions of all vectors and matrices are implicitly defined. Note that as in the hyperbolic case, the coupling between the w_1-ODE and the PDE, specifically the BC at $z = 1$, is due to the nonlinear functions f_1, B, c_1 and D.

Rearranging (53)–(56), it is apparent from

$$\dot{w}_0(t) = F_0 w_0(t) + B_0 x(0,t) \tag{57a}$$
$$\partial_z x(0,t) = Q_0 x(0,t) + C_0 w_0(t) \tag{57b}$$
$$\partial_t x(z,t) = \Lambda(z)\partial_z^2 x(z,t) + A(z)x(z,t) \tag{57c}$$
$$\partial_z x(1,t) = c_1(w_1(t), x(1,t)) + D(w_1(t), x(1,t))u(t) \tag{57d}$$
$$\dot{w}_1(t) = f_1(w_1(t), x(1,t)) + B(w_1(t), x(1,t))u(t) \tag{57e}$$

that the parabolic ODE-PDE-ODE system with the nonlinear actuator dynamics and ICs $x(z,0) = x_0(z) \in \mathbb{R}^n$, $z \in [0,1]$, $w_0(0) = w_{0,0} \in \mathbb{R}^{n_0}$ as well as $w_1(0) = w_{1,0} \in \mathbb{R}^{n_1}$ is given in strict feedback form (cf. (14) and Fig. 1). As such, assumptions analogous to A1–A3 are sufficient to design a tracking controller that stabilizes a reference trajectory (15).

- **Assumption B1:** A reference trajectory (15) is known that satisfies (57).
- **Assumption B2:** The pair (F_0, B_0) is stabilizable.
- **Assumption B3:** The w_1-dynamics (57e) has a constant vector relative degree $\{r_1, \ldots, r_n\}$ w.r.t. (57d), with $\sum_{i=1}^{n} r_i \leq n_1$. The zero dynamics of order $n_1 - \sum_{i=1}^{n} r_i$ is asymptotically stable in a neighborhood of the reference.

As in the hyperbolic case, error coordinates (27) are introduced in order to rewrite (57) as the ODE-PDE-ODE (error) system

$$\dot{e}_0(t) = F_0 e_0(t) + B_0 e_x(0,t) \tag{58a}$$

$$\partial_z e_x(0,t) = Q_0 e_x(0,t) + C_0 e_0(t) \tag{58b}$$

$$\partial_t e_x(z,t) = \Lambda(z)\partial_z^2 e_x(z,t) + A(z)e_x(z,t) \tag{58c}$$

$$\partial_z e_x(1,t) = \bar{c}_1(t, e_1(t), e_x(1,t)) + \bar{D}(t, e_1(t), e_x(1,t))u(t) \tag{58d}$$

$$\dot{e}_1(t) = \bar{f}_1(t, e_1(t), e_x(1,t)) + \bar{B}(t, e_1(t), e_x(1,t))u(t). \tag{58e}$$

with $\bar{c}_1$, $\bar{D}$, $\bar{f}_1$ and $\bar{B}$ defined similarly to (29).

3.2 Stabilization of the ODE Subsystem at $z = 0$

In the first design step, the boundary value $e_x(0,t)$ is considered as a virtual control input for the e_0-ODE (58a). It is irrelevant, which method is used to find the linear (virtual) feedback

$$e_x(0,t) = K_0 e_0(t). \tag{59}$$

The Assumption B2 of stabilizability of the pair (F_0, B_0) ensure the existence of a matrix K_0 such that $F_0 + B_0 K_0$ is Hurwitz.

Remark 6 Note that as in the hyperbolic case, the ODE (58a) could also be stabilized by a nonlinear (virtual) feedback. However, this would require a nonlinear state transformation (w.r.t. $e_0(t)$) instead of (30) or (60). Not only would that destroy the linear structure of the $(e_0, \bar{e}_x)$-dynamics, it would also make it significantly more difficult to move any impact of the ODE state $e_0(t)$ on the PDE to the actuated boundary (in order to obtain a cascaded structure for the final closed loop).

The virtual feedback (59) motivates the state transformation

$$\bar{e}_x(z,t) = e_x(z,t) - N(z)e_0(t) \tag{60}$$

with $N(z) \in \mathbb{R}^{n \times n_0}$, where at least $N(0) = K_0$ so that $e_x(0,t) \to K_0 e_0(t)$ if $\bar{e}_x(z,t)$ goes to zero. In order to move any (potentially destabilizing) impact of the ODE state $e_0(t)$ on the PDE subsystem to the actuated boundary at $z = 1$, the matrix $N(z)$ has to satisfy the IVP

$$\Lambda(z)N''(z) + A(z)N(z) - N(z)(F_0 + B_0 K_0) = 0 \tag{61a}$$
$$N(0) = K_0 \tag{61b}$$
$$N'(0) = C_0 + Q_0 K_0. \tag{61c}$$

Lemma 2 *The IVP (61) admits a unique $C^2[0,1]$-solution $N(z)$.*

Similar to the proof of Lemma 1, the solvability of (61) is verified in Appendix 2 by considering an eigenvalue problem for $F_0 + B_0 K_0$ and explicitly solving (61).

By the transformation (60) with the solution $N(z)$ of (61), the ODE-PDE-ODE system (58) is mapped into

$$\dot{e}_0(t) = (F_0 + B_0 K_0)e_0(t) + B_0 \bar{e}_x(0,t) \tag{62a}$$
$$\partial_z \bar{e}_x(0,t) = Q_0 \bar{e}_x(0,t) \tag{62b}$$
$$\partial_t \bar{e}_x(z,t) = \Lambda(z)\partial_z^2 \bar{e}_x(z,t) + A(z)\bar{e}_x(z,t) - N(z)B_0 \bar{e}_x(0,t) \tag{62c}$$
$$\partial_z \bar{e}_x(1,t) = \bar{c}_1(t, e_1(t), \bar{e}_x(1,t) + N(1)e_0(t)) - N'(1)e_0(t) + \bar{D}(t, e_1(t), \bar{e}_x(1,t) + N(1)e_0(t))u(t) \tag{62d}$$
$$\dot{e}_1(t) = \bar{f}_1(t, e_1(t), \bar{e}_x(1,t) + N(1)e_0(t)) + \bar{B}(t, e_1(t), \bar{e}_x(1,t) + N(1)e_0(t))u(t). \tag{62e}$$

In these coordinates, the parabolic system shares the same coupling structure as the hyperbolic one illustrated in Fig. 2. Specifically, the ODE (62c) is exponentially stable provided $\bar{e}_x(0,t) = 0$ and the only coupling remaining from the e_0-dynamics to the PDE is at the actuated boundary, where it will be compensated later.

3.3 Stabilization of the PDE Subsystem

The stabilization of the PDE subsystem makes use of the boundary value

$$\partial_z \bar{e}_x(1,t) = u_{\text{virtual}}(t) \tag{63}$$

as a virtual input. In [10], a backstepping controller[8] is designed for a parabolic system of the form (62b), (62c), (63). By the backstepping transformation

$$\tilde{e}_x(z,t) = \bar{e}_x(z,t) - \int_0^z K(z,\zeta)\bar{e}_x(\zeta,t)\,\mathrm{d}\zeta = \mathcal{T}[\bar{e}_x(t)](z), \tag{64}$$

where the kernel $K(z,\zeta) \in \mathbb{R}^{n\times n}$ satisfies

$$\Lambda(z)\partial_z^2 K(z,\zeta) - \partial_\zeta^2\big(K(z,\zeta)\Lambda(\zeta)\big) = K(z,\zeta)\big(A(\zeta) + \mu_c I_n\big) \tag{65a}$$

$$\partial_\zeta K(z,0)\Lambda(0) + K(z,0)\big(\Lambda'(0) - \Lambda(0)Q_0\big) = -A_0(z) - \mathcal{T}[NB_0](z) \tag{65b}$$

$$\Lambda(z)\big(\tfrac{\mathrm{d}}{\mathrm{d}z}K(z,z) + \partial_z K(z,z)\big) + K(z,z)\Lambda'(z) + \partial_\zeta K(z,z)\Lambda(z) = -\big(A(z) + \mu_c I_n\big) \tag{65c}$$

$$K(z,z)\Lambda(z) - \Lambda(z)K(z,z) = 0 \tag{65d}$$

$$K(0,0) = Q_0, \tag{65e}$$

the PDE subsystem is mapped into

$$\partial_z\tilde{e}_x(0,t) = 0 \tag{66a}$$

$$\partial_t\tilde{e}_x(z,t) = \Lambda(z)\partial_z^2\tilde{e}_x(z,t) - \mu_c\tilde{e}_x(z,t) + A_0(z)\tilde{e}_x(0,t) \tag{66b}$$

$$\partial_z\tilde{e}_x(1,t) = u_{\text{virtual}}(t) - K_I(1,1)\tilde{e}_x(1,t) - \int_0^1 \partial_z K_I(1,z)\tilde{e}_x(z,t)\,\mathrm{d}z, \tag{66c}$$

with the kernel $K_I(z,\zeta) \in \mathbb{R}^{n\times n}$ of the inverse transformation (37) (see also the reciprocity (38)).

The solution $K(z,\zeta)$ of the kernel equations (65) is shown in [10] to be piecewise $C^2[0,1]$. Moreover, by (65b), the kernel defines the strictly lower triangular matrix

$$A_0(z) = \begin{bmatrix} 0 & \cdots & \cdots & 0 \\ a_{2,1}(z) & \ddots & & \vdots \\ \vdots & \ddots & \ddots & \vdots \\ a_{n,1}(z) & \cdots & a_{n,n-1}(z) & 0 \end{bmatrix}. \tag{67}$$

By this structure and the (virtual) feedback

$$u_{\text{virtual}}(t) = K_I(1,1)\tilde{e}_x(1,t) + \int_0^1 \partial_z K_I(1,z)\tilde{e}_x(z,t)\,\mathrm{d}z, \tag{68}$$

one has a cascade of n (scalar) parabolic PDEs in (66). The design parameter μ_c can easily be chosen to ensure exponential stability of the PDE subsystem (66) with $\partial_z\tilde{e}_x(1,t) = 0$ (see [10]). However, in the end, (68) cannot be implemented.

[8] One could alternatively use any other control strategy, limited as they are for these types of systems.

Overall, the backstepping transformation (64) maps the ODE-PDE-ODE system (62) into the form

$$\dot{e}_0(t) = (F_0 + B_0 K_0)e_0(t) + B_0\tilde{e}_x(0, t) \tag{69a}$$

$$\partial_z \tilde{e}_x(0, t) = 0 \tag{69b}$$

$$\partial_t \tilde{e}_x(z, t) = \Lambda(z)\partial_z^2 \tilde{e}_x(z, t) - \mu_c \tilde{e}_x(z, t) + A_0(z)\tilde{e}_x(0, t) \tag{69c}$$

$$\begin{aligned}\partial_z \tilde{e}_x(1, t) = {} & \bar{c}_1(t, e_1(t), \mathcal{T}^{-1}[\tilde{e}_x(t)](1) + N(1)e_0(t)) - N'(1)e_0(t) \\ & - K_I(1, 1)\tilde{e}_x(1, t) - \int_0^1 \partial_z K_I(1, z)\tilde{e}_x(z, t)\,\mathrm{d}z \\ & + \bar{D}(t, e_1(t), \mathcal{T}^{-1}[\tilde{e}_x(t)](1) + N(1)e_0(t))u(t)\end{aligned} \tag{69d}$$

$$\begin{aligned}\dot{e}_1(t) = {} & \bar{f}_1(t, e_1(t), \mathcal{T}^{-1}[\tilde{e}_x(t)](1) + N(1)e_0(t)) \\ & + \bar{B}(t, e_1(t), \mathcal{T}^{-1}[\tilde{e}_x(t)](1) + N(1)e_0(t))u(t)\end{aligned} \tag{69e}$$

(cf. (66) and (37)). This representation is structurally identical to the one illustrated in Fig. 3. In particular, (69) is a cascade of the PDE subsystem and the e_0-ODE, both of which are exponentially stable provided $\partial_z \tilde{e}_x(1, t) = 0$. Therefore, in order to obtain a cascaded structure of all three subsystems in the next (and last) design step, $u(t)$ is chosen to stabilize the e_1-ODE and to remove the coupling from the (e_0, e_x)-dynamics to the ODE at $z = 1$.

3.4 Stabilization of the ODE Subsystem at $z = 1$

The stabilization of the e_1-system is basically identical to the hyperbolic case in Sect. 2.5. Instead of (44),

$$\partial_z \tilde{e}_x(1, t) = y(t) \tag{70}$$

takes the role of an output, allowing one to rewrite (69e) and (69d) in terms of the affine-input nonlinear state representation (45). The design only differs in $y(t) \in \mathbb{R}^p$ for the hyperbolic system and $y(t) \in \mathbb{R}^n$ in (70). Again, $u(t)$ is found such that $y(t) \to 0$ for $t \to \infty$ using the Byrnes-Isidori normal form and exact feedback linearization. In the end, a feedback (50) depends on the state $(e_0(t), \tilde{e}_x(z, t), e_1(t))$ of the ODE-PDE-ODE (error) system (69), as well as time derivatives of the boundary values $\tilde{e}_x(0, t)$ and $\tilde{e}_x(1, t)$. The stability of the closed loop follows from its structure in Fig. 4 and the stability properties of each of the subsystems.

4 Conclusion

The modular and systematic design of successive (virtual) controllers, that is presented in this contribution for hyperbolic and parabolic ODE-PDE-ODE systems (with nonlinear actuator dynamics) could easily be adjusted to use other control tools known for PDEs or ODEs. Especially the control strategy employed in the third design step in Sects. 2.5 and 3.4 could be replaced, e.g. by using a sliding mode or high gain controller. Moreover, as already suggested in [15], any ODE-PDE-ODE system in strict feedback form could be treated similar to the systems in Sects. 2 and 3, always provided that controllers are known for each of the subsystems. One such an example is the generalization to ODE-PIDE-ODE systems, i.e. those with a partial integro-differential equation (PIDE) instead of a PDE.[9] Moreover, the state $w_0(t)$ of the distal ODE (and $e_0(t)$ respectively) affecting the actuated boundary system would not require any essential modifications of the presented design.

One would naturally also be interested in the control of PDE-ODE-PDE systems or even more complex coupled structures of ODE and PDE subsystems. However, this generalization is not straightforward as the problem of controllability comes into play, even more so than is already indicated in the context of Assumptions A1 and A3 (see also Remark 1). In fact, a PDE-ODE-PDE system could also be regarded as an underactuated PDE-ODE system, the control of which is more challenging than the case of full boundary actuation considered here (see, e.g., [2]). As ODE systems in strict feedback form are also differentially flat, or controllable if they are linear, controllability (in some appropriate sense) is essential for the control design, not only for coupled PDE-ODE-PDE systems. The controllability aspect as well as the modular and systematic design of state observers will be the focus of future works. Specifically, the observer design for the case of entirely linear ODE-PDE-ODE systems can be derived from the controller design by duality.

Acknowledgements This work has partially been supported by the FFG, Contract No. 854184: "Pro^{2}Future is funded within the Austrian COMET Program Competence Centers for Excellent Technologies under the auspices of the Austrian Federal Ministry for Climate Action, Environment, Energy, Mobility, Innovation and Technology, the Austrian Federal Ministry for Digital and Economic Affairs and of the Provinces of Upper Austria and Styria. COMET is managed by the Austrian Research Promotion Agency FFG."

[9] A backstepping controller for linear parabolic PIDEs is given in [10].

Appendix 1

In order to solve the IVP (33), let μ_i, $i = 1, \ldots, r$ be the eigenvalues of the matrix $F_0 + B_0 K_0$. The corresponding generalized (right) eigenvectors $\varphi_{i(k)}$ of rank k satisfy

$$(F_0 + B_0 K_0)\varphi_{i(1)} = \mu_i \varphi_{i(1)} \tag{71a}$$

$$(F_0 + B_0 K_0)\varphi_{i(k)} = \mu_i \varphi_{i(k)} + \varphi_{i(k-1)}, \qquad k = 2, \ldots, l_i, \tag{71b}$$

where $\sum_{i=1}^{r} l_i = n_0$. Consequently, r denotes the number of eigenvectors (of rank 1) and l_i the dimension of the respective eigenspace.

Multiplying (33) by $\varphi_{i(k)}$ from the right and introducing $\mathbb{R}^n \ni n_{i(k)}(z) = N(z)\varphi_{i(k)}$, the matrix-valued IVP is reduced to n_0 vector-valued ones:

$$n'_{i(k)}(z) = \bar{A}_i(z) n_{i(k)}(z) + \Lambda^{-1}(z) n_{i(k-1)}(z) \tag{72a}$$

$$n_{i(k)}(0) = \begin{bmatrix} K_0 \\ C_0 + Q_0 K_0 \end{bmatrix} \varphi_{i(k)}, \tag{72b}$$

with $k = 1, \ldots, l_i$, $i = 1, \ldots, r$, $n_{i(0)}(z) = 0$ and

$$\bar{A}_i(z) = \Lambda^{-1}(z)[\mu_i I_n - A(z)]. \tag{73}$$

As $\Lambda, A \in (C^1[0, 1])^{n \times n}$ (see Sect. 2.1), the elements of $\bar{A}_i$ are $C^1[0, 1]$-functions, too. With that, well-known results for linear time-varying systems guarantee the existence of the (piecewise) $C^1[0, 1]$-solution

$$n_{i(k)}(z) = \Phi_i(z, 0) n_{i(k)}(0) + \int_0^z \Phi_i(z, \zeta) \Lambda^{-1}(\zeta) n_{i(k-1)}(\zeta)\, d\zeta, \tag{74}$$

where $\Phi_i(z, \zeta) : [0, 1] \times [0, 1] \to \mathbb{C}^{n \times n}$ denotes the fundamental matrix satisfying

$$\partial_z \Phi_i(z, \zeta) = \bar{A}_i(z) \Phi_i(z, \zeta), \qquad \Phi_i(z, z) = I_n \tag{75}$$

(e.g. [21]).

The solutions $n_{i(k)}(z)$ in (74) are evaluated for each $i = 1, \ldots, r$ with increasing $k = 1, \ldots, l_i$. Combining them in $\Xi = [n_{1(1)}, \ldots, n_{1(l_1)}, n_{2(1)}, \ldots, n_{r(l_r)}]$ and defining the matrix $\Psi = [\varphi_{1(1)}, \ldots, \varphi_{1(l_1)}, \varphi_{2(1)}, \ldots, \varphi_{r(l_r)}]$ of all (generalized) eigenvectors, the $C^1[0, 1]$-solution

$$N(z) = \Xi(z)\Psi^{-1} \tag{76}$$

of (33) is found.

Appendix 2

The solution of the IVP (61) makes use of the eigenvalues μ_i and generalized eigenvectors $\varphi_{i(k)}$ of the matrix $F_0 + B_0 K_0$, with $i = 1, \ldots, r$, $k = 1, \ldots, l_i$ and $\sum_{i=1}^{r} l_i = n_0$ (cf. Appendix 1 and in particular (71)). Then, introducing

$$m_{i(k)}(z) = \begin{bmatrix} n_{i(k)}(z) \\ n'_{i(k)}(z) \end{bmatrix} = \begin{bmatrix} N(z)\varphi_{i(k)} \\ N'(z)\varphi_{i(k)} \end{bmatrix} \in \mathbb{R}^{2n}, \tag{77}$$

the multiplication of (61) by $\varphi_{i(k)}$ from the right results in an IVP

$$m'_{i(k)}(z) = \bar{A}_i(z) m_{i(k)}(z) + D(z) m_{i(k-1)}(z) \tag{78a}$$

$$m_{i(k)}(0) = \begin{bmatrix} K_0 \\ C_0 + Q_0 K_0 \end{bmatrix} \varphi_{i(k)} \tag{78b}$$

similar to (72), with

$$\bar{A}_i(z) = \begin{bmatrix} 0 & I_n \\ \Lambda^{-1}(z)(\mu_i I_n - A(z)) & 0 \end{bmatrix} \quad \text{and} \quad D(z) = \begin{bmatrix} 0 & 0 \\ \Lambda^{-1}(z) & 0 \end{bmatrix}. \tag{79}$$

The fundamental matrix $\Phi_i(z, \zeta) : [0, 1] \times [0, 1] \to \mathbb{C}^{2n \times 2n}$ associated with $\bar{A}_i(z)$ satisfies

$$\partial_z \Phi_i(z, \zeta) = \bar{A}_i(z) \Phi_i(z, \zeta), \qquad \Phi_i(z, z) = I_{2n}. \tag{80}$$

With it, the solution of (78) reads

$$m_{i(k)}(z) = \Phi_i(z, 0) m_{i(k)}(0) + \int_0^z \Phi_i(z, \zeta) D(\zeta) m_{i(k-1)}(\zeta)\, d\zeta. \tag{81}$$

Its elements stem from $C^1[0, 1]$ as the same holds true for those of Λ and A (see Sect. 3.1). By $n'_{i(k)} \in (C^1[0, 1])^n$ (cf. (77)), the $C^2[0, 1]$-solution $N(z)$ of (61) is implied by (76), with matrices Ξ and Ψ similar to Appendix 1.

References

1. Anfinsen, H., Aamo, O.: Stabilization of a linear hyperbolic PDE with actuator and sensor dynamics. Automatica **95**, 104–111 (2018)
2. Auriol, J., Aarsnes, U., Di Meglio, F., Shor, R.: Robust control design of underactuated 2×2 PDE-ODE-PDE systems. IEEE Control Syst. Lett. **5**, 469–474 (2021)
3. Auriol, J., Bribiesca-Argomedo, F.: Delay-robust stabilization of an $n + m$ hyperbolic PDE-ODE system. In: 2019 IEEE 58th Conference on Decision and Control (CDC), pp. 4964–4970 (2019)

4. Balogh, A., Krstic, M.: Infinite dimensional backstepping-style feedback transformations for a heat equation with an arbitrary level of instability. Eur. J. Control **8**, 165–175 (2002)
5. Bastin, G., Coron, J.M.: Stability and Boundary Stabilization of 1-D Hyperbolic Systems. Springer (2016)
6. Bou Saba, D., Bribiesca-Argomedo, F., Di Loreto, M., Eberhard, D.: Backstepping stabilization of 2×2 linear hyperbolic PDEs coupled with potentially unstable actuator and load dynamics. In: Proceedings of 56th IEEE Conference on Decision and Control (CDC), pp. 2498–2503 (2017)
7. Deutscher, J., Gehring, N.: Output feedback control of coupled linear parabolic ODE-PDE-ODE systems **66**, 4668–4683 (2021)
8. Deutscher, J., Gehring, N., Kern, R.: Backstepping control of linear 2×2 hyperbolic systems with dynamic boundary conditions. In: 20th IFAC World Congress, Toulouse, France, July 9-14, 2017 (2017)
9. Deutscher, J., Gehring, N., Kern, R.: Output feedback control of general linear heterodirectional hyperbolic ODE-PDE-ODE systems. Automatica **95**, 472–480 (2018)
10. Deutscher, J., Kerschbaum, S.: Backstepping control of coupled linear parabolic PIDEs with spatially varying coefficients. IEEE Trans. Autom. Control **63**, 4218–4233 (2018)
11. Di Meglio, F., Bribiesca Argomedo, F., Hu, L., Krstic, M.: Stabilization of coupled linear heterodirectional hyperbolic PDE-ODE systems. Automatica **87**, 281–289 (2018)
12. Di Meglio, F., Lamare, P.O., Aarsnes, U.: Robust output feedback stabilization of an ODE-PDE-ODE interconnection. Automatica **119**, 109059 (2020)
13. Fliess, M., Lévine, J., Martin, P., Rouchon, P.: Flatness and defect of non-linear systems: introductory theory and examples. Int. J. Control **61**, 1327–1361 (1995)
14. Gehring, N.: Ein systematischer Backstepping-Zugang zur Regelung gekoppelter ODE-PDE-ODE-Systeme. at – Automatisierungstechnik, pp. 654–666 (2020)
15. Gehring, N.: A systematic design of backstepping-based state feedback controllers for ODE-PDE-ODE systems. IFAC-PapersOnLine **54**(9), 410–415 (2021)
16. Hu, L., Vazquez, R., Di Meglio, F., Krstic, M.: Control of homodirectional and general heterodirectional linear coupled hyperbolic PDEs. IEEE Trans. Autom. Control **61**, 3301–3314 (2016)
17. Hu, L., Vazquez, R., Di Meglio, F., Krstic, M.: Boundary exponential stabilization of 1-dimensional inhomogeneous quasi-linear hyperbolic systems. SIAM J. Control Optim. **57**, 963–998 (2019)
18. Irscheid, A., Gehring, N., Rudolph, J.: Trajectory tracking control for a class of 2×2 hyperbolic PDE-ODE systems. IFAC-PapersOnLine **54**(9), 416–421 (2021)
19. Isidori, A.: Nonlinear Control Systems. Springer (1995)
20. Jakobsen, H.: Chemical Reactor Modeling. Springer (2008)
21. Kailath, T.: Linear Systems. Prentice Hall (1980)
22. Khalil, H.: Nonlinear Systems. Prentice Hall (1996)
23. Knüppel, T., Woittennek, F.: Control design for quasi-linear hyperbolic systems with an application to the heavy rope. IEEE Trans. Autom. Control **60**, 5–18 (2015)
24. Krstic, M., Smyshlyaev, A.: Boundary Control of PDEs. SIAM (2008)
25. Mohammadi, L., Aksikas, I., Dubljevic, S., Fraser Forbes, J.: Optimal boundary control of coupled parabolic PDE-ODE systems using infinite-dimensional representation. J. Process Control, pp. 102–111 (2015)
26. Tang, S., Xie, C.: State and output feedback boundary control for a coupled PDE-ODE system. Syst. Control Lett. **60**, 540–545 (2011)
27. Wang, J., Krstic, M.: Output feedback boundary control of a heat PDE sandwiched between two ODEs. IEEE Trans. Autom. Control **64**, 4653–4660 (2019)
28. Wang, J.W., Wu, H.N., Li, H.X.: Fuzzy control design for nonlinear ODE-hyperbolic PDE-cascaded systems: a fuzzy and entropy-like Lyapunov function approach. IEEE Trans. Fuzzy Syst. pp. 1313–1324 (2014)
29. Woittennek, F., Rudolph, J.: Controller canonical forms and flatness based state feedback for 1D hyperbolic systems. In: Proceedings of 7th Vienna International Conference on Mathematical Modelling (MATHMOD), pp. 792–797 (2012)

Applications and Numerical Aspects

Cable-Operated Elevators and Deep-Sea Construction: 4×4 Hyperbolic PDE-ODE Control with Moving Boundary

Miroslav Krstic and Ji Wang

Abstract In mineral exploitation (mining), a cable-operated elevator is used to transport the cargo and miners between the ground and the working platform underground. A dual-cable mining elevator is used in the operation at a great depth, such as over 2000 m, and carrying a heavy load, where the single-cable elevator is not suitable, because a very thick cable, which is required to bear the heavy load, suffers from problems in the winding on the winder drum, due to its high bending. A deep-sea construction vessel, is used to place underwater parts of an off-shore oil drilling platform at the designated locations on the seabed, and is also used in the construction of artificial reefs for enabling the development and growth of marine life, in laying communication cables on the seabed, which is often uneven and rife with obstacles, and in other undersea applications. The dynamics of the mining cable elevator and deep-sea construction vessel are described by 4×4 coupled hyperbolic partial differential equations (PDEs), which are converted from two wave PDEs by the Riemann transformation, on a time-varying spatial domain introduced by the time-varying length of the cable, and coupled with an ordinary differential equation (ODE) which describes the dynamics of the attached payload, i.e., the cage or object to be installed/recycled/moved on the seabed. In the dual-cable mining elevator, we deal with axial oscillations in a pair of cables connected by a payload at the distal boundary. So the 4×4 system in the cable elevator is a set of two 2×2 pairs that are not coupled along the domain but are coupled at the boundary. In contrast, in the deep-sea construction vessel, we deal with a single cable but with axial and lateral oscillations which bring domain-wide coupling into plant and, therefore, a fully coupled 4×4 hyperbolic system. The boundary control inputs, i.e., control forces applied at the head sheaves of the mining cable elevator and the ship-mounted crane of the deep-sea construction vessel, are designed by using the backstepping method. The simulations verify that the designed controllers achieve suppression of axial vibrations and tension oscillations in the cables, as well as balancing the cage roll, in the dual-cable mining elevator, and also effectively reduce the undesired

M. Krstic · J. Wang (✉)
Department of Mechanical and Aerospace Engineering, University of California, San Diego, CA, USA
e-mail: krstic@ucsd.edu

J. Auriol et al. (eds.), *Advances in Distributed Parameter Systems*, Advances in Delays and Dynamics 14, https://doi.org/10.1007/978-3-030-94766-8_9

longitudinal-lateral vibrations in the deep-sea construction vessel with ocean current disturbances, even though the design is developed in the absence of such real-world effects.

1 Introduction

1.1 Cable-Operated Elevators

The cable plays a vital role in the mining elevators because its advantages of low bending and torsional stiffness, resisting relatively large axial loads, are helpful to the heavy load and large depth transportation. However, the compliance property or stretch and contract abilities of cables tend to cause mechanical vibrations, especially when the elevator is running at high speed, which leads to tension oscillations and premature fatigue fracture. For the operation at a greater depth, such as over 2000 m, and carrying a heavier load, the single-cable elevator [17] is not suitable, because a very thick cable is required to bear the heavy load, and such a thick cable, at high bending, suffers from problems in the winding on the winder drum. A dual-cable mining elevator, shown in Fig. 1, is proposed to solve this problem, where the requirement of a very thick cable can be removed because two cables tow the cage. However, the imbalance problem, such as cage roll, frequently appears in the dual-cable elevator, as shown in Fig. 1 where taut cables are used as flexible guide rails because traditional steel rails are with high cost of manufacture and installment in deep mines. The cage roll would increase the differences of oscillation tensions between two cables, and enlarge the oscillation amplitude of the tension in cables, which accelerates premature fatigue and requires inspections and costly repairs. Therefore, the importance of balancing the cage roll and suppressing the vibrations and tension oscillations in cables of the dual-cable mining elevator cannot be overestimated, for the safety of personnel and profitability. One feasible arrangement is designing and applying additional control forces through actuators at floating head sheaves shown in Fig. 1. The control design for suppression of the axial vibrations and tension oscillations, and balancing the cage roll in a dual-cable mining elevator is presented in Sect. 2.

1.2 Deep-Sea Construction

Deep-sea construction vessels (DCVs) are often used for installation of underwater parts for an offshore drilling platform, such as a subsea manifold, a subsea pump station, a subsea distribution unit along with associated foundations, flowlines and umbilicals, at designated locations around the drill center on the seafloor [14, 15]. The DCV is shown in Fig. 11 in Sect. 3, where the top of the cable is attached to a crane on a vessel at the ocean surface and the cable's bottom is attached to equipment

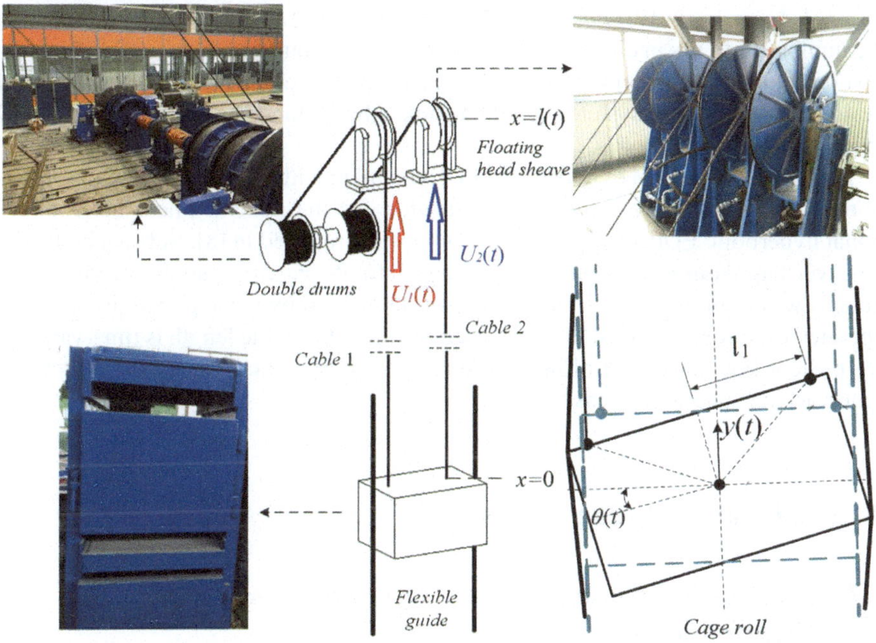

Fig. 1 Diagram and prototype of a dual-cable mining elevator with flexible guide rails (the prototype was built by CITIC Heavy Industries Co. Ltd.)

to be installed at the sea floor. The traditional method in underwater installation by DCVs is regulating the vessel dynamics position and manipulating the crane to obtain the desired heading for the payload. Such a method is not suitable for the deeper water construction (more than a thousand meter), because the long cable oscillations would cause large offset between the payload and the desired heading position of the crane, namely the designated installation location. In Sect. 3, the control forces at the onboard crane are designed to reduce the cable oscillations and place the equipment in the target area on the seafloor.

1.3 Control of Coupled Hyperbolic PDEs

The cable dynamics in the mining cable elevator and deep-sea construction vessel are described by 4×4 coupled first-order hyperbolic partial differential equations (PDEs), which are converted from two wave PDEs by the Riemann transformation [13]. Some theoretical results on boundary control of coupled first-order hyperbolic PDEs have emerged over the last decade. Basic boundary stabilization problem of 2×2 coupled linear transport PDEs by backstepping was addressed in [4, 16]. Boundary control of the 2×2 transport PDE system was further extended to that of a $n+1$ system in [9]. For a more general coupled transport PDE system where the number of PDEs in either direction is arbitrary, a boundary stabilization law was first designed by backstepping in [10], which is a systematic framework for control design for this class of systems and then some other extended results were proposed in [1, 2, 5, 6, 11].

For boundary control of hyperbolic PDEs coupled with ordinary differential equations (ODEs), the state-feedback stabilization for the general linear heterodirectional hyperbolic PDE-ODE coupled system was proposed in [8]. Subsequently, the observer-based output-feedback control design for the case of spatially varying coefficients was presented in [7]. In this chapter, motivated by the applications in cable-operated elevators and deep-sea construction where the cable length is time-varying, we present boundary control designs for 4×4 hyperbolic PDE-ODE systems on time-varying spatial domains.

2 Dual-Cable Mining Elevator

2.1 *Dual-Cable Mining Elevator Dynamics and Reformulation*

(1) Dynamics of Dual-Cable Mining Elevators

Following the modeling process in [17], and with the inclusion of the material damping of the cables, the vibration dynamics of a dual-cable mining elevator are modeled as

$$\dot{X}(t) = AX(t) + B[u_x(0,t), v_x(0,t)]^T, \tag{1}$$
$$u_{tt}(x,t) = qu_{xx}(x,t) - cu_t(x,t), \tag{2}$$
$$v_{tt}(x,t) = qv_{xx}(x,t) - cv_t(x,t), \tag{3}$$
$$C_3X(t) + l_1C_4X(t) = u_t(0,t), \tag{4}$$
$$C_3X(t) - l_1C_4X(t) = v_t(0,t), \tag{5}$$
$$u_x(l(t),t) = U_1(t), \tag{6}$$
$$v_x(l(t),t) = U_2(t), \tag{7}$$

$x \in [0, l(t)]$, $t \in [0, \infty)$, where $u(x,t)$, $v(x,t)$ are axial vibrations in two cables in a moving coordinate system associated with the elevator axial motion $l(t)$ where the origin is located at the cage. Matrices A and B are

$$A = \begin{pmatrix} 0 & 0 & 1 & 0 \\ 0 & 0 & 0 & 1 \\ 0 & 0 & \frac{-c_d}{M} & 0 \\ 0 & 0 & 0 & \frac{-c_a}{J_c} \end{pmatrix}, \; B = E_A \begin{pmatrix} 0 & 0 \\ 0 & 0 \\ \frac{-1}{M} & \frac{-1}{M} \\ \frac{-l_1}{J_c} & \frac{l_1}{J_c} \end{pmatrix} = [\bar{B}_1, \bar{B}_2] \tag{8}$$

where M denotes the mass of the load. The matrices C_3, C_4 are

$$C_3 = [0, 0, 1, 0], \quad C_4 = [0, 0, 0, 1]. \tag{9}$$

The function $l(t)$ describes the time-varying length of the cables. The parameter J_c is the moment of inertia of the cage, and $q = \frac{E \times A_a}{\rho}$, where E, A_a, and ρ are Young's modulus, the cross-sectional area, and the linear density of the cables, respectively. The parameters c_d and c_a denote the damping coefficients of cage axial and roll motion, respectively, and $c = \frac{\bar{c}}{\rho}$, where $\bar{c}$ is the material damping coefficient of the steel cables. The parameter l_1 is the cage dimension shown in Fig. 1.

The PDE states $u(x, t)$, $v(x, t)$ in (2), (3) describe the axial vibration displacements of the distributed points in cable 1 and cable 2, respectively. The functions $u_x(x, t)$ and $v_x(x, t)$ denote the distributed strain in the cables, and the tension oscillations are represented as $E_A u_x(x, t)$ and $E_A v_x(x, t)$, where the constant $E_A = E \times A_a$. The functions $E_A u_x(0, t)$ and $E_A v_x(0, t)$ that denote forces acting on the cage drive the ODE dynamics (1), where the ODE state $X(t) = [y(t), \theta(t), \dot{y}(t), \dot{\theta}(t)]^T$ describes the cage dynamics. The functions $y(t)$, $\dot{y}(t)$ are the axial vibration displacement and velocity of the centroid of the cage, and $\theta(t)$, $\dot{\theta}(t)$ are the cage roll angle and roll rate around the axis which is vertical to the door and through the centroid of the cage. Equations (4), (5) describe the velocity relationship between the cage and the bottom boundaries of the two cables. Equations (6), (7) come from $E_A u_x(l(t), t) = U_{1v}(t)$ and $E_A v_x(l(t), t) = U_{2v}(t)$ with the definition of $U_{1v}(t) = E_A U_1(t)$ and $U_{2v}(t) = E_A U_2(t)$, from which the two actual control forces at the two floating sheaves $U_{1v}(t)$, $U_{2v}(t)$ can be obtained by $U_1(t)$, $U_2(t)$ to be designed in this section.

We pursue the control design for a general model where the damping coefficients c_a, c_d, c in (1)–(7) are arbitrary. In other words, these damping coefficients can be damped (> 0), undamped ($= 0$), or even anti-damped (< 0).

Assumption 1 The varying length $l(t)$ is bounded, i.e., $0 < l(t) \leq L, \forall t \geq 0$, where the positive constant L is the maximal length. □

Assumption 2 The velocity $\dot{l}(t)$ of the moving boundary is bounded by $|\dot{l}(t)| \leq \bar{v}_{\max} < \sqrt{q}$, where $\bar{v}_{\max}$ is the maximum velocity of the mining cable elevator. □

The assumptions are fully satisfied in the mining cable elevator. In addition to the suppression of oscillations of tension in each cable, suppression of tension oscillation discrepancy between two cables is required as well. Therefore, the error between tension oscillations in two cables should be a part of the model used to design a controller. Toward that end we introduce the new error and mean variables

$$e(x, t) = v(x, t) - u(x, t), \quad s(x, t) = v(x, t) + u(x, t), \tag{10}$$

where $E_A e_x(x, t) = E_A v_x(x, t) - E_A u_x(x, t)$ is the tension oscillation discrepancy between two cables, and $E_A s_x(x, t)$ is the total tension oscillations in the two cables.

The control objectives are formulated in terms of the following physical behavior and theoretical results:

- Physically, suppress the oscillations of tension in each cable as fast as possible. Theoretically, make the (u, v)-system exponentially stable in the sense of terms including $\|u_{xx}(\cdot, t)\| + \|v_{xx}(\cdot, t)\|$, where the exponential decay rate can be chosen.
- Physically, reduce the error of oscillations of tension between two cables as fast as possible. Theoretically, make the (e, s)-system exponentially stable in the sense of terms including $\|e_{xx}(\cdot, t)\| + \|s_{xx}(\cdot, t)\|$, where the exponential decay rate can be chosen.
- Physically, suppress the axial vibration displacement $y(t)$ and the roll angle $\theta(t)$ of the cage as fast as possible. Theoretically, make $|X(t)|$ exponentially convergent to zero when $t \to \infty$, where the exponential decay rate can be chosen.

This may be a convenient place to remind the reader that $\|u(\cdot, t)\|$ is a compact notation for $\sqrt{\int_0^{l(t)} u(x, t)^2 dx}$, and the single bars $|\cdot|$ denotes the Euclidean norm.
(2) Reformulation in Riemann Coordinates
In order to convert the wave models to first-order systems, for the sake of enabling the applicability of backstepping designs for first-order hyperbolic PDE-ODE systems, we introduce the following Riemann coordinates,

$$z_1(x, t) = s_t(x, t) - \sqrt{q} s_x(x, t), \quad w_1(x, t) = s_t(x, t) + \sqrt{q} s_x(x, t), \tag{11}$$
$$z_2(x, t) = e_t(x, t) - \sqrt{q} e_x(x, t), \quad w_2(x, t) = e_t(x, t) + \sqrt{q} e_x(x, t). \tag{12}$$

Recalling (10), the system (1)–(7) is rewritten as

$$\dot{X}(t) = \bar{A} X(t) + \sum_{i=1}^{2} \frac{B_i}{\sqrt{q}} w_i(0, t), \tag{13}$$
$$z_{it}(x, t) = -\sqrt{q} z_{ix}(x, t) - \frac{c}{2}(z_i(x, t) + w_i(x, t)), \tag{14}$$
$$w_{it}(x, t) = \sqrt{q} w_{ix}(x, t) - \frac{c}{2}(z_i(x, t) + w_i(x, t)), \tag{15}$$
$$z_i(0, t) = D_i X(t) - w_i(0, t), \tag{16}$$
$$w_i(l(t), t) = z_i(l(t), t) + 2\sqrt{q} U_{ei}(t), \tag{17}$$

for $x \in [0, l(t)], t \in [0, \infty), i = 1, 2$, where

$$B_1 = \frac{(\bar{B}_1 + \bar{B}_2)}{2} = \left[0, 0, \frac{E_A}{M}, 0\right]^T, \; B_2 = \frac{(\bar{B}_2 - \bar{B}_1)}{2} = \left[0, 0, 0, \frac{E_A l_1}{J_c}\right]^T, \tag{18}$$

$$\bar{A} = A - \frac{2B_1}{\sqrt{q}} C_3 + \frac{2B_2}{\sqrt{q}} C_4 l_1, \tag{19}$$

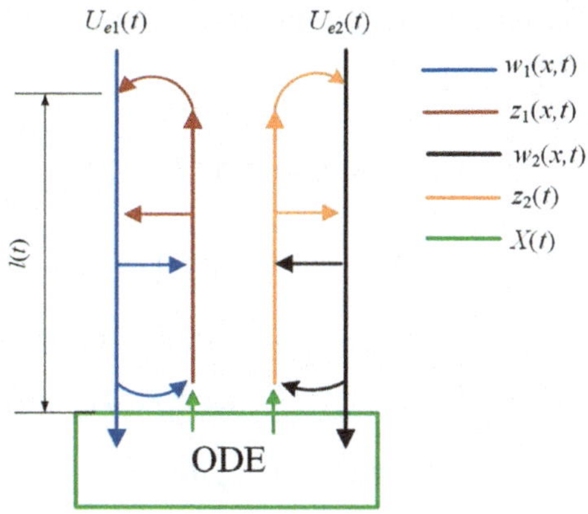

Fig. 2 Diagram of the plant dynamics (13)–(17)

$$D_1 = 4C_3 = [0, 0, 4, 0]^T, \quad D_2 = -4l_1 C_4 = [0, 0, 0, -4l_1]^T, \tag{20}$$

and where

$$U_{e1}(t) = U_2(t) + U_1(t), \quad U_{e2}(t) = U_2(t) - U_1(t). \tag{21}$$

The PDEs (14)–(17) are two pairs of heterodirectional transport equations with states $w_i(x,t), z_i(x,t), i = 1, 2$ and with source terms that result from in-domain damping. The PDEs governing $w_i(x,t), z_i(x,t)$ are coupled with an ODE (13) at the distal boundary. The diagram describing these plant dynamics is shown in Fig. 2.

2.2 *Observer for Cable Tension*

In this subsection we pursue observer design to estimate the distributed states $z_i(x,t)$, $w_i(x,t)$, which are associated with tension oscillations $u_x(x,t)$, $v_x(x,t)$ in the two cables according to (10)–(12). Given that distributed tension usually cannot be measured in practice, though it is needed in the controller, we design an observer of distributed tension using only the available boundary measurements.

(1) Observer Structure

The available measurements in the mining cable elevator are

- Axial vibration acceleration $\ddot{y}(t)$ and roll angular acceleration $\ddot{\theta}(t)$ in the cage where accelerometers are placed. Measuring acceleration is the prevalent method in control of vibrating mechanical systems because accelerations are easier to measure with accelerometers than displacements or velocities.
- Force $E_A u_x(l(t),t)$, $E_A v_x(l(t),t)$ and velocity $u_t(l(t),t)$, $v_t(l(t),t)$ feedback signals of the actuators at two floating sheaves.

The observer is constructed as a copy of the plant (13)–(17) with output error injections,

$$\begin{aligned}\dot{\hat{X}}(t) =&\bar{A}\hat{X}(t) + \sum_{i=1}^{2} \frac{B_i}{\sqrt{q}}\hat{w}_i(0,t)\\&+\bar{L}\left((y(t)+\theta(t)) - (C_1+C_2)\hat{X}(t)\right), \end{aligned}\tag{22}$$

$$\begin{aligned}\hat{z}_{it}(x,t) =& -\sqrt{q}\hat{z}_{ix}(x,t) - \frac{c}{2}(\hat{z}_i(x,t)+\hat{w}_i(x,t))\\&+\bar{\Gamma}_i(x,t)(z_i(l(t),t)-\hat{z}_i(l(t),t)),\end{aligned}\tag{23}$$

$$\begin{aligned}\hat{w}_{it}(x,t) =&\sqrt{q}\hat{w}_{ix}(x,t) - \frac{c}{2}(\hat{z}_i(x,t)+\hat{w}_i(x,t))\\&+\Gamma_i(x,t)(z_i(l(t),t)-\hat{z}_i(l(t),t)),\end{aligned}\tag{24}$$

$$\hat{z}_1(0,t) =D_1X(t) - \hat{w}_1(0,t) = 4\dot{y}(t) - \hat{w}_1(0,t),\tag{25}$$

$$\hat{z}_2(0,t) =D_2X(t) - \hat{w}_2(0,t) = -4l_1\dot{\theta}(t) - \hat{w}_2(0,t),\tag{26}$$

$$\hat{w}_i(l(t),t) =z_i(l(t),t) + 2\sqrt{q}U_{ei}(t),\tag{27}$$

where $z_i(l(t),t)$ can be obtained by using $s_x(l(t),t), s_t(l(t),t), e_x(l(t),t), e_t(l(t),t)$, which are computed by the measurements $u_x(l(t),t), v_x(l(t),t), u_t(l(t),t), v_t(l(t),t)$. Because it is the acceleration (rather than displacement) sensor that is more convenient to place at the cage, $y(t), \theta(t), \dot{y}(t), \dot{\theta}(t)$ in $X(t)$ are calculated by integrating the measured accelerations $\ddot{y}(t), \ddot{\theta}(t)$ with known initial conditions $y(0), \theta(0), \dot{y}(0)$, $\dot{\theta}(0)$ (this is only the explanation of an implementable signal acquisition method in practice, and no restrictions are imposed on any initial conditions in the design and theory). The ODE measurements $y(t), \theta(t)$ and the matrix $\bar{A}$ in (22) form an observable matrix pair $(\bar{A}, C_1+C_2)$ where $C_1 = [1,0,0,0]$, $C_2 = [0,1,0,0]$.

(2) Observer Error Dynamics and Backstepping

Defining the observer error

$$(\tilde{z}_i(x,t), \tilde{w}_i(x,t), \tilde{X}(t)) = (z_i(x,t), w_i(x,t), X(t)) - (\hat{z}_i(x,t), \hat{w}_i(x,t), \hat{X}(t)),$$

we have the error dynamics between the plant and the observer as

$$\dot{\tilde{X}}(t) = \hat{A}\tilde{X}(t) + \sum_{i=1}^{2}\frac{B_i}{\sqrt{q}}\tilde{w}_i(0,t),\tag{28}$$

$$\tilde{z}_{it}(x,t) = -\sqrt{q}\tilde{z}_{ix}(x,t) - \frac{c}{2}(\tilde{z}_i(x,t)+\tilde{w}_i(x,t)) - \bar{\Gamma}_i(x,t)\tilde{z}_i(l(t),t),\tag{29}$$

$$\tilde{w}_{it}(x,t) = \sqrt{q}\tilde{w}_{ix}(x,t) - \frac{c}{2}(\tilde{z}_i(x,t)+\tilde{w}_i(x,t)) - \Gamma_i(x,t)\tilde{z}_i(l(t),t),\tag{30}$$

$$\tilde{z}_i(0,t) = -\tilde{w}_i(0,t),\quad \tilde{w}_i(l(t),t) = 0,\tag{31}$$

for $i = 1, 2$, where the system matrix $\hat{A} = \bar{A} - \bar{L}(C_1 + C_2)$ in (28) is made Hurwitz by choosing $\bar{L}$.

Introduce the backstepping transformation

$$\tilde{z}_i(x,t) = \tilde{\alpha}_i(x,t) - \int_x^{l(t)} \bar{\phi}_i(x,y)\tilde{\alpha}_i(y,t)dy, \tag{32}$$

$$\tilde{w}_i(x,t) = \tilde{\beta}_i(x,t) - \int_x^{l(t)} \bar{\psi}_i(x,y)\tilde{\alpha}_i(y,t)dy, \tag{33}$$

where the kernels $\bar{\phi}_i(x,y), \bar{\psi}_i(x,y)$ on $\mathcal{D}_1 = \{(x,y)|0 \leq x \leq y \leq l(t)\}$ satisfy the following well-posed equations:

$$-\sqrt{q}\bar{\phi}_{ix}(x,y) - \sqrt{q}\bar{\phi}_{iy}(x,y) = \frac{c}{2}\bar{\psi}_i(x,y), \tag{34}$$

$$-\sqrt{q}\bar{\psi}_{iy}(x,y) + \sqrt{q}\bar{\psi}_{ix}(x,y) = \frac{c}{2}\bar{\phi}_i(x,y), \tag{35}$$

$$\bar{\phi}_i(0,y) = -\bar{\psi}_i(0,y), \quad \bar{\psi}_i(x,x) = \frac{c}{4\sqrt{q}}. \tag{36}$$

Through (32), (33), with choosing the observer gains $\Gamma_i(x,t), \bar{\Gamma}_i(x,t)$ as

$$\Gamma_i(x,t) = \dot{l}(t)\bar{\psi}_i(x,l(t)) - \sqrt{q}\bar{\psi}_i(x,l(t)), \tag{37}$$

$$\bar{\Gamma}_i(x,t) = \dot{l}(t)\bar{\phi}_i(x,l(t)) - \sqrt{q}\bar{\phi}_i(x,l(t)), \tag{38}$$

we convert the error dynamics (28)–(31) into the target error system as

$$\dot{\tilde{X}}(t) = \hat{A}\tilde{X}(t) + \sum_{i=1}^{2} \frac{B_i}{\sqrt{q}}\left(\tilde{\beta}_i(0,t) + \int_0^{l(t)} \bar{\psi}_i(0,y)\tilde{\alpha}_i(y,t)dy\right), \tag{39}$$

$$\tilde{\alpha}_{it}(x,t) = -\sqrt{q}\tilde{\alpha}_{ix}(x,t) + \int_x^{l(t)} \bar{M}_i(x,y)\tilde{\beta}_i(y,t)dy - \frac{c}{2}(\tilde{\alpha}_i(x,t) + \tilde{\beta}_i(x,t)), \tag{40}$$

$$\tilde{\beta}_{it}(x,t) = \sqrt{q}\tilde{\beta}_{ix}(x,t) - \frac{c}{2}\tilde{\beta}_i(x,t) + \int_x^{l(t)} \bar{N}_i(x,y)\tilde{\beta}_i(y,t)dy, \tag{41}$$

$$\tilde{\alpha}_i(0,t) = -\tilde{\beta}_i(0,t), \quad \tilde{\beta}_i(l(t),t) = 0, \tag{42}$$

where $\bar{M}_i(x,y), \bar{N}_i(x,y)$ in (40) and (41) satisfy $\bar{M}_i(x,y) = -\frac{c}{2}\bar{\phi}_i(x,y) + \int_x^y \bar{\phi}_i(x,z)\bar{M}_i(z,y)dz$, $\bar{N}_i(x,y) = -\frac{c}{2}\bar{\psi}_i(x,y) + \int_x^y \bar{\psi}_i(x,z)\bar{M}_i(z,y)dz$.

Remark that the heterodirectional coupled hyperbolic PDEs (29), (30) which include in-domain unstable sources, especially the coupling terms $\tilde{z}_i(x,t), \tilde{z}_i(l(t),t)$ in (30) whose propagate direction is from the right boundary to the left boundary and the ODE, i.e., the "forward" direction, are intentionally converted into the target system where there is no coupling term (the states form the transport PDE in the reverse direction) in the "forward" transport PDE (41). This motivation is also used to construct the target system for (22)–(27), which will be shown in Sect. 2.3.

We can prove exponential stability of the observer error dynamics (28)–(31) with the designed gains $\bar{\Gamma}_i(x,t)$, $\Gamma_i(x,t)$ in (37), (38) in the sense of the norm $(\sum_{i=1}^{2}(\|\tilde{z}_i(\cdot,t)\|^2 + \|\tilde{w}_i(\cdot,t)\|^2 + \|\tilde{z}_{ix}(\cdot,t)\|^2 + \|\tilde{w}_{ix}(\cdot,t)\|^2) + |\tilde{X}(t)|^2)^{\frac{1}{2}}$, by using the exponential stability of the target observer error system (39)–(42) and the invertibility of the backstepping transformations (32), (33). The proof is omitted due to the space limit.

2.3 *Controller for Cable Tension Oscillation Suppression and Cage Balance*

In Sect. 2.2, we have obtained an observer which can exponentially track the distributed states of the system (13)–(17). Now, we design output-feedback control laws $U_1(t)$, $U_2(t)$ by using the states recovered from the observer via the backstepping method.

The design of the observer-based output-feedback controller is based on the observer (22)–(27). Introduce the backstepping transformation

$$\alpha_i(x,t) \equiv \hat{z}_i(x,t), \tag{43}$$

$$\beta_i(x,t) = \hat{w}_i(x,t) - \int_0^x \psi_i(x,y)\hat{z}_i(y,t)dy - \int_0^x \phi_i(x,y)\hat{w}_i(y,t)dy - \gamma_i(x)\hat{X}(t), \tag{44}$$

where the kernels $\psi_i(x,y)$, $\phi_i(x,y)$ on $\mathcal{D} = \{(x,y)|0 \le y \le x \le l(t)\}$ and $\gamma_i(x)$, $i = 1, 2$, satisfy

$$\sqrt{q}\phi_{ix}(x,y) + \sqrt{q}\phi_{iy}(x,y) = -\frac{c}{2}\psi_i(x,y), \tag{45}$$

$$\sqrt{q}\psi_{ix}(x,y) - \sqrt{q}\psi_{iy}(x,y) = -\frac{c}{2}\phi_i(x,y), \tag{46}$$

$$\phi_i(x,0) = \frac{1}{q}\gamma_i(x)B_i - \psi_i(x,0), \quad \psi_i(x,x) = \frac{c}{4\sqrt{q}}, \tag{47}$$

$$\gamma_i{}'(x) - \frac{1}{\sqrt{q}}\gamma_i(x)\left(\bar{A} + \frac{c}{2}I_4\right) = \frac{1}{\sqrt{q}}\psi_i(x,0)D_i, \quad \gamma_i(0) = \sqrt{q}\kappa_i. \tag{48}$$

The matrix I_4 is an identity matrix with dimension 4. The row vectors κ_i are defined as $\kappa_1 = [\hat{k}_{11}, 0, \hat{k}_{13}, 0]$, $\kappa_2 = [0, \hat{k}_{22}, 0, \hat{k}_{24}]$. By choosing the design parameters $\hat{k}_{11}, \hat{k}_{13}, \hat{k}_{22}, \hat{k}_{24}$ in κ_i, a Hurwitz matrix $\bar{A} + \sum_{i=1}^{2} B_i\kappa_i$ is made according to (8), (9), (18), (19). It follows from (20), (48) that $\gamma_1(x)$ and $\gamma_2(x)$ are in the forms $\gamma_1(x) = [\hat{\gamma}_{11}(x), 0, \hat{\gamma}_{13}(x), 0]$, $\gamma_2(x) = [0, \hat{\gamma}_{22}(x), 0, \hat{\gamma}_{24}(x)]$. Recalling (18), we know that

$$\gamma_i(x)B_j = 0, \quad i \neq j. \tag{49}$$

With extending $\mathcal{D}$ to a fixed domain $\{(x, y)|0 \le y \le x \le L\}$, and applying the results in [8], the well-posedness of the kernel equations (45)–(48) can be proved.

Through (43), (44), with recalling (49), we convert the observer system (22)–(27) to the following target system:

$$\dot{\hat{X}}(t) = \left(\bar{A} + \sum_{i=1}^{2} B_i\kappa_i\right)\hat{X}(t) + \sum_{i=1}^{2}\frac{B_i}{\sqrt{q}}\beta_i(0,t) + \bar{L}(C_1 + C_2)\tilde{X}(t), \tag{50}$$

$$\begin{aligned}\alpha_{it}(x,t) =& -\sqrt{q}\alpha_{ix}(x,t) - \frac{c}{2}\beta_i(x,t) - \frac{c}{2}\alpha_i(x,t)\\ &- \frac{c}{2}\int_0^x \hat{M}_i(x,y)\alpha_i(y,t)dy - \frac{c}{2}\int_0^x \hat{N}_i(x,y)\beta_i(y,t)dy\\ &- \frac{c}{2}\hat{\vartheta}_i(x)\hat{X}(t) + \bar{\Gamma}_i(x,t)\tilde{z}_i(l(t),t),\end{aligned} \tag{51}$$

$$\beta_{it}(x,t) = \sqrt{q}\beta_{ix}(x,t) - \frac{c}{2}\beta_i(x,t) - \mathcal{N}_i(x,t)\tilde{z}_i(l(t),t) - \mathcal{N}_{1i}(x)\tilde{X}(t), \tag{52}$$

$$\alpha_i(0,t) = \bar{D}_i\hat{X}(t) - \beta_i(0,t) + D_i\tilde{X}(t), \tag{53}$$

$$\beta_i(l(t),t) = 0, \tag{54}$$

where $\mathcal{N}_i(x,t) = \int_0^x(\phi_i(x,y)\Gamma_i(y,t) + \psi_i(x,y)\bar{\Gamma}_i(y,t))dy - \Gamma_i(x,t)$ and $\mathcal{N}_{1i}(x) = \gamma_i(x)\bar{L}(C_1 + C_2) + \sqrt{q}\psi_i(x,0)D_i$, and where $\hat{N}_i(x,y)$, $\hat{M}_i(x,y)$, $\hat{\vartheta}_i(x)$ satisfy that $\hat{N}_i(x,y) = \int_y^x \hat{N}_i(x,\delta)\phi_i(\delta,y)d\delta + \phi_i(x,y)$, $\hat{M}_i(x,y) = \int_y^x \hat{N}_i(x,\delta)\psi_i(\delta,y)d\delta + \psi_i(x,y)$, $\hat{\vartheta}_i(x) = \int_0^x \hat{N}_i(x,y)\gamma_i(y)dy + \gamma_i(x)$. The matrices $\bar{D}_i$ are $\bar{D}_i = D_i - \gamma_i(0)$.

The boundary condition (54) is ensured by choosing the control inputs as

$$\begin{aligned}U_{ei}(t) =& \frac{-1}{2\sqrt{q}}\Bigg(z_i(l(t),t) - \int_0^{l(t)} \psi_i(l(t),y)\hat{z}_i(y,t)dy\\ &- \int_0^{l(t)} \phi_i(l(t),y)\hat{w}_i(y,t)dy - \gamma_i(l(t))\hat{X}(t)\Bigg),\end{aligned} \tag{55}$$

for $i = 1, 2$. The signals $\hat{z}_i(x,t)$, $\hat{w}_i(x,t)$, $\hat{X}(t)$ are obtained from the observer (22)–(27) constructed by measurements $\ddot{y}(t)$, $\ddot{\theta}(t)$ and $u_x(l(t),t)$, $v_x(l(t),t)$, $u_t(l(t),t)$, $v_t(l(t),t)$ which are used to calculate $z_i(l(t),t)$. The gains $(\psi_i(x,y), \phi_i(x,y), \gamma_i(x))$ are the solution of (45)–(48). Recalling (21), the two control inputs $U_1(t)$, $U_2(t)$ in (1)–(7) are derived as

$$U_1(t) = U_{e1}(t) - U_{e2}(t), \; U_2(t) = U_{e1}(t) + U_{e2}(t) \tag{56}$$

where only readily available measurements on the mining cable elevator are used.

2.4 Stability Result

The following theorem establishes the exponential stability of the $(\hat{z}_i, \hat{w}_i)$-system in (22)–(27) under the control (55). The proof is omitted due to the space limit.

Theorem 1 *For all initial values* $(\hat{z}_i(x, t_0), \hat{w}_i(x, t_0), z_i(x, t_0), w_i(x, t_0)) \in H^1(0, L_0)$, *where* $L_0 = l(t_0)$ *is the initial length of the cable, the closed-loop system including the plant* (13)–(17), *the controller* (55) *and the observer* (22)–(27) *has the following properties.*

(1) The closed-loop system is uniformly exponentially stable in the sense of the norm

$$\Bigg(\sum_{i=1}^{2}\left(\|\hat{z}_i(\cdot,t)\|^2 + \|\hat{w}_i(\cdot,t)\|^2 + \|z_i(\cdot,t)\|^2 + \|w_i(\cdot,t)\|^2 + \|\hat{z}_{ix}(\cdot,t)\|^2\right.$$
$$\left. + \|\hat{w}_{ix}(\cdot,t)\|^2 + \|z_{ix}(\cdot,t)\|^2 + \|w_{ix}(\cdot,t)\|^2\right) + \left|\hat{X}(t)\right|^2 + |X(t)|^2\Bigg)^{1/2}, \quad (57)$$

with a decay rate which can be adjusted by the choices of the design parameters κ_i, $\bar{L}$.

(2) The control inputs $U_{ei}(t)$ *are bounded and exponentially convergent to zero.*

The term "uniformly" refers to uniformity in the initial time t_0. According to Theorem 1, recalling (10)–(12), (56), applying the Cauchy-Schwarz inequality, the control objectives formulated in Sect. 2.1 are achieved. Moreover, the control inputs $U_1(t)$, $U_2(t)$ realized by hydraulic actuators at the floating sheaves are bounded and exponentially convergent to zero.

2.5 Simulation Test in a Dual-Cable Mining Elevator

Details of conducting the simulation are contained in Sect. 3.5 of [18], including model parameters, design parameters, initial conditions, and numerical methods. We only present the simulation results here due to the space limit.

For the sake of comparison, we also apply the boundary damper which is traditionally utilized in industry at the head sheaves, to compare with the proposed output-feedback controller. The boundary dampers feedback laws are given as $U_{d1}(t) = k_{d1}\dot{u}(l(t), t)$, $U_{d2}(t) = k_{d2}\dot{v}(l(t), t)$, where k_{d1}, k_{d2} are tuned to attain the efficient control performance. We have tested different values of k_{d1}, k_{d2}, and the best regulating performance is achieved with $k_{d1} = -0.33$, $k_{d2} = -0.31$.

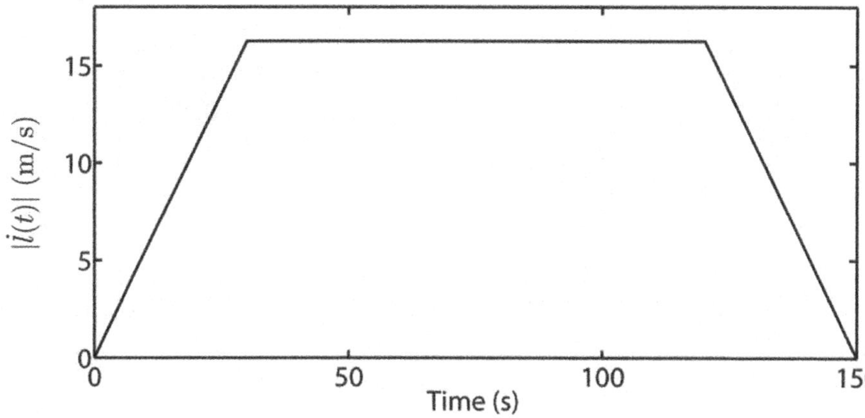

Fig. 3 The hoisting velocity $\dot{l}(t)$

The designed reference of the hoisting velocity $\dot{l}(t)$ is plotted in Fig. 3. The responses of tension oscillations at the midpoint of cable 1 and cable 2 are shown in Figs. 4 and 5 respectively, with the open loop, the boundary damper, and the proposed control law. The red dashed lines, depicting the open loop responses in Figs. 4 and 5, show that large tension oscillations would be caused in the accelerated ascending process for the hoisting velocity curve $\dot{l}(t)$ in Fig. 3. To suppress the oscillations of tension, the proposed output-feedback controller and the boundary damper are applied at the head sheaves, respectively, and the responses are shown in black and blue lines in Figs. 4 and 5. It can be seen that tension oscillations are suppressed by both the proposed control law and the boundary damper. However, we observe that the responses with the proposed control law have faster convergence and less overshoot than the responses with the boundary damper.

The errors of tension oscillations between cables in the open loop, under the boundary damper, and under the proposed control law are shown in Fig. 6. It shows that the tension oscillation error increases in 0–30 s, which is the accelerated ascending process, in the open-loop case, while it is reduced and convergent to zero under the proposed output-feedback controller, and moreover, with a faster convergence than the boundary damper.

The responses of the axial and roll vibrations of the cage are shown in Figs. 7 and 8 in the cases without control, with the boundary damper, and with the proposed output-feedback controller. It can be observed that the large axial vibrations and roll of the cage in the open loop are suppressed to zero more efficiently under the proposed output-feedback controller than the traditional boundary damper, with the faster convergence and less overshoot.

Figure 9 shows the convergent observer errors between $\hat{z}_2$, $\hat{w}_2$ and the plant states z_2, w_2 at the midpoint of the domain $[0, l(t)]$, which indicates that the observer (22)–(27) can reconstruct the actual distributed states in (13)–(17). Because the locations of the actuator and the sensor are at the boundaries, the estimation of the states at the mid-

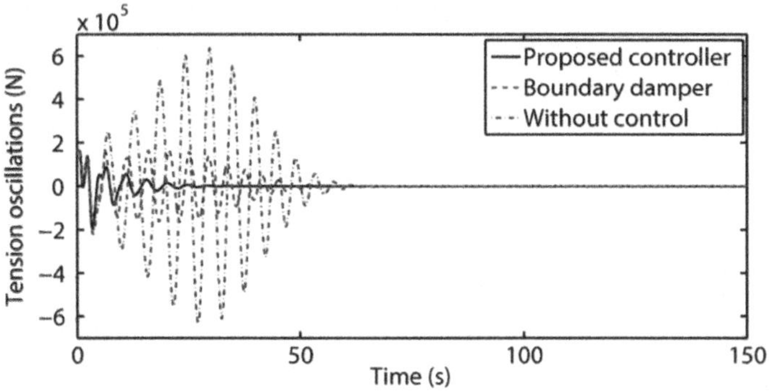

Fig. 4 Tension oscillations $E_A \times u_x(l(t)/2, t)$ at the midpoint of cable 1

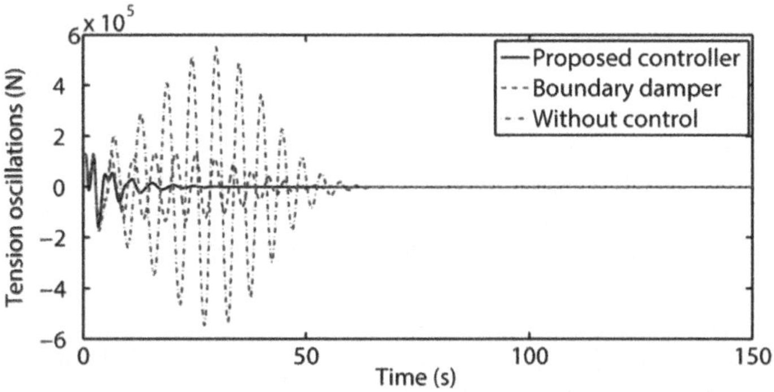

Fig. 5 Tension oscillations $E_A \times v_x(l(t)/2, t)$ at the midpoint of cable 2

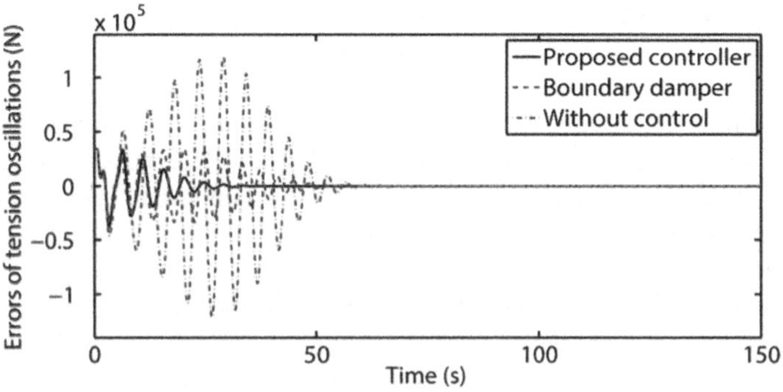

Fig. 6 Errors $E_A \times (u_x(l(t)/2, t) - v_x(l(t)/2, t))$ of tension oscillations between cable 1 and cable 2

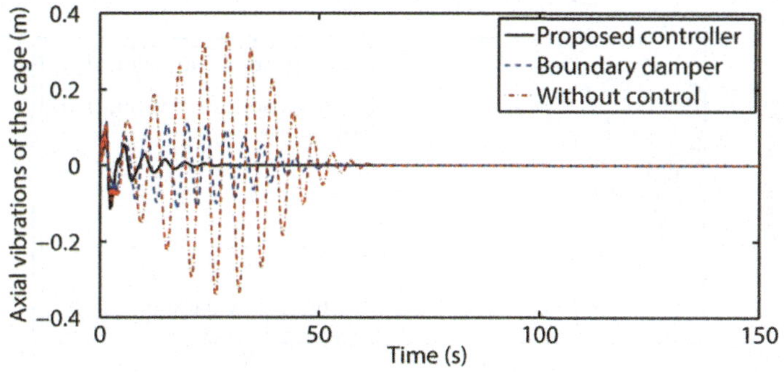

Fig. 7 Axial vibration displacements $y(t)$ of the cage

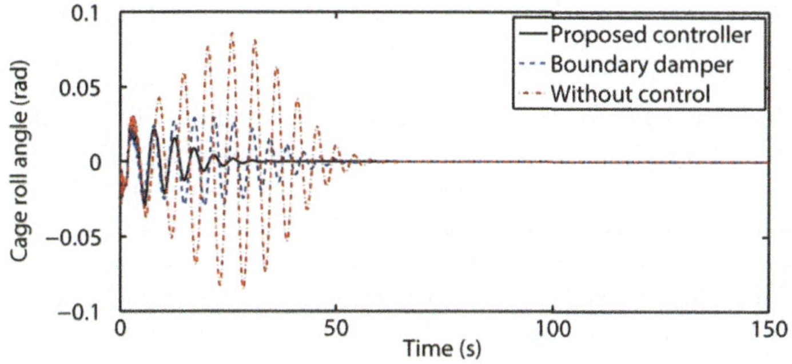

Fig. 8 Cage roll angles $\theta(t)$

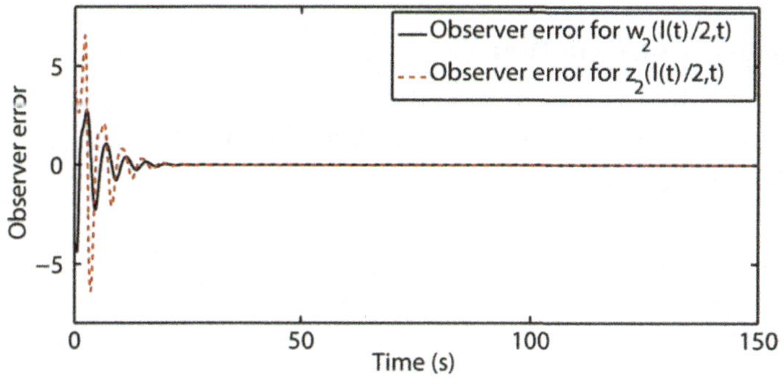

Fig. 9 Observer errors $z_2(l(t)/2, t) - \hat{z}_2(l(t)/2, t)$ and $w_2(l(t)/2, t) - \hat{w}_2(l(t)/2, t)$ between the plant (13)–(17) and the observer (22)–(27)

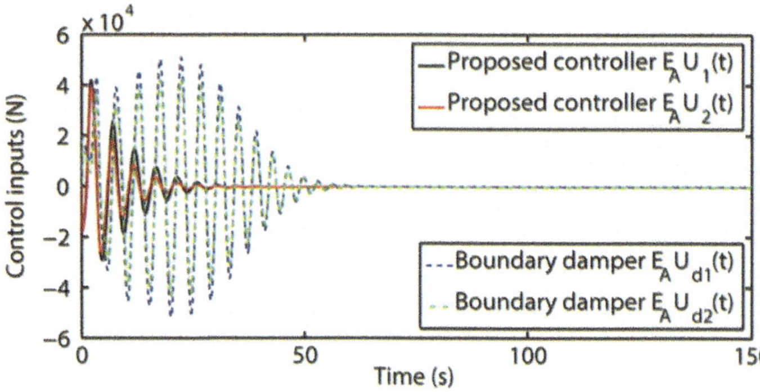

Fig. 10 The proposed observer-based output-feedback control forces $E_A U_1(t) = E_A(U_{e2} + U_{e1})$, $E_A U_2(t) = E_A(U_{e2} - U_{e1})$ and the boundary dampers $E_A U_{d1}(t)$, $E_A U_{d2}(t)$ at two head sheaves

point $x = l(t)/2$ is the most challenging due to its accessibility. The initial conditions of the observer $\hat{z}_1(x, 0)$, $\hat{w}_1(x, 0)$ are the same as the plant $z_1(x, 0)$, $w_1(x, 0)$ in the simulation, the observer errors of z_1, w_1 are thus at a very small magnitude 10^{-14} and convergent to zero as well. We only show the observer errors of z_2, w_2 here to avoid repetitiousness and excessive use of space. The control forces $E_A U_1(t)$, $E_A U_2(t)$ obtained from (55), (56) at the two head sheaves in the closed-loop system are given in Fig. 10, which shows that the control forces $E_A U_1(t)$, $E_A U_2(t)$ are bounded and convergent to zero.

3 Deep-Sea Construction

3.1 Dynamics of a Deep-Sea Construction Vessel and Reformulation

(1) Dynamics of a deep-sea construction vessel
A DCV is depicted in Fig. 11, where a crane mounted on a ship regulates a cable to install a piece of equipment (attached payload) at the target position on the seabed. The attached payload is subject to a constant drag force caused by a constant water stream velocity. We consider only the water-stream-caused drag force at the payload because the diameter of the cable is much smaller than that of the payload. For suppression of cable oscillation/vibration, two-directional control forces implemented by two actuators (hydraulic cylinder and hydraulic motor) and measurements by accelerometers are applied/placed at the ship-mounted crane. The varying cable length, driven by an additional winch wound with the cable on the ship, can be considered as a pre-defined time-varying function $l(t)$ in the cable vibration dynamics regulated by the cranes.

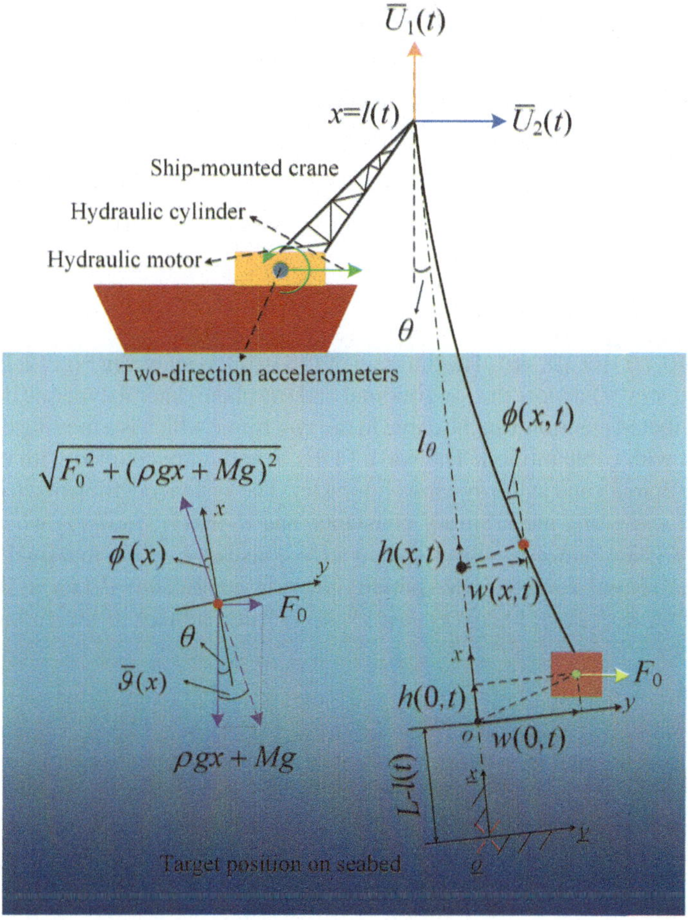

Fig. 11 Diagram of a deep-sea construction vessel

Through a modeling process (please refer to Sect. 6.1 in [18] for details) where Hamilton's principle and the moving frame approach [12] are used, a nonlinear mathematical model of the two-dimensional vibrations of the DCV is obtained as (6.13)–(6.18) in [18]. The nonlinear PDE system is then linearized around a steady state as the procedure in [3], generating a linear model as (6.21)–(6.26) in [18].

In the control design, we consider a general form of (6.21)–(6.26) in [18], where more couplings between two wave PDEs are included, in both the domain and at the dynamic boundary. The obtained theoretical result is then applied back to the specific application problem of the DCV in the simulation. The class of plants we consider is

$$w_{tt}(x,t) = d_1(x)w_{xx}(x,t) + d_2(x)w_x(x,t) + d_3(x)u_x(x,t) + d_4(x)w_t(x,t) + d_5(x)u_t(x,t), \tag{58}$$
$$u_{tt}(x,t) = d_6(x)u_{xx}(x,t) + d_7(x)w_x(x,t) + d_8(x)u_x(x,t) + d_9(x)w_t(x,t) + d_{10}(x)u_t(x,t), \tag{59}$$
$$w_{tt}(0,t) = d_{11}w_t(0,t) + d_{12}w_x(0,t) + d_{13}u_t(0,t) + d_{14}u_x(0,t), \tag{60}$$
$$u_{tt}(0,t) = d_{15}u_t(0,t) + d_{16}u_x(0,t) + d_{17}w_t(0,t) + d_{18}w_x(0,t), \tag{61}$$
$$u_x(l(t),t) = d_{19}(l(t))U_1(t), \tag{62}$$
$$w_x(l(t),t) = d_{20}(l(t))U_2(t), \tag{63}$$

$\forall (x,t) \in [0, l(t)] \times [0, \infty)$. The diagram of (58)–(63) is shown in Fig. 12. Physically $u(x,t)$ and $w(x,t)$ denote the longitudinal and (in-plane) lateral dynamic deflections in the distributed position x in the cable in the xoy frame which is a moving coordinate associated with cable motion. The wave PDEs w and u are coupled with each other both in the domain and at the dynamic boundary. The system coefficients d_{11}, d_{12}, d_{13}, d_{14}, d_{15}, d_{16}, d_{17}, d_{18} are arbitrary constants, and $d_{19}(l(t))$, $d_{20}(l(t))$ which belong to $C^0([0, L])$ are nonzero, where the positive constant L is the maximal length of the varying spatial domain. The spatially varying coefficients $d_1(x)$, $d_2(x)$, $d_3(x)$, $d_4(x)$, $d_5(x)$, $d_6(x)$, $d_7(x)$, $d_8(x)$, $d_9(x)$, $d_{10}(x)$ are subject to the following two assumptions.

Assumption 3 The spatially varying coefficients $d_2(x)$, $d_3(x)$, $d_4(x)$, $d_5(x)$, $d_7(x)$, $d_8(x)$, $d_9(x)$, $d_{10}(x)$ belong to $C^0([0, L])$. □

Assumption 4 The spatially varying coefficients $d_1(x), d_6(x)$, which belong to $C^1([0, L])$, are positive, and $d_1(x) \neq d_6(x)$, $\forall x \in [0, L]$. □

The time-varying domain, i.e., the moving boundary $l(t)$, is subject to the following two assumptions.

Assumption 5 The function $l(t)$ is bounded, i.e., $0 < l(t) \leq L, \forall t \geq 0$. □

Assumption 6 The function $\dot{l}(t)$ is bounded as

$$|\dot{l}(t)| < \min_{0\leq x\leq L}\{\sqrt{d_1(x)}, \sqrt{d_6(x)}\}.$$ □

Assumptions 3–6 about the spatially varying coefficients and the time-varying spatial domain of (58)–(63) are fully satisfied in the application of the DCV.

From a practical point of view, the available measurements are the acceleration signals $u_{tt}(l(t),t)$, $w_{tt}(l(t),t)$ obtained by accelerometers placed at the crane, because, as mentioned before, measuring vibrational acceleration rather than velocity/displacement is more viable in vibrational mechanical systems. The velocity signals $u_t(l(t),t)$, $w_t(l(t),t)$ can then be obtained by integrations of the measured acceleration signals under known initial conditions (this is only the explanation of an implementable signal acquisition method in practice, and no restrictions are imposed on any initial conditions in the design and theory).

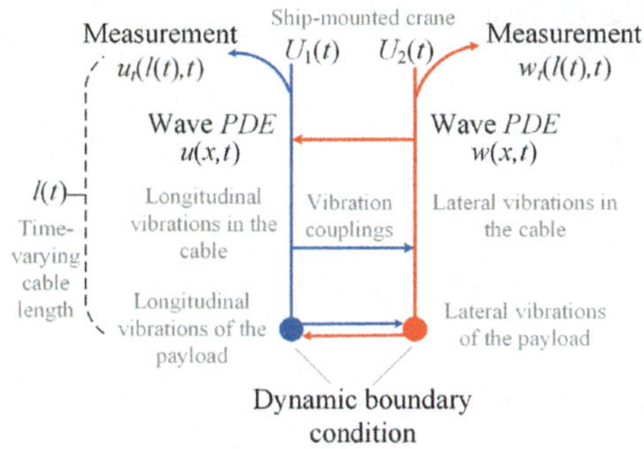

Fig. 12 Diagram of the plant (58)–(63). The text in gray describes the according physical meanings of (58)–(63) in the specific application of the DCV

Our objective is to exponentially stabilize (58)–(63) through designing the control inputs $U_1(t), U_2(t)$ in (62), (63) using only the boundary values $u_t(l(t), t)$, $w_t(l(t), t)$, i.e., a collocated type output-feedback control system.

(2) Reformulated Plant

A preliminary transformation, which allows one to convert the original plant (58)–(63) to a reformulated plant, where backstepping is readily applicable, is utilized. We introduce a set of the Riemann transformations

$$z(x,t) = w_t(x,t) + \sqrt{d_1(x)}w_x(x,t), \tag{64}$$

$$v(x,t) = w_t(x,t) - \sqrt{d_1(x)}w_x(x,t), \tag{65}$$

$$k(x,t) = u_t(x,t) + \sqrt{d_6(x)}u_x(x,t), \tag{66}$$

$$y(x,t) = u_t(x,t) - \sqrt{d_6(x)}u_x(x,t) \tag{67}$$

and define new variables as

$$X(t) = [w(0,t), w_t(0,t)], \quad Y(t) = [u(0,t), u_t(0,t)], \tag{68}$$

to reformulate (58)–(63), in a matrix representation, as

$$p_t(x,t) + Q(x)p_x(x,t) = T_a(x)r(x,t) + T_b(x)p(x,t), \tag{69}$$

$$r_t(x,t) - Q(x)r_x(x,t) = T_a(x)r(x,t) + T_b(x)p(x,t), \tag{70}$$

$$p(0,t) = C_3W(t) - r(0,t), \tag{71}$$

$$\dot{W}(t) = (\bar{A} - \bar{B}C_3)W(t) + 2\bar{B}r(0,t), \tag{72}$$

$$r(l(t),t) = R(l(t))U(t) + p(l(t),t), \tag{73}$$

where

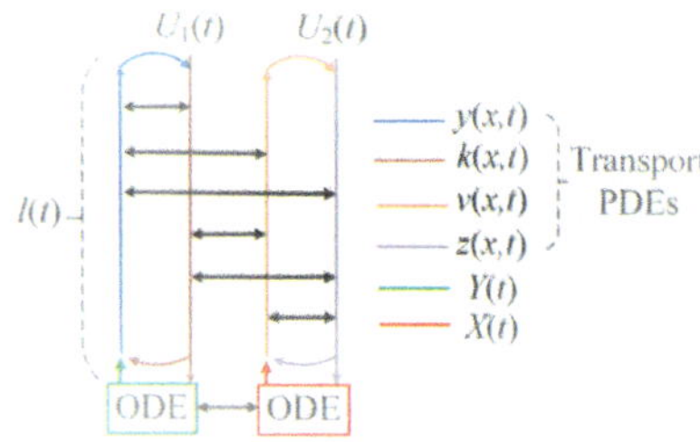

Fig. 13 Diagram of the system (69)–(73)

$$p(x,t) = [y(x,t), v(x,t)]^T, \tag{74}$$

$$r(x,t) = [k(x,t), z(x,t)]^T, \quad W(t) = [X(t), Y(t)]^T \tag{75}$$

and where $U(t) = [U_1(t), U_2(t)]^T$, and $R(l(t)) = 2\mathrm{diag}(\sqrt{d_6(l(t))}d_{19}(l(t)), \sqrt{d_1(l(t))}d_{20}(l(t)))$, $Q(x) = \mathrm{diag}\{Q_1(x), Q_2(x)\} = \mathrm{diag}\{\sqrt{d_6(x)}, \sqrt{d_1(x)}\}$. The matrices $\bar{A}$, C_3, $\bar{B}$ and $T_a(x) = \{T_{aij}(x)\}_{1\le i,j\le 2}$, $T_b(x) = \{T_{bij}(x)\}_{1\le i,j\le 2}$ are given in (6.68)–(6.71) with (6.54)–(6.57) in [18].

The following assumption is required for the stabilization of the ODE subsystem (72) in the state-feedback and observer design, and is also satisfied in the DCV model.

Assumption 7 The pair $(\bar{A} - \bar{B}C_3, \bar{B})$ is controllable and $(\bar{A} - \bar{B}C_3, C_3)$ is observable. □

The diagram of the system (69)–(73) is depicted in Fig. 13. The system (69)–(73) is 4×4 coupled linear heterodirectional hyperbolic PDE-ODE system, where $y(x,t)$, $v(x,t)$ and $k(x,t)$, $z(x,t)$ propagate in opposite directions, and where the four transport PDEs are coupled with each other in the time-varying spatial domain $[0, l(t)]$ and are coupled with two ODEs $X(t)$, $Y(t)$, which are coupled with each other as well, at the uncontrolled boundary $x = 0$.

In the following subsections, we develop an observer-based output-feedback controller for (69)–(73) with only collocated boundary states assumed measurable. Then the control input and the stability result will be formulated back for the original plant (58)–(63).

3.2 Basic Control Design Using Full States

We seek an invertible transformation that converts the system (69)–(73), with variables $(p(x,t), r(x,t), W(t))$, into a target system whose exponential stability is straightforward to prove.

We postulate the backstepping transformation in the form

$$\alpha(x,t) = p(x,t) - \int_0^x K(x,y)p(y,t)dy - \int_0^x J(x,y)r(y,t)dy - \gamma(x)W(t), \tag{76}$$

$$\beta(x,t) = r(x,t) - \int_0^x F(x,y)p(y,t)dy - \int_0^x N(x,y)r(y,t)dy - \lambda(x)W(t), \tag{77}$$

where $K(x,y) = \{K_{ij}(x,y)\}_{1\le i,j\le 2}$, $J(x,y) = \{J_{ij}(x,y)\}_{1\le i,j\le 2}$, $F(x,y) = \{F_{ij}(x,y)\}_{1\le i,j\le 2}$, $N(x,y) = \{N_{ij}(x,y)\}_{1\le i,j\le 2}$, on the triangular domain $\mathcal{D} = \{0 \le y \le x \le l(t)\}$, and $\gamma(x) = \{\gamma_{ij}(x)\}_{1\le i\le 2, 1\le j\le 4}$, $\lambda(x) = \{\lambda_{ij}(x)\}_{1\le i\le 2, 1\le j\le 4}$ are to be determined.

The target system is designed as

$$\dot{W}(t) = \hat{A}W(t) + 2\bar{B}\beta(0,t), \tag{78}$$

$$\alpha_t(x,t) = -Q(x)\alpha_x(x,t) + \bar{T}_b(x)\alpha(x,t) + g_1(x)\beta(0,t), \tag{79}$$

$$\beta_t(x,t) = Q(x)\beta_x(x,t) + \bar{T}_a(x)\beta(x,t) + g(x)\beta(0,t), \tag{80}$$

$$\alpha(0,t) = -\beta(0,t), \tag{81}$$

$$\beta(l(t),t) = 0, \tag{82}$$

where $\hat{A} = \bar{A} - \bar{B}C_3 + 2\bar{B}\kappa$ is a Hurwitz matrix, by choosing the design parameter $\kappa = \{\kappa_{ij}\}_{1\le i\le 2, 1\le j\le 4}$, recalling Assumption 7. The matrices $\bar{T}_a(x)$, $\bar{T}_b(x)$ are diagonal ones. Thus all coupling terms in the PDE domain (69), (70) are removed. The functions $g(x)$, $g_1(x)$ are in the form of

$$g(x) = \begin{pmatrix} 0 & 0 \\ g_a(x) & 0 \end{pmatrix}, \quad g_1(x) = \begin{pmatrix} 0 & 0 \\ g_b(x) & 0 \end{pmatrix} \tag{83}$$

where

$$g_a(x) = \sqrt{d_6(0)}F_{21}(x,0) - \frac{\lambda_{22}(x)d_{14}}{\sqrt{d_6(0)}} - \frac{\lambda_{24}(x)d_{16}}{\sqrt{d_6(0)}} + \sqrt{d_6(0)}N_{21}(x,0),$$

$$g_b(x) = \sqrt{d_6(0)}J_{21}(x,0) - \frac{\gamma_{22}(x)d_{14}}{\sqrt{d_6(0)}} - \frac{\gamma_{24}(x)d_{16}}{\sqrt{d_6(0)}} + \sqrt{d_6(0)}K_{21}(x,0).$$

By matching the systems (69)–(73) and (78)–(82) through (76), (77), a lengthy but straightforward calculation leads to the conditions on the kernels $F(x,y)$, $N(x,y)$, $\lambda(x)$, $K(x,y)$, $J(x,y)$, $\gamma(x)$ in (76), (77), which are coupled transport PDE-ODE systems, whose details and well-posedness are given in (6.87)–(6.100) and Lemma 6.1 in [18].

The control input is obtained as

$$U(t) =- R(l(t))^{-1}\Big[p(l(t), t) - \int_0^{l(t)} F(l(t), y)p(y, t)dy \\ - \int_0^{l(t)} N(l(t), y)r(y, t)dy - \lambda(l(t))W(t)\Big] \tag{84}$$

by matching boundary conditions (73) and (82) via (77).

3.3 *Observer for Two-Dimensional Oscillations of the Cable*

We consider only sensors placed at the actuated boundary (ship-mounted crane). An observer should then be designed to estimate the states w, u in the domain and at the uncontrolled boundary, i.e., the lateral-longitudinal coupled vibration states of the cable and the attached payload. These estimates are required in the state-feedback control laws.

The available measurements are $u_t(l(t), t)$, $w_t(l(t), t)$, which makes $p(l(t), t)$ known through the conversion

$$p(l(t), t) =[u_t(l(t), t) - \sqrt{d_6(l(t))}d_{19}(l(t))U_1(t), \\ w_t(l(t), t) - \sqrt{d_1(l(t))}d_{20}(l(t))U_2(t)], \tag{85}$$

by recalling (62), (63), (65), (67), and (74). Using the known signal $p(l(t), t)$, the observer for the coupled wave PDE plant (58)–(63) is constructed as

$$\hat{w}_t(x, t) =\frac{1}{2}(\hat{z}(x, t) + \hat{v}(x, t)), \tag{86}$$

$$\hat{w}_x(x, t) =\frac{1}{2\sqrt{d_1(x)}}(\hat{z}(x, t) - \hat{v}(x, t)), \tag{87}$$

$$\hat{u}_t(x, t) =\frac{1}{2}(\hat{k}(x, t) + \hat{y}(x, t)), \tag{88}$$

$$\hat{u}_x(x, t) =\frac{1}{2\sqrt{d_6(x)}}(\hat{k}(x, t) - \hat{y}(x, t)), \tag{89}$$

$$\hat{p}_t(x, t) + Q(x)\hat{p}_x(x, t) =T_a(x)\hat{r}(x, t) + T_b(x)\hat{p}(x, t) \\ + \Gamma_1(x, t)(p(l(t), t) - \hat{p}(l(t), t)), \tag{90}$$

$$\hat{r}_t(x, t) - Q(x)\hat{r}_x(x, t) =T_a(x)\hat{r}(x, t) + T_b(x)\hat{p}(x, t) \\ + \Gamma_2(x, t)(p(l(t), t) - \hat{p}(l(t), t)), \tag{91}$$

$$\hat{p}(0, t) =C_3\hat{W}(t) - \hat{r}(0, t), \tag{92}$$

$$\dot{\hat{W}}(t) =(\bar{A} - \bar{B}C_3)\hat{W}(t) + 2\bar{B}\hat{r}(0, t) \\ + \Gamma_3(t)(p(l(t), t) - \hat{p}(l(t), t)), \tag{93}$$

$$\hat{r}(l(t), t) =R(l(t))U(t) + p(l(t), t) \tag{94}$$

where $\hat{p} = [\hat{y}(x,t), \hat{v}(x,t)]^T$, $\hat{r} = [\hat{k}(x,t), \hat{z}(x,t)]^T$, $\hat{W}(t) = [\hat{X}(t), \hat{Y}(t)]^T = [\hat{w}(0,t), \hat{w}_t(0,t), \hat{u}(0,t), \hat{u}_t(0,t)]^T$.

The observer consists of two parts: 1. The system (90)–(94), as a copy of plant (69)–(73) with the addition of output injection, is to estimate $p(x,t), r(x,t)$; 2. Once $p(x,t), r(x,t)$ are estimated successfully by (90)–(94), the estimates of the original plant are simply obtained as (86)–(89), by virtue of the Riemann transformations (64)–(67).

The observer gains $\Gamma_1(x,t)$, $\Gamma_2(x,t)$, and $\Gamma_3(t)$ are determined, by using the backstepping method whose details are omitted here due to the space limit, as

$$\Gamma_1(x,t) = \dot{l}(t)\bar{\varphi}(x,l(t)) - \bar{\varphi}(x,l(t))Q(l(t)), \tag{95}$$

$$\Gamma_2(x,t) = \dot{l}(t)\bar{\psi}(x,l(t)) - \bar{\psi}(x,l(t))Q(l(t)), \tag{96}$$

$$\Gamma_3(t) = \dot{l}(t)\bar{K}(l(t)) - \bar{K}(l(t))Q(l(t)), \tag{97}$$

where $\bar{\varphi}(x,y) = \{\bar{\varphi}_{ij}(x,y)\}_{1\le i,j\le 2}$, $\bar{\psi}(x,y) = \{\bar{\psi}_{ij}(x,y)\}_{1\le i,j\le 2}$ on a triangular domain $\mathcal{D}_1 = \{0 \le x \le y \le l(t)\}$, and $\bar{K}(y) = \{\bar{K}_{ij}(y)\}_{1\le i\le 4, 1\le j\le 2}$ satisfy a coupled transport PDE-ODE system, whose details and well-posedness are given in (6.178)–(6.184) and Lemma 6.3 in [18].

Under the observer gains $\Gamma_1(x,t)$ in (95), $\Gamma_2(x,t)$ in (96), $\Gamma_3(t)$ in (97), it can be shown that the estimates $\hat{p}, \hat{r}, \hat{W}$ in (90)–(94) track the states p, r, W in (74), (75) in the sense that $\|p(\cdot,t) - \hat{p}(\cdot,t)\|^2 + \|r(\cdot,t) - \hat{r}(\cdot,t)\|^2 + |W(t) - \hat{W}(t)|^2$ exponentially converge to zero. Recalling (64)–(67) and (86)–(89), we then obtain that the observer can track the actual states in the original plant (58)–(63) in the sense that there exist positive constants Υ_1, σ_1 such that $\tilde{\Theta}(t) \le \Upsilon_1\tilde{\Theta}(0)e^{-\sigma_1 t}$, where $\tilde{\Theta}(t) = \big(\|u_t(\cdot,t) - \hat{u}_t(\cdot,t)\|^2 + \|u_x(\cdot,t) - \hat{u}_x(\cdot,t)\|^2 + \|w_t(\cdot,t) - \hat{w}_t(\cdot,t)\|^2 + \|w_x(\cdot,t) - \hat{w}_x(\cdot,t)\|^2 + (w(0,t) - \hat{w}(0,t))^2 + (w_t(0,t) - \hat{w}_t(0,t))^2 + (u(0,t) - \hat{u}(0,t))^2 + (u_t(0,t) - \hat{u}_t(0,t))^2\big)^{\frac{1}{2}}$.

3.4 Controller with Collocated Boundary Sensing

The output-feedback control law, employing axial and lateral control forces at the ship-mounted crane with sensors placed only at that location, is obtained by combining the state-feedback controller and the observer from Sects. 3.2 and 3.3.

Applying (64)–(67), (68), (74), (75), then the state-feedback controller (84) can be expanded and rewritten in terms of the original states, where inserting (62), (63) to replace $u_x(l(t),t)$, $w_x(l(t),t)$ is used. After inserting the observer states into the state-feedback controller to replace the unmeasurable states, the output-feedback form denoted as $U_{o1}(t)$, $U_{o2}(t)$ are obtained as

$$
\begin{aligned}
U_{oi}(t) = & -\frac{1}{h_i(l(t))}\Big[H_i(l(t),t) - \int_0^{l(t)} \big[(F_{i1}(l(t),y) + N_{i1}(l(t),y))\hat{u}_t(y,t) \\
& + (F_{i2}(l(t),y) + N_{i2}(l(t),y))\hat{w}_t(y,t) \\
& + (N_{i1}(l(t),y) - F_{i1}(l(t),y))\sqrt{d_6(y)}\hat{u}_x(y,t) \\
& + (N_{i2}(l(t),y) - F_{i2}(l(t),y))\sqrt{d_1(y)}\hat{w}_x(y,t)\big]dy \\
& - \lambda_{i1}(l(t))\hat{w}(0,t) - \lambda_{i2}(l(t))\hat{w}_t(0,t) \\
& - \lambda_{i3}(l(t))\hat{u}(0,t) - \lambda_{i4}(l(t))\hat{u}_t(0,t)\Big] \qquad (98)
\end{aligned}
$$

for $i = 1, 2$, where $h_1(l(t)) = d_{19}(l(t))\sqrt{d_6(l(t))}$, $h_2(l(t)) = d_{20}(l(t))\sqrt{d_1(l(t))}$, $H_1(l(t), t) = u_t(l(t), t)$, and $H_2(l(t), t) = w_t(l(t), t)$.

The control inputs $U_{o1}(t)$, $U_{o2}(t)$ are implemented only based on the boundary measurements $u_t(l(t), t)$, $w_t(l(t), t)$. To be exact, $u_t(l(t), t)$, $w_t(l(t), t)$ directly act as the first term in the expressions (98), and are also used to obtain the solutions of the observer system (86)–(94) which are required in the remaining terms in (98), through the conversion (85).

The following theorem physically implies that the vibrational energy of the cable, including kinetic energy and potential energy in two directions is exponentially convergent to zero. The proof is omitted due to the space limit.

Theorem 2 *For the closed-loop system consisting of the plant* (58)–(63)*, the observer* (86)–(94)*, and the output-feedback controller* (98)*, with arbitrary initial values* $(w(x,0), w_t(x,0)) \in H^2(0,L) \times H^1(0,L)$, $(u(x,0), u_t(x,0)) \in H^2(0,L) \times H^1(0,L)$, $(\hat{w}(x,0), \hat{w}_t(x,0)) \in H^2(0,L) \times H^1(0,L)$, $(\hat{u}(x,0), \hat{u}_t(x,0)) \in H^2(0,L) \times H^1(0,L)$, *one obtains*

(1) There exist positive constants Υ_2 *and* σ_2 *such that* $\Xi(t) \le \Upsilon_2 \Xi(0) e^{-\sigma_2 t}$, *where* $\Xi(t)$ *is defined as* $\Xi(t) = \big(\|u_x(\cdot,t)\|^2 + \|u_t(\cdot,t)\|^2 + \|w_x(\cdot,t)\|^2 + \|w_t(\cdot,t)\|^2 + u(0,t)^2 + u_t(0,t)^2 + w(0,t)^2 + w_t(0,t)^2 + \|\hat{u}_x(\cdot,t)\|^2 + \|\hat{u}_t(\cdot,t)\|^2 + \|\hat{w}_x(\cdot,t)\|^2 + \|\hat{w}_t(\cdot,t)\|^2 + \hat{u}(0,t)^2 + \hat{u}_t(0,t)^2 + \hat{w}(0,t)^2 + \hat{w}_t(0,t)^2\big)^{\frac{1}{2}}$.

(2) The output-feedback control signals (98) *are bounded and exponentially convergent to zero.*

3.5 Simulation for Deep-Sea Construction System

The simulation is conducted based on the actual nonlinear model (6.13)–(6.18) in [18] with external disturbances. The plant (and the observer) on the time-varying domain with pre-determined time-varying functions $l(t)$ and $\dot{l}(t)$, which are shown in Fig. 14, is converted to a $u(\iota, t)$ plant on the fixed spatial domain $\iota \in [0, 1]$, with time-varying coefficients related to $l(t)$, $\dot{l}(t)$, $\ddot{l}(t)$, by introducing $\iota = \frac{x}{l(t)}$. Please see Sect. 6.5 in [18] for more details, which are omitted here due to space limit, in conducting the

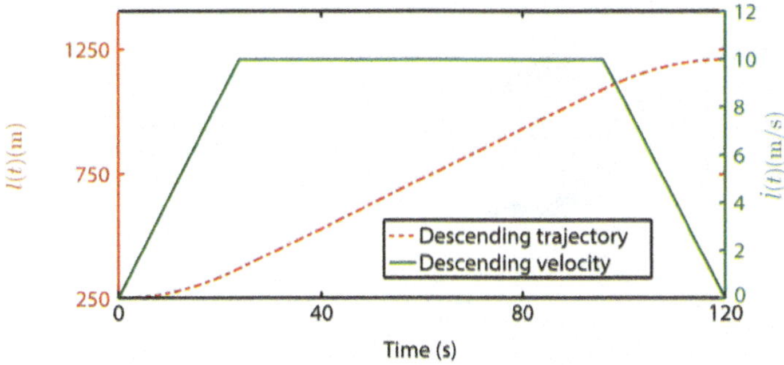

Fig. 14 Descending trajectory and velocity, i.e., the time-varying cable length $l(t)$ and the changing rate $\dot{l}(t)$

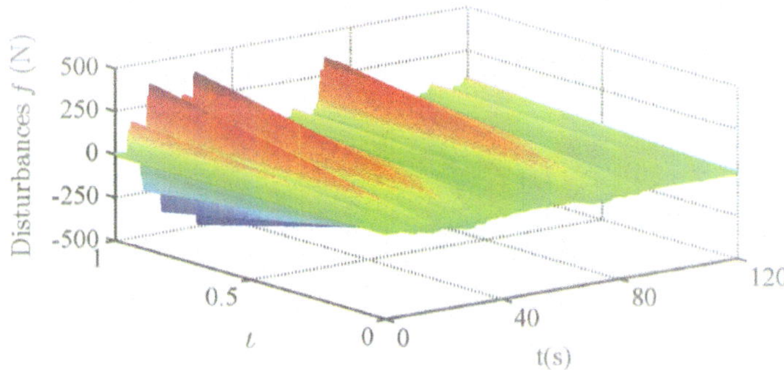

Fig. 15 Lateral oscillation drag forces from ocean current disturbances

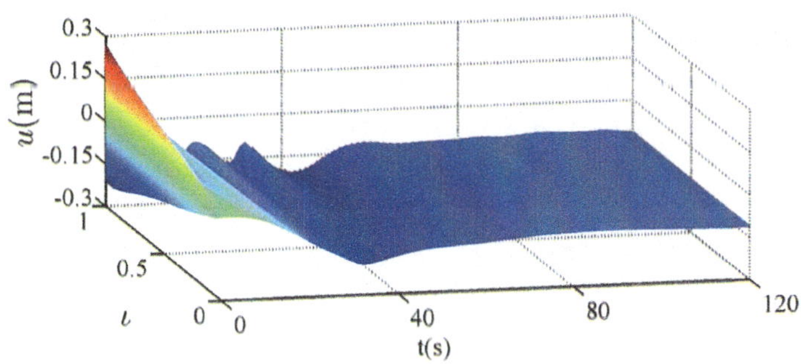

Fig. 16 Closed-loop responses of longitudinal vibrations $u(\iota, t)$ in the actual nonlinear model with unmodeled disturbances

simulation. Lateral oscillation drag forces from ocean current disturbances, whose formulation is given in (6.223), (6.224) in [18], are shown in Fig. 15.

We apply the proposed output-feedback controller to the actual nonlinear model with the ocean current disturbances. It is shown in Figs. 16, 17 that the longitudinal vibrations and lateral vibrations are reduced as time goes on. It is particularly remarkable in Fig. 17, that, while the control law is designed only for suppressing

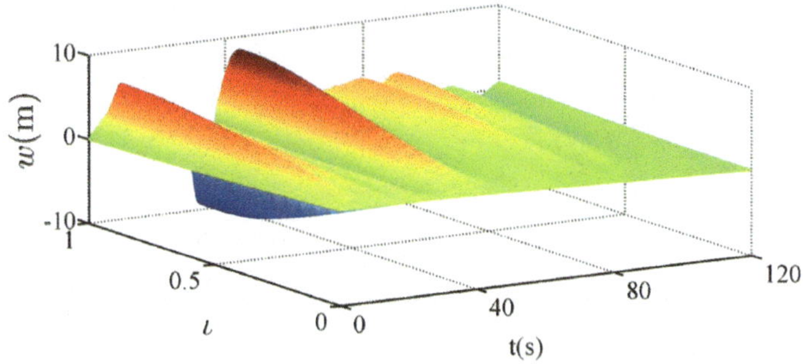

Fig. 17 Closed-loop responses of lateral vibrations $w(\iota, t)$ in the actual nonlinear model with unmodeled disturbances

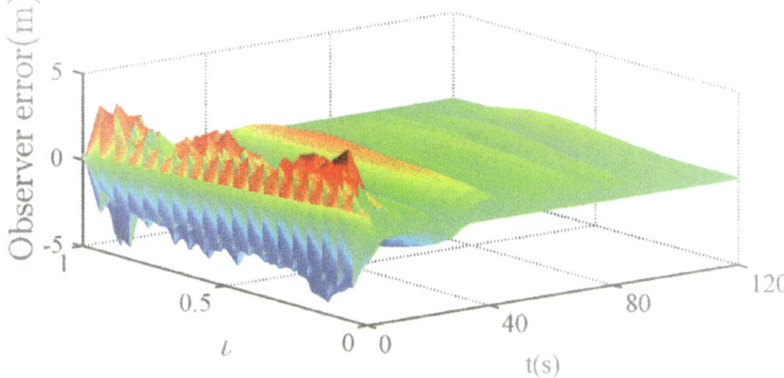

Fig. 18 Observer errors $\tilde{u}(\iota, t) + \tilde{w}(\iota, t)$ in the actual nonlinear model with unmodeled disturbances

Fig. 19 Control forces in the actual nonlinear model

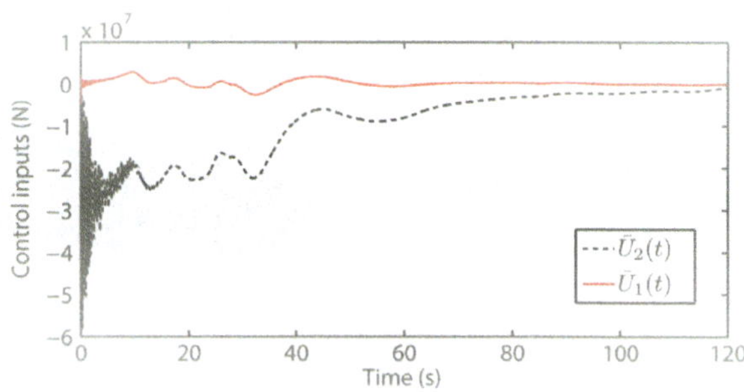

oscillations in response to initial lateral displacements, and not for attenuating a lateral disturbance, like the disturbance in Fig. 15, the controller is effective in disturbance attenuation after an initial transient of about 40 s. From Fig. 18, the observer errors converge to a small range around zero. Simulation results illustrate the effectiveness of the proposed control design when applied into vibration suppression of the DCV, where the two-directional output-feedback control inputs are shown in Fig. 19.

References

1. Anfinsen, H., Aamo, O.M.: Disturbance rejection in general heterodirectional 1-D linear hyperbolic systems using collocated sensing and control. Automatica **76**, 230–242 (2017)
2. Auriol, J., Di Meglio, F.: Minimum time control of heterodirectional linear coupled hyperbolic PDEs. Automatica **71**, 300–307 (2016)
3. Bohm, M., Krstic, M., Kuchler, S., Sawodny, O.: Modeling and boundary control of a hanging cable immersed in water. J. Dyn. Syst. Meas. Control **136**, 011006 (2014)
4. Coron, J.M., Vazquez, R., Krstic, M., Bastin, G.: Local exponential H^2 stabilization of a 2×2 quasilinear hyperbolic system using backstepping. SIAM J. Control Optim. **51**(3), 2005–2035 (2013)
5. Deutscher, J.: Output regulation for general linear heterodirectional hyperbolic systems with spatially-varying coefficients. Automatica **85**, 34–42 (2017)
6. Deutscher, J., Gabrie, J.: Minimum time output regulation for general linear heterodirectional hyperbolic systems. Int. J. Control **93**, 1826–1838 (2020)
7. Deutscher, J., Gehring, N., Kern, R.: Output feedback control of general linear heterodirectional hyperbolic PDE-ODE systems with spatially-varying coefficients. Int. J. Control **92**(10), 2274–2290 (2019)
8. Di Meglio, F., Bribiesca, F., Hu, L., Krstic, M.: Stabilization of coupled linear heterodirectional hyperbolic PDE-ODE systems. Automatica **87**, 281–289 (2018)
9. Di Meglio, F., Vazquez, R., Krstic, M.: Stabilization of a system of $n + 1$ coupled first-order hyperbolic linear PDEs with a single boundary input. IEEE Trans. Autom. Control **58**, 3097–3111 (2013)
10. Hu, L., Di Meglio, F., Vazquez, R., Krstic, M.: Control of homodirectional and general heterodirectional linear coupled hyperbolic PDEs. IEEE Trans. Autom. Control **61**(11), 3301–3314 (2016)
11. Hu, L., Vazquez, R., Di Meglio, F., Krstic, M.: Boundary exponential stabilization of 1-dimensional inhomogeneous quasi-linear hyperbolic systems. SIAM J. Control Optim. **57**(2), 963–998 (2019)
12. Kaczmarczyk, S., Ostachowicz, W.: Transient vibration phenomena in deep mine hoisting cables-part 1: mathematical model. J. Sound Vib. **262**(2), 219–244 (2003)
13. Roman, C., Bresch-Pietri, D., Prieur, C., Sename, O.: Robustness to in-domain viscous damping of a collocated boundary adaptive feedback law for an antidamped boundary wave PDE. IEEE Trans. Autom. Control **64**(8), 3284–3299 (2019)
14. Standing, R.G., Mackenzie, B.G., Snell, R.O.: Enhancing the technology for deepwater installation of subsea hardware. In: Offshore Technology Conference, pp. 1–6 (2002)
15. Stensgaard, T., White, C., Schiffer, K.: Subsea hardware installation from a FDPSO. In: Offshore Technology Conference, pp. 1–6 (2010)
16. Vazquez, R., Krstic, M., Coron, J.M.: Backstepping boundary stabilization and state estimation of a 2×2 linear hyperbolic system. In: 50th IEEE Conference on Decision and Control and European Control Conference, pp. 4937–4942 (2011)
17. Wang, J., Koga, S., Pi, Y., Krstic, M.: Axial vibration suppression in a PDE model of ascending mining cable elevator. J. Dyn. Syst. Meas. Control **140**, 111003 (2018)
18. Wang, J., Krstic, M.: PDE Control of String-Actuated Motion. Princeton University Press (2022)

Ramp Metering: Modeling, Simulations and Control Issues

Cédric Join, Hassane Abouaïssa, and Michel Fliess

Abstract The aim of ramp metering is to improve the highway traffic conditions by an appropriate regulation of the inflow from the on-ramps to the highway mainstream. Our presentation rests on several improvements: (1) Our simulation techniques do not need contrarily to other approaches any heuristic fundamental law. (2) There is no need of crucial time-varying quantities, like the critical density, which is most difficult to estimate correctly online. (3) Our feedback loop, which is stemming from model-free control, is easy to implement and shows an excellent robustness with respect to model mismatch. Several computer experiments are displayed and discussed.

1 Introduction

Ramp metering is about the use of traffic signals at highway on-ramps in order to control the rate of vehicles entering the highway. The signals can be set for different metering rates to optimize freeway flow and minimize congestion. The social and ecological damages due to traffic jams justify the huge academic literature which has been devoted to traffic flow dynamics and its regulation. See, e.g., [20–23, 28, 29,

C. Join (✉)
CRAN (CNRS, UMR 7039), Université de Lorraine, BP 239, 54506 Vandœuvre-lès-Nancy, France
e-mail: cedric.join@univ-lorraine.fr; cedric.join@alien-sas.com

H. Abouaïssa
Univ. Artois, UR 3926, Laboratoire de Génie Informatique et d'Automatique de l'Artois (LGI2A), 62400 Béthune, France
e-mail: hassane.abouaissa@univ-artois.fr

M. Fliess
LIX (CNRS, UMR 7161), École polytechnique, 91128 Palaiseau, France
e-mail: Michel.Fliess@polytechnique.edu; michel.fliess@alien-sas.com

C. Join · M. Fliess
AL.I.E.N., 7 rue Maurice Barrès, 54330 Vézelise, France

J. Auriol et al. (eds.), *Advances in Distributed Parameter Systems*, Advances in Delays and Dynamics 14, https://doi.org/10.1007/978-3-030-94766-8_10

32, 39], and references therein. Lack of space prevents us from a careful analysis of the various standpoints which are often antagonist.

Our contributions may be summarized like follows:

1. **No fundamental diagram for simulation purposes.** Before being implemented in practice any ramp metering procedure ought to be tested via computers. The similarities between highway traffic and hydrodynamics explain the importance of the Lighthill-Whitham-Richards (LWR) first order partial differential equation [26, 37]. Its numerical integration necessitates to know the relationship between the traffic density and the traffic speed. Several empirical laws, called *fundamental diagrams*, have been proposed. The calibration of those diagrams is far from being obvious. That is why contrarily to other approaches (see, e.g., [6]) we ignore them. Fundamental diagrams are replaced by traffic data which are collected during various situations. Note moreover that our viewpoint, contrarily to other ones like METANET [25], does not necessitate higher order partial differential equations.
2. **No difficult estimation technique.** Implementing any control law requires crucial quantities like the *critical density*, which are time-varying and therefore almost impossible to estimate in real time. They are replaced by ad hoc quantities which are deduced at once from easily measurable data.
3. **Model-free control.** ALINEA[1] [18, 19, 33, 34] is perhaps the most popular feedback control algorithm for traffic regulation. After first trials in Paris it has been quite often employed in many different places. Following [1], we are using here *model-free control* in the sense of [12, 13] and, more precisely, *intelligent proportional controllers*. This setting, which has been successfully tested in many concrete situations, has already been illustrated via various questions about intelligent transportation systems (see, e.g., [5, 9, 17, 30, 31, 36, 40–42]). Concrete experiments show today that model-free control not only yields better traffic regulation than ALINEA but is also simpler to implement.[2]

Some remarks might be useful:

- Let us emphasize that the LWR partial differential equation is only related to simulation purposes and not to traffic control. Such a control would necessitate a real-time calibration, which seems today beyond all reasonable hope.
- Achieving an efficient traffic regulation is simpler than obtaining reliable computer simulations. This ascertainment remains valid in most applications where model-free control plays a key rôle.

Our paper is organized as follows. Section 2 reviews model-free control and the corresponding *intelligent proportional controller*. Following [3], Sect. 2.3 investigates the relationship with proportional controllers, which should be considered as the backbone of ALINEA. Modeling issues are discussed in Sect. 3: The partial differential equation stemming from elementary hydrodynamic conservation laws; some

---

[1] It is the acronym of Asservissement *LIN*éaire d'*E*ntrée Autoroutière.

[2] There are many variants of ALINEA in the literature. We have selected for our computer comparisons the version [2].

fundamental empirical laws; the space discretization and its use without any fundamental empirical law; the ALINEA and model-free regulation without the critical density. Computer simulations are displayed and commented in Sect. 4. Section 4.2 in particular exhibits the superiority of model-free control in the presence of sudden changes. Section 5 contains some concluding remarks.

2 Model-Free Control

2.1 *A Short Review*

2.1.1 Ultra-Local Model

Consider only, for simplicity's sake, SISO (single-input single-output) systems. Elementary functional analysis and differential algebra as used in [12] show as well as practical experiments that most, or at least many, concrete systems may be approximated by the *ultra-local*

$$\boxed{\dot{y} = F + \alpha u} \tag{1}$$

where

- the control and output variables are respectively u and y,
- the constant $\alpha \in \mathbb{R}$ is chosen by the practitioner such that αu and $\dot{y}$ are of the same magnitude. Therefor α does not need to be precisely estimated.

The following comments might be useful:

- F is estimated via the knowledge of the control and output variables u and y,
- F subsumes not only the unknown structure of the system, but also any external disturbance.

2.1.2 Intelligent Controllers

Close the loop with the following *intelligent proportional controller*, or *iP*,

$$\boxed{u = -\frac{F - \dot{y}^* + K_P e}{\alpha}} \tag{2}$$

where:

- y^* is the reference trajectory,
- $e = y - y^*$ is the tracking error,
- K_P is a tuning gain.

Combining Eqs. (1) and (2) yields:

$$\dot{e} + K_P e = 0$$

where F does not appear anymore. Thus $\lim_{t \to +\infty} e(t) = 0$ iff $K_P > 0$. This local stability property proves that the tuning of K_P is straightforward. This is a major difference with the classic gain tuning for PIs and PIDs (see, e.g., [4] and the references therein).

2.1.3 Estimation of *F*

Under a weak integrability condition, F in Eq. (1) may be "well" approximated by a piecewise constant function F_{est} (see, e.g., [8]). The estimation techniques below are borrowed from [14, 15, 38].[3] Let us summarize two types of computations:

1. Rewrite Eq. (1) in the operational domain (see, e.g., [43]):

$$sY = \frac{\Phi}{s} + \alpha U + y(0)$$

   where Φ is a constant. We get rid of the initial condition $y(0)$ by deriving both sides with respect to s:

$$Y + s\frac{dY}{ds} = -\frac{\Phi}{s^2} + \alpha \frac{dU}{ds}$$

   Noise attenuation is achieved by multiplying both sides on the left by s^{-2}, since integration with respect to time is a lowpass filter (see [11] for further details). It yields in the time domain the realtime estimate, thanks to the equivalence between $\frac{d}{ds}$ and the multiplication by $-t$,

$$F_{\text{est}}(t) = -\frac{6}{\tau^3} \int_{t-\tau}^{t} [(\tau - 2\sigma)y(\sigma) + \alpha\sigma(\tau - \sigma)u(\sigma)]\, d\sigma \tag{3}$$

   where $\tau > 0$ might be quite small. This integral may of course be replaced in practice by a classic digital filter.
2. Close the loop with the iP (2). It yields:

$$F_{\text{est}}(t) = \frac{1}{\tau}\left[\int_{t-\tau}^{t} \left(\dot{y}^* - \alpha u - K_P e\right) d\sigma\right]$$

[3] They were often used in practice for parameter identification (see, e.g., [27]).

2.2 PI and iP

Consider the classic proportional-integral controller, or PI,

$$\boxed{u(t) = k_p e(t) + k_i \int e(\tau)d\tau} \tag{4}$$

where $k_p, k_i \in \mathbb{R}$ are constants. A crude sampling of the integral $\int e(\tau)d\tau$ through a Riemann sum $\mathcal{I}(t)$ leads to

$$\int e(\tau)d\tau \simeq \mathcal{I}(t) = \mathcal{I}(t-h) + he(t)$$

where h is the sampling interval. The corresponding discrete form of Eq. (4) reads:

$$u(t) = k_p e(t) + k_i \mathcal{I}(t) = k_p e(t) + k_i \mathcal{I}(t-h) + k_i he(t)$$

Combining the above equation with

$$u(t-h) = k_p e(t-h) + k_i \mathcal{I}(t-h)$$

yields

$$u(t) = u(t-h) + k_p (e(t) - e(t-h)) + k_i he(t) \tag{5}$$

Replace in Eq. (2) F by $\dot{y}(t) - \alpha u(t-h)$ and therefore by

$$\frac{y(t) - y(t-h)}{h} - \alpha u(t-h) \tag{6}$$

It yields

$$u(t) = u(t-h) - \frac{e(t) - e(t-h)}{h\alpha} - \frac{K_P}{\alpha} e(t) \tag{7}$$

Equations (5) and (7) become **identical** if we set

$$k_p = -\frac{1}{\alpha h}, \quad k_i = -\frac{K_P}{\alpha h} \tag{8}$$

Let us emphasize that this important result, which was first stated in [10, 12], is only valid in discrete time: Formulae (8) become meaningless if $h \downarrow 0$.

2.3 Application to Integral Controllers

Set in Eq. (4) $k_p = 0$. It yields the integral controller, or I,

$$\boxed{u(t) = k_i \int e(\tau)d\tau} \tag{9}$$

Derive both sides in Eq. (4):

$$\dot{u} = k_p \dot{e} + k_i e$$

It shows that the PI and I controllers are "close" if $\dot{e}$ remains "small." It implies of course that the initial condition $y(0)$ is "close" to the initial point $y^*(0)$ of the reference trajectory. Let us try to explain why:

- the feedback loop is supposed to render $y - y^*$ quickly as small as possible,
- the derivative $\dot{e}$ might therefore be large.

It follows from Eqs. (5) and (6) that the sampled versions of the iP (2) and the I (9) are "close" (see [3] for more details) if

- the reference trajectory y^* is "slowly" varying, and starts at the initial condition $y(0)$ or, at least, at a point which is quite close to it,
- the disturbances and the corrupting noises are rather mild.

3 Application to Ramp Metering

3.1 Traffic Flow Modeling

3.1.1 The LWR Partial Differential Equation

The Lighthill-Whitham-Richards, or LWR, partial differential equation [26, 37], which was derived 65 years ago, is mimicking hydrodynamic conservation laws. It is the simplest model for a macroscopic traffic flow model

$$\frac{\partial}{\partial t}\rho(t, x) + \frac{\partial}{\partial x} f(t, x) = 0 \tag{10}$$

- ρ is the traffic density,
- the traffic flux $f = \rho \times v$ is the product of the traffic density and of the (mean) traffic speed v.

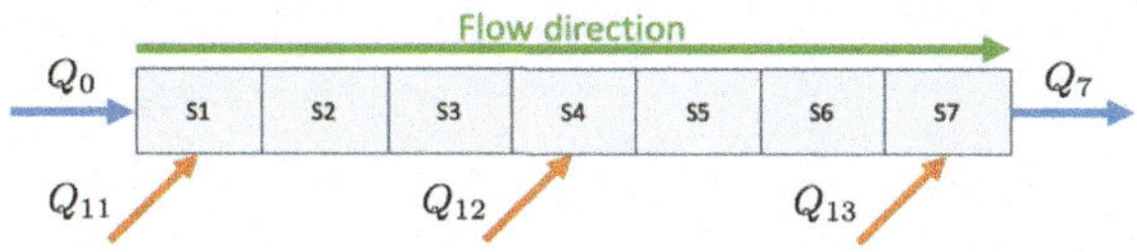

Fig. 1 Segments and ramps of motorway portion

3.1.2 Fundamental Laws

There are many empirical laws relating ρ and v. The Greenshield model [16]

$$v(\rho) = v_f \left(1 - \frac{\rho}{\rho_m}\right) \tag{11}$$

and the May formula [29]:

$$v(\rho) = v_f \exp\left(-\frac{1}{a}\left(\frac{\rho}{\rho_c}\right)^a\right) \tag{12}$$

for instance are quite popular. We will not try here to define the corresponding parameters. Note however that they depend on t and x. Their estimation is therefore rather cumbersome.

3.1.3 A Space Discretization for Numerical Simulations

Figure 1 displays a portion of a motorway in France, with 7 segments $S\iota, \iota = 1, \ldots, 7$. There are 3 lanes on $S\iota, \iota = 1, \ldots, 6$, and 4 on S_7. The lengths of the various segments are respectively 4.7, 0.6, 1.4, 1.7, 3.7, 0.6, 0.9 ([km]). Write Q_0 (resp. Q_7) the inlet (resp. output) flow. The traffic flow on $S\iota$ is defined by $(Q_\iota, To_\iota, V_\iota)$, where Q_ι, To_ι and V_ι are respectively the flow rate, the occupation rate and the speed. Our aim is to regulate the flow rates Q_{11}, Q_{12}, Q_{13} on the 3 access ramps in order to ease the traffic flow on the motorway. The control variables are the green light durations GD_κ, $\kappa = 1, 2, 3$, on the 3 on-ramps. The duration of a cycle of the ramp signals is 40s. Set $15\text{s} \leq GD_\kappa \leq 29\text{s}$.

The vehicle conservation principle for each segment ι should be understood as a space discretization of Eq. (10). It reads

$$\dot{\rho}_\iota = \frac{1}{L_\iota}(Q_{\iota-1} - Q_\iota)$$

- ρ_ι is the density ([Veh/km]),
- L_ι is the length ([km]),
- $Q_{\iota-1}$ is the inlet flow ([Veh/min]): it is the sum of the upstream flow and of the ramp flow if any,
- Q_ι is the output flow ([Veh/min]).

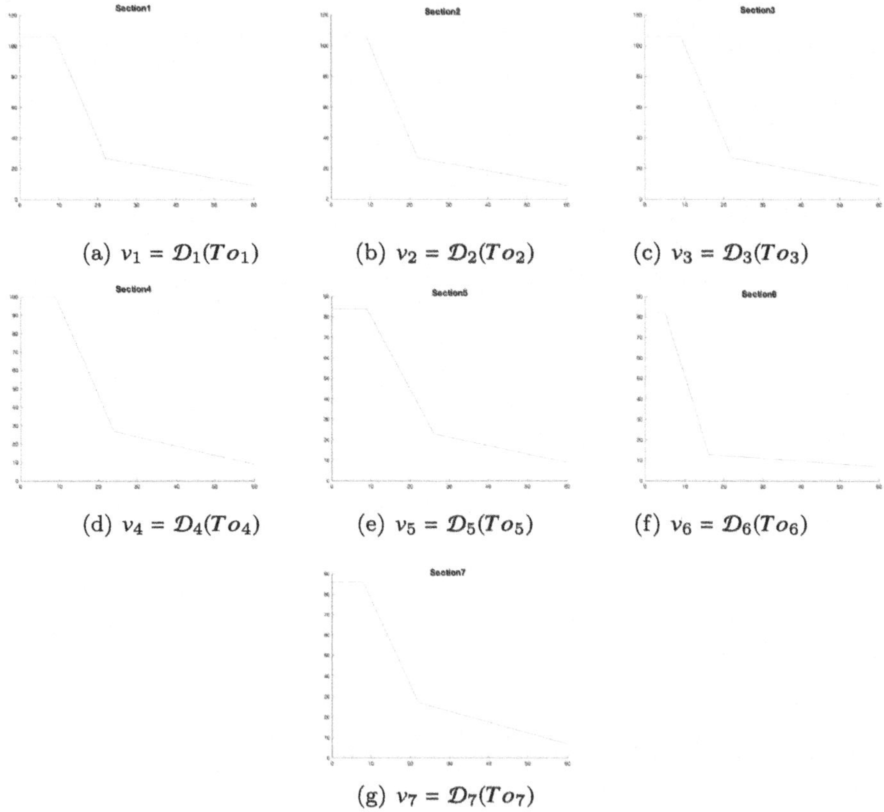

Fig. 2 Diagrams

Define the *occupancy* rate ([%]) by

$$To_\iota = \frac{\rho_\iota}{\rho_{\iota,\max}} \times 100$$

where $\rho_{\iota,\max} = \frac{\lambda_\iota}{\texttt{size}}$, λ_ι is the number of lanes, `size` is the mean length ([km]) of a vehicle (here $5.5.10^{-3}$).

The fundamental diagrams in Fig. 2 relate the (mean) speed ([km/h]) and the occupancy, i.e., $v_i = \mathcal{D}_\iota(To_\iota)$. They are derived from real data which are, therefore, confidential.

3.2 Feedback Loops

3.2.1 ALINEA Without Calibration

Equation (9) yields ALINEA in ramp metering. For $\kappa = 2$ it reads[4]

$$GD_2 = -K_{I,2} \int_0^t (To_4 - To_4^*) d\tau$$

where the gain $K_{I,2} \in \mathbb{R}$ is set equal to 1. A classic *anti-windup* setting (see, e.g., [4]) is mandatory. The *critical* occupancy rate To_4^* corresponds to the *critical* density ρ_c in Formula (12). A "good" real-time calibration of To_4^* seems today out of reach. We therefore set as in [2]

$$\begin{cases} To_4^*(t+1) = To_4^*(t) + \delta^+ \text{ if } V_4 > V_{4,\text{threshold}} \\ To_4^*(t+1) = To_4^*(t) - \delta^- \text{ if not} \end{cases}$$

where

- V_4 is the mean speed on the 4th segment,
- $V_{4,\text{threshold}} = V_{4,f} - 10$,
- $V_{4,f}$ is the *free* speed, i.e., the maximum speed when the traffic is light,
- $\delta^+ = 0.15$, $\delta^- = 0.3$.

3.2.2 iP

Equation (1) reads here

$$\begin{aligned} \dot{To}_1 &= F_1 + \alpha_1 GD_1 \\ \dot{To}_4 &= F_2 + \alpha_2 GD_2 \\ \dot{To}_7 &= F_3 + \alpha_3 GD_3 \end{aligned}$$

The analogous feedback loops of Eq. (2) become

$$\begin{aligned} GD_1 &= \frac{\dot{To}_1^* - F_1 - \mathcal{K}_1(To_1 - To_1^*)}{\alpha_1} \\ GD_2 &= \frac{\dot{To}_4^* - F_2 - \mathcal{K}_2(To_4 - To_4^*)}{\alpha_2} \\ GD_3 &= \frac{\dot{To}_7^* - F_3 - \mathcal{K}_3(To_7 - To_7^*)}{\alpha_3} \end{aligned}$$

[4] The cases $\kappa = 1$ and $\kappa = 3$ are similar.

Set $\alpha_j = 30$, $K_{p,j} = 0.5$. To^*_κ, $\kappa = 1, 2, 3$ is replaced as above.

4 Simulations

4.1 Generalities

Figures 3, 4, 5, 6, 7, 8, 9, 10, 11, 12 and 13 display convincing results via the setting of Sect. 3.2.2:

- The reference trajectory is a decreasing time function when the traffic is dense (see Fig. 10). This property does not hold anymore if the traffic becomes fluid.
- Figure 11 shows that the green light duration is set to the whole cycle of 40s when the queue on the ramp is large (see Fig. 12).
- Figure 13 indicates that the output flow on the ramps are reduced when there is a congestion.

Remark 10.1 Those results are only slightly better than those obtained via ALINEA with the same calibration of the setpoint. Let us emphasize however that ALINEA when implemented with the critical density ρ_c gives results which are disapointly similar to those obtained without any control, i.e., without ramp metering.

4.2 What Is Happening with a Sudden Change?

Assume only 2 lanes, instead of 3, on Segment S_2: it might be due to an accident or to some work on the highway. Figures 14 shows that the iP, without any new tuning, behave much better than ALINEA, especially outside the congestion hours when ramp metering is more or less useless. Those results agree with Sect. 2.3.

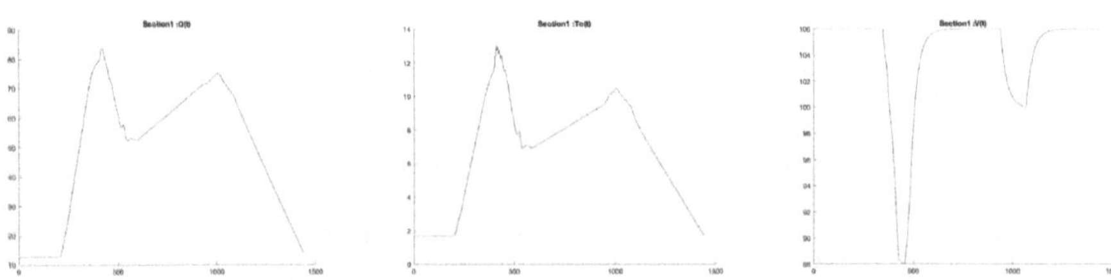

(a) Flow in *[Veh/min]* (b) Occupancy rate in *%* (c) Speed in *[km/h]*

Fig. 3 Section 1

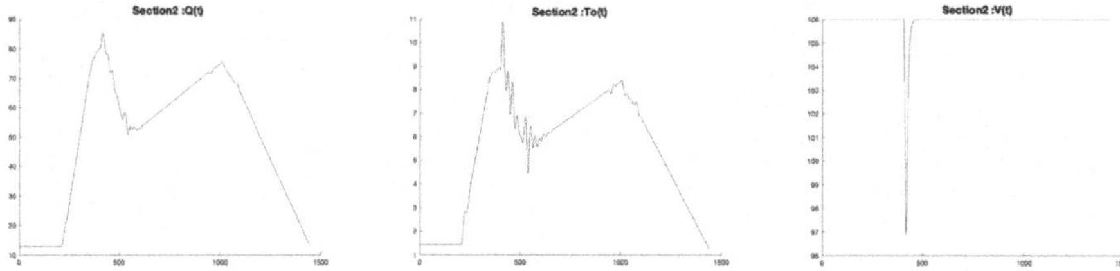

(a) Flow in *[Veh/min]* (b) Occupancy rate in % (c) Speed in *[km/h]*

Fig. 4 Section 2

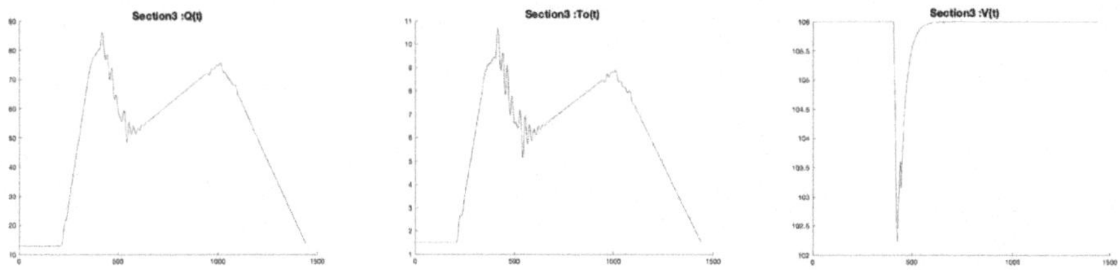

(a) Flow in *[Veh/min]* (b) Occupancy rate in % (c) Speed in *[km/h]*

Fig. 5 Section 3

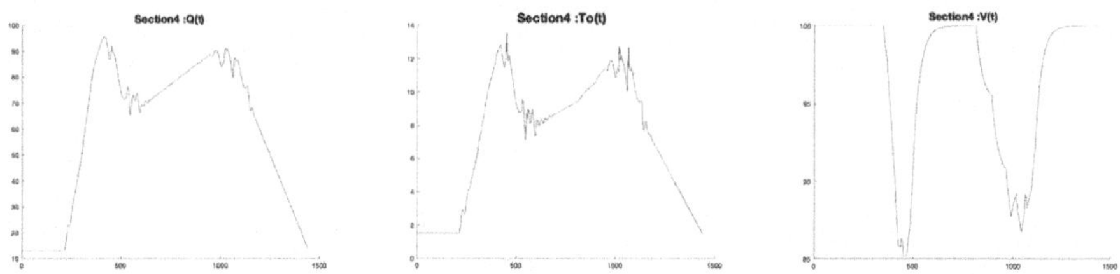

(a) Flow in *[Veh/min]* (b) Occupancy rate in % (c) Speed in *[km/h]*

Fig. 6 Section 4

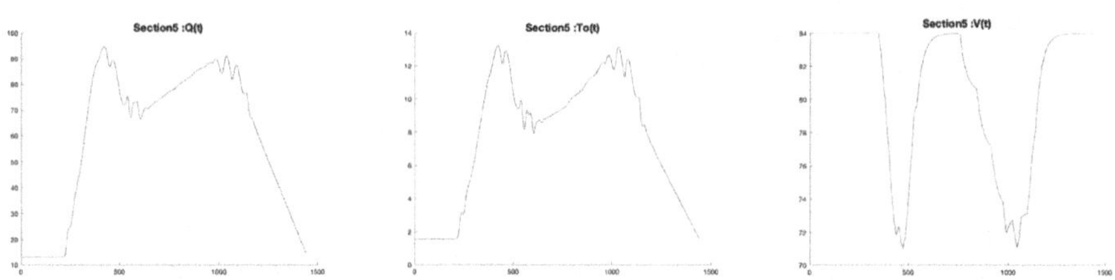

(a) Flow in *[Veh/min]* (b) Occupancy rate in % (c) Speed in *[km/h]*

Fig. 7 Section 5

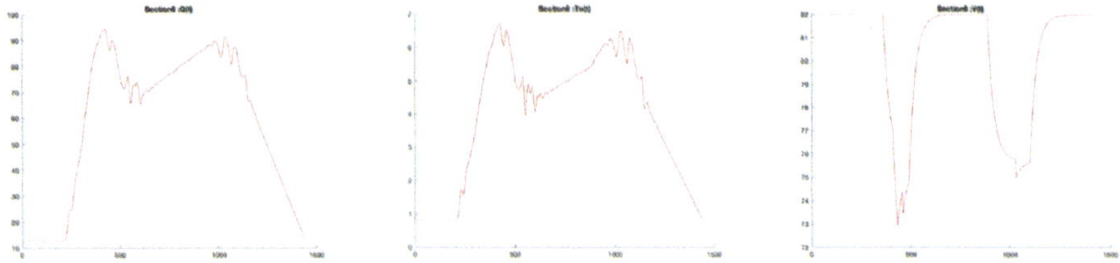

(a) Flow in *[Veh/min]* (b) Occupancy rate in % (c) Speed in *[km/h]*

Fig. 8 Section 6

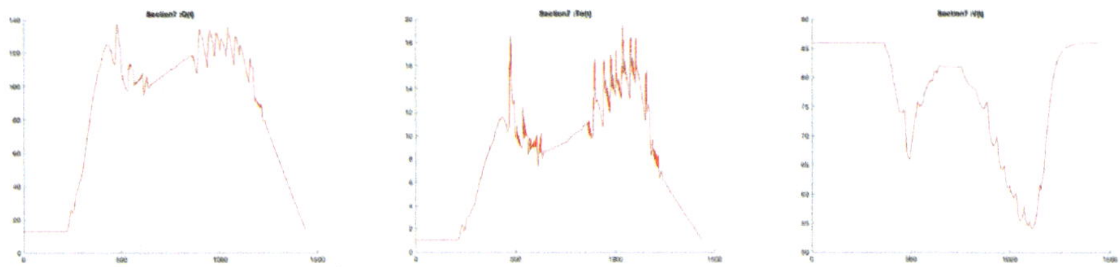

(a) Flow in *[Veh/min]* (b) Occupancy rate in % (c) Speed in *[km/h]*

Fig. 9 Section 7

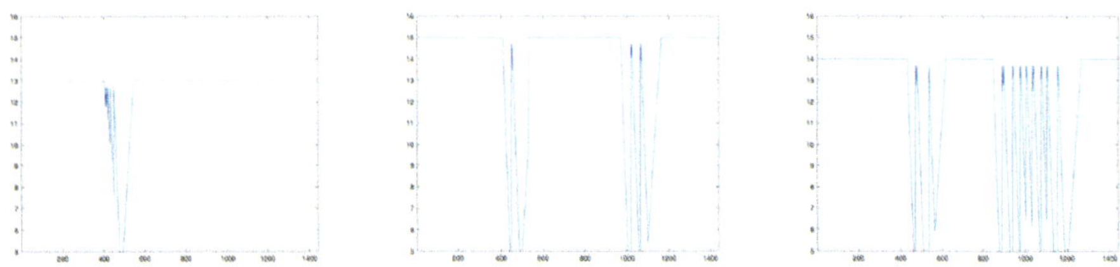

(a) Ramp 1 (b) Ramp 2 (c) Ramp 3

Fig. 10 Setpoint time evolution

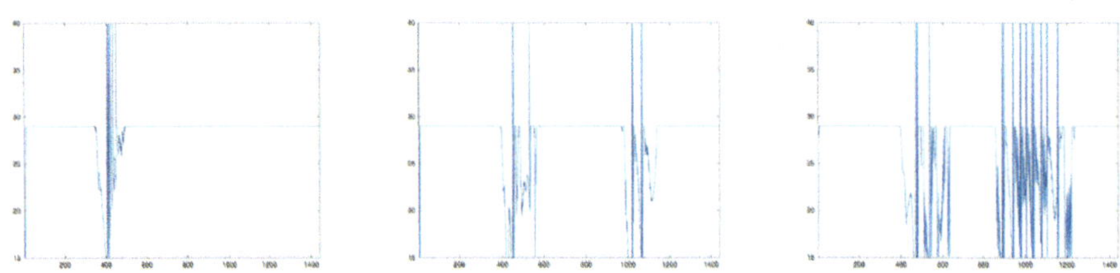

(a) Ramp 1 (b) Ramp 2 (c) Ramp 3

Fig. 11 Time evolution of green lights according to the control inputs GD_1, GD_2, GD_3

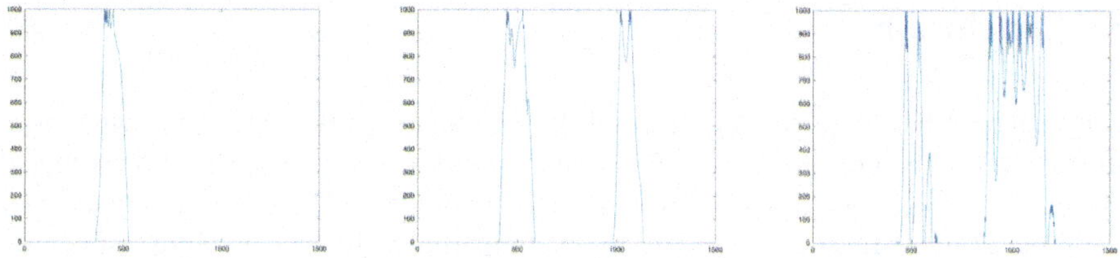

(a) Ramp 1 (b) Ramp 2 (c) Ramp 3

Fig. 12 Queue length in [m]

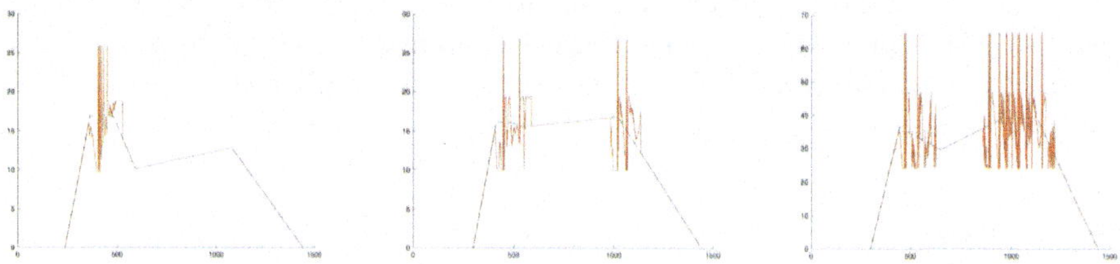

(a) Ramp 1 (b) Ramp 2 (c) Ramp 3

Fig. 13 Demand D_1, j (- -) and flow Q_1, j (–) in *[Veh/min]*

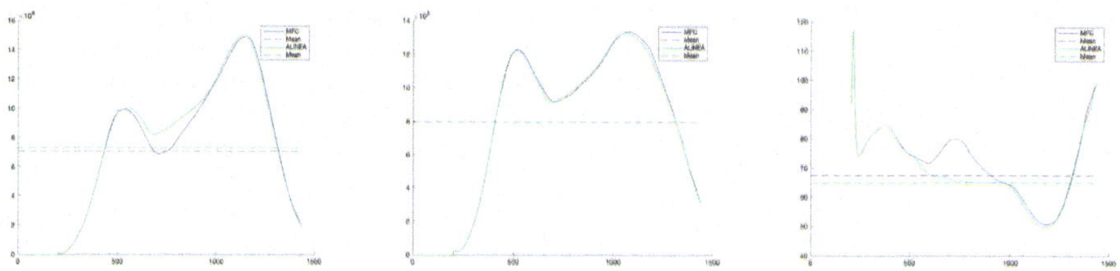

(a) TTS (b) TTP (c) MS

Fig. 14 ALINEA and MFC (TTS: Travel Time Spent, TTD: Total Travel Distance, MS: Mean Speed)

5 Conclusion

Our approach has already been successfully employed on several French highways. Its adaptative features explains its excellent robustness with respect to unexpected events. Its implementation moreover is quite elementary. There are of course other ramp metering settings which ought to be compared with our viewpoint, especially now those stemming from artificial intelligence. Le us notice that *deep reinforcement learning* methods do not seem today to perform better than ALINEA (see, e.g., [7]).

The *coordination* of several ramp-metering actions on highway networks is being investigated. Coordinated ALINEA techniques have been used in Australia some time ago [35]. It should be clear that any such coordination ought to be connected to a *variable speed limit* (see, e.g., [24]) in order to be more efficient.

References

1. Abouaïssa, H., Fliess, M., Iordanova, V., Join, C.: Freeway ramp metering control made easy and efficient. In: 13th IFAC Symposium on Control in Transportation Systems, Sofia (2012). https://hal.archives-ouvertes.fr/hal-00711847/en/
2. Abouaïssa, H., Fliess, M., Join, C.: A fresh look at ramp metering control: ALINEA without any tedious calibration. Res. Report. https://hal.archives-ouvertes.fr/hal-01117068/en/
3. Abouaïssa, H., Fliess, M., Join, C.: On ramp metering: towards a better understanding of ALINEA via model-free control. Int. J. Control **90**, 1018–1026 (2017)
4. Åström, K.J., Murray, R.M.: Feedback Systems: An Introduction for Scientists and Engineers. Princeton University Press (2008)
5. Baciu, A., Lazar, C.: Model-free iPD control design for a complex nonlinear automotive system. In: 24th International Conference on System Theory, Control, and Computing, Sinaia (2020)
6. Barceló, J. (Ed.): Fundamental of Traffic Simulation. Springer (2010)
7. Belletti, F., Haziza, D., Gomes, G., Bayen, A.M.: Expert level control of ramp metering based on multi-task deep reinforcement learning. IEEE Trans. Intel. Transport. Syst. **19**, 1198–1207 (2018)
8. Bourbaki, N.: Fonctions d'une variable réelle. Hermann (1976). English translation: Functions of a Real Variable. Springer (1994)
9. d'Andréa-Novel, B.: Model-free control of longitudinal and lateral dynamics for automated vehicles. JTEKT Engin. J. **1015E** (2018)
10. d'Andréa-Novel, B., Fliess, M., Join, C., Mounier, H., Steux, B.: A mathematical explanation via "intelligent" PID controllers of the strange ubiquity of PIDs. In: 18th Mediterranean Conference on Control and Automation, Marrakech (2010). https://hal.inria.fr/inria-00480293/en/
11. Fliess, M.: Analyse non standard du bruit. C.R. Acad. Sci. Paris Ser. I **342**, 797–802 (2006)
12. Fliess, M., Join, C.: Model-free control. Int. J. Control **86**, 2228–2252 (2013)
13. Fliess, M., Join, C.: An alternative to proportional-integral and proportional-integral-derivative regulators: intelligent proportional-derivative regulators. Int. J. Robust Nonlinear Control (2021). https://doi.org/10.1002/rnc.5657
14. Fliess, M., Sira-Ramírez, H.: An algebraic framework for linear identification. ESAIM Contr. Optimiz. Calc. Variat. **9**, 151–168 (2003)
15. Fliess, M., Sira-Ramírez, H.: Closed-loop parametric identification for continuous-time linear systems via new algebraic techniques. In: Garnier, H., Wang, L. (eds.) Identification of Continuous-time Models from Sampled Data, pp. 362–391. Springer (2008)
16. Greenshields, B.D.: A study of traffic capacity. Highw. Res. Board **14**, 448–477 (1935)

17. Haddar, M., Chaari, R., Baslamisli, S.C., Chaari, F., Haddar, M.: Intelligent PD controller design for active suspension system based on robust model-free control strategy. Proc. Inst. Mech. Engin. Part C: J. Mech. Engin. Sci., **233**, 4863–4880 (2019)
18. Haj-Salem, H., Blosseville, J.-M., Davée, M.M., Papageorgiou, M.: ALINEA: Un outil de régulation d'accès isolé sur autoroute – Étude comparative sur site réel. Rapport INRETS n° 80, Arcueil (1988)
19. Haj-Salem, H., Blosseville, J.-M., Papageorgiou, M.: ALINEA - a local feedback control law for on-ramp metering: a real life study. In: 3rd IEE International Conference on Road Traffic Control, London (1990)
20. Kachroo, P., Özbay, K.: Feedback Ramp Metering in Intelligent Transportation Systems. Springer (2003)
21. Kachroo, P., Özbay, K.: Feedback Control Theory for Dynamic Traffic Assignment, 2nd edn. Springer (2018)
22. Kerner, B.S.: Introduction to Modern Traffic Flow Theory and Control. Springer (2009)
23. Kerner, B.S. (Ed.): Dynamics of Traffic Management. Springer (2019)
24. Khondaker, B., Kattan, L.: Variable speed limit: an overview. Transport. Lett. **7**, 264–278 (2015)
25. Kotsialos, A., Papageorgiou, M., Diakaki, C., Pavlis, Y., Middelham, F.: Traffic flow modeling of large-scale motorway networks using the macroscopic modeling tool METANET. IEEE Trans. Intel. Transport. Syst. **3**, 282–292 (2002)
26. Lighthill, M.J., Whitham, G.B.: On kinematic waves II. A theory of traffic flow on long crowded roads, Proc. Royal Soc. **229-1178**, 317–345 (1955)
27. Liu, X., Xun, J., Ning, B., Wang, C.: Braking process identification of high-speed trains for automatic train stop control. ISA Trans. **111**, 171–179 (2021)
28. Mammar, S (Ed.): Systèmes de transport intelligents : modélisation, information et contrôle. Hermes - Lavoisier (2007)
29. May, A.D.: Traffic Flow Fundamentals. Prentice-Hall (1990)
30. Menhour, L., d'Andréa-Novel, B., Fliess, M., Gruyer, D., Mounier, H.: An efficient model-free setting for longitudinal and lateral vehicle control: validation through the interconnected Pro-SiVIC/RTMaps prototyping platform. IEEE Trans. Intel. Transport. Syst. **19**, 461–475 (2018)
31. Milanés, V., Villagrá, J., Godoy, J., González, C.: Comparing fuzzy and intelligent PI controllers in stop-and-go manoeuvres. IEEE Trans. Control Syst. Technol. **20**, 770–778 (2011)
32. Papageorgiou, M.: Applications of Automatic Control Concepts to Traffic Flow Modeling and Control. Springer (1983)
33. Papageorgiou, M., Blosseville, J.-M., Hadj-Salem, H.: Modelling and real-time control of traffic flow on the southern part of boulevard périphérique in Paris: Part II: Coordinated on-ramp metering. Transport. Res. A **24A**, 361–370 (1990)
34. Papageorgiou, M., Blosseville, J.-M., Hadj-Salem, H.: ALINEA: a local feedback control law for on-ramp metering. Transport. Res. Rec. **1320**, 58–64 (1991)
35. Papamichail, I., Papageorgiou, M., Vong, V., Gaffney, J.: Heuristic ramp-metering coordination strategy implemented at Monash freeway. Aust. Transp. Res. Rec. **2178**, 10–20 (2010)
36. Polack, P., Delprat, S., d'Andréa-Novel, B.: Brake and velocity model-free control on an actual vehicle. Control Engin. Pract. **92**, 104072 (2019)
37. Richards, P.I.: Shock waves on the highways. Oper. Res. **4**, 42–51 (1956)
38. Sira-Ramírez, H., García-Rodríguez, C., Cortès-Romero, J., Luviano-Juárez, A.: Algebraic Identification and Estimation Methods in Feedback Control Systems. Wiley (2013)
39. Treiber, M., Kesting, A.: Verkehrsdynamik und -simulation – Daten, Modelle und Anwendungen der Verkehrsflussdynamik. Springer (2010). English translation: Traffic Flow Dynamics – Data, Models and Simulation. Springer (2013)
40. Villagra, J., Herrero-Pérez, D.: A comparison of control techniques for robust docking maneuvers of an AVG. IEEE Trans. Control Syst. Technol. **20**, 1116–1123 (2012)
41. Wang, Z., Wang, J.: Ultra-local model predictive control: A model-free approach and its application on automated vehicle trajectory tracking. Control Engin. Pract. **101**, 104482 (2020)

42. Yang, H., Liu, C., Shi, J., Zheng, G.: Development and control of four-wheel independent driving and modular steering electric vehicles for improved maneuverability limits, SAE Techn. Paper, 2019-01-0459 (2019)
43. Yosida, K.: Operational Calculus (translated from the Japanese). Springer (1984)

Event-Triggered Output Feedback Control of Traffic Flow on Cascaded Roads

Nicolas Espitia, Jean Auriol, Huan Yu, and Miroslav Krstic

Abstract In this chapter, we develop an event-triggered boundary output feedback controller that guarantees the simultaneous stabilization of traffic flow on connected roads. The density and velocity traffic dynamics are described with the linearized Aw-Rascle-Zhang (ARZ) macroscopic traffic partial differential equation (PDE) model, which results in a coupled hyperbolic system. The control objective is to simultaneously stabilize the upstream and downstream traffic to a given spatially uniform constant steady-state that is in the congested regime. To suppress stop-and-go traffic oscillations on the cascaded roads, we consider a ramp metering strategy that regulates the traffic flow rate entering from the on-ramp to the mainline freeway. The ramp metering is located at the outlet with only boundary measurements of flow rate and velocity. The main idea is that the control signal is only updated when an event triggering condition is satisfied. Compared with the continuous input signal, the event-triggered boundary output control presents a more realistic setting to implement by ramp metering on a digital platform. The event-triggered boundary output control design relies on the emulation of the backstepping boundary output feedback and on a dynamic event-triggered strategy to determine the time instants at which the control value must be updated. We prove that there is a uniform minimal dwell-time (independent of initial conditions), thus avoiding the so-called Zeno phenomenon.

N. Espitia
Univ. Lille, CNRS, Centrale Lille, UMR 9189 - CRIStAL - Centre de Recherche en Informatique Signal et Automatique de Lille, 59000 Lille, France
e-mail: nicolas.espitia-hoyos@univ-lille.fr

J. Auriol (✉)
Laboratoire des Signaux et Systèmes, Université Paris-Saclay, CNRS, CentraleSupélec, 91190 Gif-sur-Yvette, France
e-mail: jean.auriol@centralesupelec.fr

H. Yu · M. Krstic
Department of Mechanical and Aerospace Engineering, University of California, San Diego, La Jolla, CA 92093-0411, USA
e-mail: huy015@eng.ucsd.edu

M. Krstic
e-mail: krstic@ucsd.edu

J. Auriol et al. (eds.), *Advances in Distributed Parameter Systems*, Advances in Delays and Dynamics 14, https://doi.org/10.1007/978-3-030-94766-8_11

We guarantee the exponential convergence of the closed-loop system under the proposed event-triggered boundary control. A numerical example illustrates the results.

1 Introduction

Freeway traffic modeling and management have been intensively investigated due to the increasing demand of traffic mobility over the past decades. Various traffic control methods have been studied to regulate freeway traffic systems and mitigate traffic congestion. In particular, we focus on the stop-and-go traffic, a common phenomenon appearing on congested freeways. In congested traffic, drivers are forced into the acceleration-and-deceleration cycles. The stop-and-go traffic is characterized by such oscillations and causes increased consumption of fuel and unsafe driving conditions. We are interested in developing control strategies to mitigate stop-and-go oscillations.

Among different models for freeway traffic, macroscopic modeling is particularly suitable to describe the stop-and-go traffic since the propagation of traffic waves is described in the temporal and spatial domain. The macroscopic models predict the evolution of continuous traffic states by employing hyperbolic PDEs to govern traffic density and velocity dynamics. Among the models, the second-order Aw-Rascle-Zhang (ARZ) model [3, 20] for the stop-and-go traffic stands out. Indeed, the ARZ PDE model describes the traffic density and velocity of a freeway segment with two coupled hyperbolic PDEs. For more complex road network structures, the traffic network PDE model is developed in [10, 11] based on the family of ARZ models. Traffic control strategies are mainly developed and implemented on the traffic management infrastructures, that is, ramp metering and varying speed limits (VSL). Ramp metering controls the traffic lights on a ramp such that the inflow traffic is regulated for the mainline traffic. The VSL regulates traffic velocity by displaying driving velocities that are time-varying and dependent on real-time traffic. Boundary control algorithms have been developed for traffic control of a single freeway segment in [4, 12, 19, 21]. If we consider a traffic control problem on cascaded freeway segments, the application of these control laws needs to assume road homogeneity. In this chapter, we solve the control problem of stop-and-go traffic congestion on cascaded freeway segments. We adopt the state-of-art second-order macroscopic traffic network models in [11], and we build on the linearized ARZ model around a steady-state in the congested regime. The model results in a coupled hyperbolic system with boundary input. We design the boundary control by the backstepping method and perform its emulation towards an event-triggered implementation. Hence, the boundary control signal is updated according to some policy while accounting for the information of estimated states (obtained from a suitable observer) and the error of the sampling scheduling. The triggering strategy, in turn, relies on a dynamic triggering condition which determines when the control action has to be updated. In this chapter, we only provide the results on avoidance of Zeno phenomenon, through the explicit characterization of a minimal inter-sampling time, and on the stability result for the closed-loop system under the event-triggered output control strategy.

2 Preliminaries and Problem Description

The evolution of traffic density $\rho_1(t, x)$ and velocity $v_1(t, x)$ (with $(t, x) \in [0, \infty) \times [0, L]$) on the downstream road segment and traffic density $\rho_2(t, s)$ and velocity $v_2(t, s)$ ($(t, s) \in [0, \infty) \times [-L, 0]$) on the upstream road segment are modeled by the following ARZ model.

$$\partial_t \rho_i + \partial_x (\rho_i v_i) = 0, \tag{1}$$

$$\partial_t (\rho_i (v_i + p_i)) + \partial_x (\rho_i v_i (v_i + p_i)) = -\frac{\rho_i (v_i - V_i (\rho_i))}{\tau_i}, \tag{2}$$

where $i \in \{1, 2\}$ represents downstream and upstream road respectively. The labeling of freeway segments is chosen as the reverse direction of traffic flow but same as the propagation direction of the control signal, which will be explained later. The traffic pressure $p_i (\rho_i)$ is defined as an increasing function of the density $p_i (\rho_i) = c_i \rho_i^{\gamma_i}$, where $\gamma_i, c_i \in \mathbb{R}^+ = [0, \infty)$ is defined as $c_i = v_m / \rho_{m,i}^{\gamma_i}$. The coefficient γ_i represents the overall drivers' property, reflecting their change of driving behavior to the increase of density. The changes in γ_i may reflect changes in the road properties (e.g. number of lanes). The positive constant v_m represents the maximum velocity and the positive constant $\rho_{m,i}$ is the maximum density defined as the number of vehicles per unit length. The equilibrium density-velocity relation $V_i (\rho_i)$ is given by $V_i (\rho_i) = v_m - p_i (\rho_i)$ for both segments, which assumes the same maximum velocity for the two segments when there are no vehicles on the road $\rho_i = 0$. We define the following variable

$$w_i = v_i + p_i (\rho_i), \tag{3}$$

which is interpreted as traffic "friction" or drivers' property [9]. This property transports in the traffic flow with vehicle velocity, representing the heterogeneity of individual driver with respect to the equilibrium density-velocity relation $V_i (\rho_i)$. The maximum velocity v_m is assumed to be the same for the two road segments while the maximum density $\rho_{m,i}$ and coefficient γ_i are allowed to vary. The positive constant τ_i is the relaxation time that represents the time scale for traffic velocity v_i adapting to the equilibrium density velocity relation $V_i (\rho_i)$. We denote the traffic flow rate on each road as $q_i = \rho_i v_i$ The equilibrium flow and density relation, also known as the fundamental diagram, is then given by $Q_i (\rho_i) = \rho_i V (\rho_i) = \rho_i v_m \left(1 - (\rho_i / \rho_{m,i})^{\gamma_i}\right)$. We assume that the equilibrium traffic relation is different for the two segments due to the change of road situations and access to road junction. The illustration is given in Fig. 1. The critical density ρ_c segregates the free and congested regimes

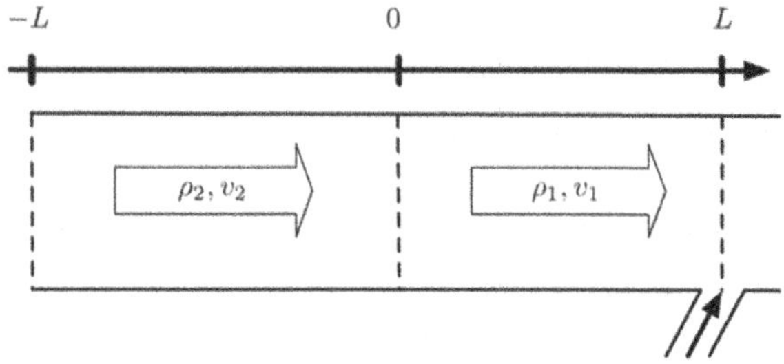

Fig. 1 Traffic flow on an incoming road and an outgoing road connected with a junction, actuation is implemented at the outlet with ramp metering

of traffic states. The critical density is given by $\rho_{c,i} = \rho_{m,i}/\left(1+\gamma_i\right)^{1/\gamma_i}$ such that $Q_i'(\rho)\big|_{\rho=\rho_c} = 0$. The traffic is free when the density satisfies $\rho < \rho_{c,i}$. The traffic is defined as the congested one when the density satisfies $\rho > \rho_{c,i}$. For the free traffic, oscillations around the steady states will be damped out fast. For the congested traffic, there are two directional waves on road with one being the velocity oscillation propagating upstream and the other one being the density oscillation propagating downstream with the traffic. The congested traffic can become unstable [18]. We consider the situation that the upstream road segment 2 for $s \in [-L, 0]$ has more lanes than the downstream road segment for $x \in [0, L]$, in which congested traffic is usually formed up from downstream to upstream. Therefore, the maximum density $\rho_{m,2} > \rho_{m,1}$. The maximum driving speed v_m is assumed to be the same for the two segments. The maximum flow rate of the upstream road $Q_2\left(\rho_c\right)$ is reduced in the downstream to $Q_1\left(\rho_c\right)$, due to the change of road conditions from segment 2 to segment 1.

2.1 *Actuated Boundary*

Regarding the boundary conditions connecting the two PDE systems, the Rankine-Hugoniot condition is satisfied at the junction such that the weak solution exists for the network (1)–(2) [14]. This condition implies piecewise smooth solutions and corresponds to the conservation of the mass and of the drivers' properties defined in (3) at the junction. Thus the flux and drivers' property are assumed to be continuous across the boundary conditions at $x = 0$, that is

$$\rho_1(t,0)v_1(t,0) = \rho_2(t,0)v_2(t,0), \tag{4}$$

$$w_2(t,0) = w_1(t,0). \tag{5}$$

For the open-loop system, we assume a constant inflow $q^\star$ entering the inlet boundary $s = -L$ and a constant outflow $q^\star$ at the outlet boundary for $x = L$:

$$q_2(t,-L) = q^\star \tag{6}$$

$$q_1(t,L) = q^\star \tag{7}$$

The control problem we solve consists in stabilizing on events the traffic flow in both the upstream and downstream road segments with a single actuator. Three possible locations for implementing a ramp metering control input are either at the inlet $x = -L$, at the junction $x = 0$ or at the outlet $x = L$ as in [17]. However, in this chapter we only present the observer-based event triggered control results for control input acting on the outlet and that is updated according to a suitable event-triggering condition. Note that the other cases could be solved adjusting the proposed techniques.

Ramp metering control $U_{\text{nom}}(t)$ from the outlet $x = L$: The downstream outflow at $x = L$ is actuated by $U_{\text{nom}}(t)$

$$q_1(t, L) = q^\star + U_{\text{nom}}(t), \tag{8}$$

where the outflow rate equals the summation of the onramp metering flow and the constant mainline flow. It should be noted that the designed controller U_{nom} is the flow rate perturbation around a nominal flow rate. We assume that the steady-state flow rate consists of a nominal onramp flow rate $q_r \geq 0$, which is a component of the steady-state flow rate $q^\star$. Then the actual ramp flow input at an onramp is given by

$$q_{\text{ramp}}(t) = q_r + U_{\text{nom}}(t) \geq 0. \tag{9}$$

In practice, we only need to guarantee that $q_{\text{ramp}}(t)$ is nonnegative so that $U_{\text{nom}}(t) \geq -q_r$. The value of q_r depends on the road configuration and real-time traffic conditions. We assume that there exists $q_r > 0$ such that (9) always holds. Combining the proposed control law with a saturation could guarantee that the proposed condition is satisfied.

2.2 Congested Steady States

We are concerned with the congested traffic and assume that the equilibrium of both segments $(\rho_1^\star, v_1^\star)$, $(\rho_2^\star, v_2^\star)$ are in the congested regime, which is the only one of theoretical control interest among all four traffic scenarios including free and free, free and congested, congested and free, congested and congested. If the traffic of both segments is free, there is no need for ramp metering control. If the upstream segment 2 is in the free regime and the downstream segment 1 is congested, then we only need to control the congested downstream traffic with $U_{\text{nom}}(t)$ as presented in [19]. The oscillations propagated from the congested segment to the free regime segment will be damped out soon. The same applies to the scenario of free traffic in downstream segment 1 and congested traffic in upstream segment 2. The control objective is to stabilize the traffic flow in the two segments around the steady states. In practice, the steady states represent the equilibrium state values of the traffic flow when oscillations are successfully suppressed by our control design.

The steady states $\left(\rho_1^\star, v_1^\star\right)$, $\left(\rho_2^\star, v_2^\star\right)$ are considered to be in the congested regime and the boundary conditions (4) and (5) are satisfied, i.e.,

$$\rho_1^\star v_1^\star = \rho_2^\star v_2^\star = q^\star, \tag{10}$$

$$w_1^\star = w_2^\star = v_m, \tag{11}$$

where the steady state velocities satisfy the equilibrium density-velocity relation $v_i^\star = V_i\left(\rho_i^\star\right)$. According to (3) the constant driver's property in (11) implies that we have the same maximum velocity v_m for the two segments (which corresponds to our initial assumption):

$$v_1^\star + p_1^\star = v_2^\star + p_2^\star = v_m, \tag{12}$$

where $p_i^\star = p_i\left(\rho_i^\star\right)$. The steady states can be solved from the above nonlinear equations (10), (12) however there are no explicit solutions. Therefore we show the derivation process for obtaining the steady state values when $\rho_1^\star$ and the model parameters v_m, $\rho_{m,i}$ and γ_i are given. The functions $V_i(\rho)$, $Q_i(\rho)$ and $p_i(\rho)$ are also known. The steady state flow rate in (10) is obtained as $q^\star = Q_1\left(\rho_1^\star\right)$, and the constant flux $Q_1\left(\rho_1^\star\right) = Q_2\left(\rho_2^\star\right)$, yields a relation for the steady state densities of the two segments $\frac{\rho_1^\star \rho_{m,1}^{\gamma_1} - \left(\rho_1^\star\right)^{\gamma_1+1}}{\rho_2^\star \rho_{m,2}^{\gamma_2} - \left(\rho_2^\star\right)^{\gamma_2+1}} = \frac{\rho_{m,1}^{\gamma_1}}{\rho_{m,2}^{\gamma_2}}$. Knowing $\rho_1^\star$, $\rho_2^\star$ and $q^\star$, the steady states velocities are obtained as $v_i^\star = q^\star/\rho_i^\star$.

2.3 Linearized ARZ Model in Riemann Coordinates

We linearize the ARZ based traffic network model (ρ_i, v_i) in (1), (2) with the boundary conditions (4)–(7) around the steady states $\left(\rho_i^\star, v_i^\star\right)$ defined in the previous section. In order to simplify the control design, the linearized model is then rewritten into the Riemann variables to which we apply an invertible spatial transformation

$$\bar{w}_i = \exp\left(\frac{x}{\tau_i v_i^\star}\right)\left(\frac{\gamma_i p_i^\star}{q^\star}\left(\rho_i v_i - \rho_i^\star v_i^\star\right) + \frac{1}{r_i}\left(v_i - v_i^\star\right)\right), \tag{13}$$

$$\bar{v}_i = v_i - v_i^\star, \tag{14}$$

where $p_i^\star = p_i\left(\rho_i^\star\right)$ and the constant coefficients r_i are defined as

$$r_i = -\frac{v_i^\star}{\gamma_i p_i^\star - v_i^\star}. \tag{15}$$

For the congested regime we have $\rho_i^\star > \frac{\rho_{m,i}}{(1+\gamma_i)^{1/\gamma_i}}$ so that the characteristic speed $\gamma_i p_i^\star - v_i^\star > 0$. The velocity variations $\bar{v}_1(t,x)$ $\bar{v}_2(t,x)$ transport upstream which means the action of velocity acceleration or deceleration is repeated from the leading vehicle to the following vehicle. More precisely, we have $p_i^\star = v_m - V\left(\rho_i^\star\right) =$

$v_m\left(\frac{\rho_i^\star}{\rho_{m,i}}\right)^{\gamma_i} > \frac{v_m}{1+\gamma_i}$, since $\rho_i^\star > \frac{\rho_{m,i}}{(1+\gamma_i)^{1/\gamma_i}}$. The more congested the traffic, the lower the absolute value of this ratio. The linearized system with the controlled boundary condition (8) is written as

$$\partial_t \bar{w}_1(t,x) + v_1^\star \partial_x \bar{w}_1(t,x) = 0, \tag{16}$$

$$\partial_t \bar{v}_1(t,x) - (\gamma_1 p_1^\star - v_1^\star)\partial_x \bar{v}_1(t,x) = \bar{c}_1(x)\bar{w}_1(t,x), \tag{17}$$

$$\partial_t \bar{w}_2(t,s) + v_2^\star \partial_s \bar{w}_2(t,s) = 0, \tag{18}$$

$$\partial_t \bar{v}_2(t,s) - (\gamma_2 p_2^\star - v_2^\star)\partial_s \bar{v}_2(t,s) = \bar{c}_2(x)\bar{w}_2(t,s), \tag{19}$$

$$\bar{w}_1(t,0) = \bar{w}_2(t,0), \tag{20}$$

$$\bar{v}_1(t,L) = r_1 \exp\left(\frac{-L}{\tau_1 v_1^\star}\right)\bar{w}_1(t,L) + \frac{1-r_1}{\rho_1^\star}U_{\mathrm{nom}}(t), \tag{21}$$

$$\bar{w}_2(t,-L) = \exp\left(\frac{-L}{\tau_2 v_2^\star}\right)\frac{1}{r_2}\bar{v}_2(t,-L), \tag{22}$$

$$\bar{v}_2(t,0) = \delta\frac{r_2}{r_1}\bar{v}_1(t,0) + (1-\delta)r_2\bar{w}_2(t,0), \tag{23}$$

where $s \in [-L, 0]$, $x \in [0, L]$, and where the spatially varying coefficient $\bar{c}_i(x)$ are defined by $\bar{c}_i(x) = -\frac{1}{\tau_i}\exp\left(-\frac{x}{\tau_i v_i^\star}\right)$. The constant coefficient δ (ratio related to the traffic pressure of the segments) is defined by $\delta = \frac{\gamma_2 p_2^\star}{\gamma_1 p_1^\star}$. Although the cascade structure of the network presents some advantages for the design of a stabilizing control law (see [17]), it is more convenient for the design of an event-triggered algorithm to have all the states defined on the same spatial domain. The control diagram is shown in Fig. 2.

To rewrite the states $\bar{w}_2$ and $\bar{v}_2$ as functions defined on $[0, L]$, we consider the folding transformation $\bar{x} = -s$. The variable $\bar{x}$ belongs to $[0, L]$. For the sake of simplicity, we will omit the bar and abusively denote $w_2(x) = w_2(x)$. With this transformation, the previous system rewrites

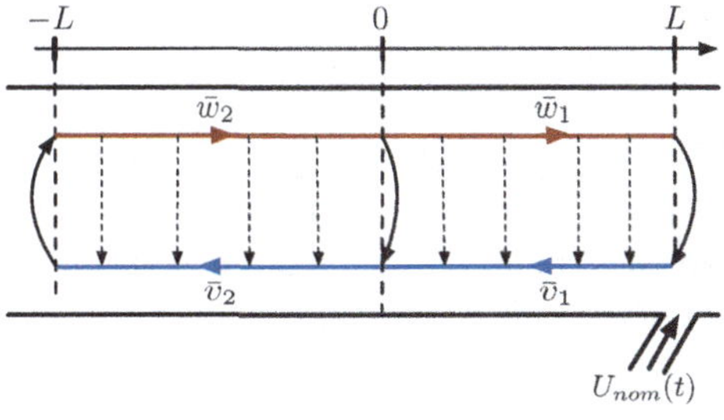

Fig. 2 Traffic flow on an incoming road and an outgoing road connected with a junction. Actuation is implemented at the outlet with ramp metering

$$\partial_t \bar{w}_1(t,x) + v_1^\star \partial_x \bar{w}_1(t,x) = 0, \tag{24}$$

$$\partial_t \bar{v}_1(t,x) - (\gamma_1 p_1^\star - v_1^\star)\partial_x \bar{v}_1(t,x) = c_1(x)\bar{w}_1(t,x), \tag{25}$$

$$\partial_t \bar{w}_2(t,x) - v_2^\star \partial_x \bar{w}_2(t,x) = 0, \tag{26}$$

$$\partial_t \bar{v}_2(t,x) + (\gamma_2 p_2^\star - v_2^\star)\partial_x \bar{v}_2(t,x) = c_2(x)\bar{w}_2(t,x), \tag{27}$$

$$\bar{w}_1(t,0) = \bar{w}_2(t,0), \tag{28}$$

$$\bar{v}_1(t,L) = r_1 \exp\left(\frac{-L}{\tau_1 v_1^\star}\right)\bar{w}_1(t,L) + \tfrac{1-r_1}{p_1^\star} U_{\mathrm{nom}}(t), \tag{29}$$

$$\bar{w}_2(t,L) = \exp\left(\frac{-L}{\tau_2 v_2^\star}\right)\frac{1}{r_2}\bar{v}_2(t,L), \tag{30}$$

$$\bar{v}_2(t,0) = \delta\frac{r_2}{r_1}\bar{v}_1(t,0) + (1-\delta)r_2\bar{w}_2(t,0), \tag{31}$$

where $x \in [0, L]$, and $c_1(x) = \bar{c}_1(x)$ and $c_2(x) = \bar{c}_2(-x)$. The open-loop system (24)–(31) (for which $U_{\mathrm{nom}} \equiv 0$) is well-posed in the sense of the L^2 norm (weak solutions) by [4, Theorem A.4], that is, for any initial conditions $(\bar{v}_0)_i, (\bar{w}_0)_i \in (L^2([0, L]))^2$, there is only one L^2-solution. It is shown in [19] that only marginal linear stability holds for the open-loop system of one segment. The control operator is admissible (i.e. it verifies the so-called admissibility condition as stated in [5]). Consequently, for any $U_L \in L^2([0, T])$, and for any initial conditions $(\bar{v}_0)_i, (\bar{w}_0)_i \in (L^2([0, L]))^2$ there is only one L^2-solution to (24)–(31). We assume that the available measurement corresponds to the values of $\bar{q}_i$ and $\bar{v}_i$ at the left side of the outlet $x = L$. Since $\bar{w}_1(t,L) = \exp\left(\frac{L}{\tau_1 v_1^\star}\right)\left(\frac{\gamma_1 p_1^\star}{q^\star}\bar{q}_1(t,L) - \frac{1}{r_i}\bar{v}_1(t,L)\right)$, we can consider that

$$Y_L(t) = \bar{w}_1(t,L). \tag{32}$$

We make the following non-restrictive assumption so that the proposed feedback laws have some (delay)-robustness margins.

Assumption 1 The boundary couplings of the system (24)–(31) are such that

$$\delta < \frac{1+\exp(\frac{L}{\tau_2 v_2^\star})}{1+\exp(\frac{-L}{\tau_1 v_1^\star})} \ \text{ if } \delta > 1, \quad \delta < \frac{1-\exp(\frac{L}{\tau_2 v_2^\star})}{1-\exp(\frac{-L}{\tau_1 v_1^\star})} \ \text{ if } \delta \leq 1, \tag{33}$$

If this assumption is not satisfied, then it is not possible to robustly stabilize the system (24)–(31) when there are input delays (as the open-loop transfer function would have an infinite chain of poles in the complex right half-plane) see [1, 17] for details. Moreover, it can be shown (see [1]) that this condition implies that the system (24)–(31) is exponentially stable in the absence of in-domain couplings and actuation. Thus, this assumption means that in the absence of in-domain couplings, the system is naturally dissipative.

The control objective is to simultaneously stabilize the upstream and downstream traffic to a given spatially uniform constant steady-state. We propose an output feedback controller located at the outlet of the downstream traffic with collocated sensing of flow rate and velocity at the outlet. The state feedback and observer designs are based on the PDE backstepping methodology. The exponential stability in the sense of L^2-norm of the under-actuated network of two systems of two hyperbolic PDEs is guaranteed. Considering the continuous boundary control and estimation designs need to be implemented into digital platforms, we develop event-triggered boundary controllers which stabilize the system on events. The proposed event-triggered controllers are piecewise constant, and the control value is updated based on a dynamic triggering condition only when needed i.e. once a given deviation term (that will be defined below) becomes larger that a Lyapunov functional.

3 Output-Feedback Stabilization and Emulation of the Control Law

An output-feedback law $U_{\text{nom}}(t)$ has been proposed in [17] to stabilize the system (24)–(31). Here, we aim at stabilizing the closed-loop system (24)–(31) on events while updating the continuous-time controller $U_{\text{nom}}(t)$ at certain sequence of time instants $(t_k)_{k\in\mathbb{N}}$, that will be characterized later on. The control value is held constant between two successive time instants and it is updated when some triggering condition is verified. This procedure is referred to as *event-triggering*. It is an efficient way to suitably sample the control value, thus avoiding useless actuation solicitations. To that end, we need to modify the control law proposed in [17]. More precisely, the control law $U_{\text{nom}}(t)$ will be replaced by $U_{\text{nom}}(t_k)$ for all $t \in [t_k, t_{k+1})$, $k \geq 0$. Consequently, we have $U_{\text{nom}}(t_k) = U_{\text{nom}}(t) + d(t)$, where d can be seen as a deviation of actuation. In what follows, we recall the observer/controller design proposed in [17] using the backstepping transformations since these transformations will be crucial to prove that our emulated control law still stabilizes the system (24)–(31). We define the sets $\bar{\mathcal{T}}_1, \bar{\mathcal{T}}_2$ as follows

$$\bar{\mathcal{T}}_1 = \{(x, \xi) \in [0, L]^2, \xi \geq x\}, \; \bar{\mathcal{T}}_2 = \{(x, \xi) \in [0, L]^2, \xi \leq x\}. \tag{34}$$

Finally, the set $\bar{\mathcal{T}}_3$ is defined as the unit square $[0, L]^2$: $\bar{\mathcal{T}}_3 = \{(x, \xi) \in [0, L]^2\}$. The set $\bar{\mathcal{T}}_1$ is the upper-part of this square while $\bar{\mathcal{T}}_2$ corresponds to its lower part.

3.1 Observer Design

The first step for the design of an output-feedback law is the design of a state estimator. Consider an arbitrary control law $U_{\text{nom}}(t_k)$ acting on the system (24)–(31), for all $t \in [t_k, t_{k+1})$. We will defined $\hat{w}_i$ and $\hat{v}_i$ as the observer states. The error states are defined as the difference between the real states and their estimations: $\tilde{w}_i = \bar{w}_i - \hat{w}_i$ and $\tilde{v}_i = \bar{v}_i - \hat{v}_i$. A suitable observer system is defined in [17] by

$$\partial_t \hat{w}_1(t,x) + v_1^\star \partial_x \hat{w}_1(t,x) = -\mu_1(x)\tilde{w}_1(t,L), \tag{35}$$
$$\partial_t \hat{v}_1(t,x) - (\gamma_1 p_1^\star - v_1^\star)\partial_x \hat{v}_1(t,x) = c_1(x)\hat{w}_1(t,x) - \nu_1(x)\tilde{w}_1(t,L), \tag{36}$$
$$\partial_t \hat{w}_2(t,x) - v_2^\star \partial_x \hat{w}_2(t,x) = -\mu_2(x)\tilde{w}_1(t,L), \tag{37}$$
$$\partial_t \hat{v}_2(t,x) + (\gamma_2 p_2^\star - v_2^\star)\partial_x \hat{v}_2(t,x) = c_2(x)\hat{w}_2(t,x) - \nu_2(x)\tilde{w}_1(t,L), \tag{38}$$
$$\hat{w}_1(t,0) = \hat{w}_2(t,0), \tag{39}$$
$$\hat{v}_1(t,L) = r_1 \exp\left(\frac{-L}{\tau_1 v_1^\star}\right)\hat{w}_1(t,L) + \tfrac{1-r_1}{\rho_1^\star} U_{\text{nom}}(t_k), \tag{40}$$
$$\hat{w}_2(t,L) = \exp\left(\frac{-L}{\tau_2 v_2^\star}\right)\frac{1}{r_2}\hat{v}_2(t,L), \tag{41}$$
$$\hat{v}_2(t,0) = \delta\frac{r_2}{r_1}\hat{v}_1(t,0) + (1-\delta)r_2\hat{w}_2(t,0), \tag{42}$$

where $\hat{w}_i, \hat{v}_i$ are the estimates of the state variables $\bar{w}_i, \bar{v}_i$ in (24)–(31). The corresponding initial conditions are L^2 functions. The terms μ_i and ν_i are the output injection terms that are given as follows [17]:

$$\mu_1(x) = v_1^\star N_1^{\alpha\alpha}(x,L), \quad \nu_1(x) = v_1^\star N_1^{\beta\alpha}(x,L), \tag{43}$$
$$\mu_2(x) = v_1^\star N^{\alpha}(x,L), \quad \nu_2(x) = v_1^\star N^{\beta}(x,L), \tag{44}$$

where the kernels $N_1^{\alpha\alpha}$, $N_1^{\alpha\beta}$ are bounded functions defined on $\bar{\mathcal{T}}_1$, the kernels $N_2^{\alpha\alpha}$, $N_2^{\alpha\beta}$ are bounded functions defined on $\bar{\mathcal{T}}_2$, and the kernels N^{α}, N^{β} are bounded functions defined on $\bar{\mathcal{T}}_3$. They satisfy the following set of equations for all (x,ξ) that belong to their respective domain of definition

$$\partial_x N_i^{\alpha\alpha}(x,\xi) + \partial_\xi N_i^{\alpha\alpha}(x,\xi) = 0, \tag{45}$$
$$(\gamma_1 p_1^\star - v_1^\star)\partial_x N_1^{\beta\alpha}(x,\xi) - v_1^\star \partial_\xi N_1^{\beta\alpha}(x,\xi) = -c_1(x)N_1^{\alpha\alpha}(x,\xi), \tag{46}$$
$$(\gamma_2 p_2^\star - v_2^\star)\partial_x N_2^{\beta\alpha}(x,\xi) - v_2^\star \partial_\xi N_2^{\beta\alpha}(x,\xi) = c_2(x)N_2^{\alpha\alpha}(x,\xi), \tag{47}$$
$$v_2^\star \partial_x N^{\alpha}(x,\xi) - v_1^\star \partial_\xi N^{\alpha}(x,\xi) = 0, \tag{48}$$
$$(\gamma_2 p_2^\star - v_2^\star)\partial_x N^{\beta}(x,\xi) + v_1^\star \partial_\xi N^{\beta}(x,\xi) = c_2(x)N^{\alpha}(x,\xi), \tag{49}$$

with the boundary conditions

$$N_i^{\beta\alpha}(x,x) = -\frac{c_i(x)}{\gamma_i p_i^\star}, \; N_1^{\alpha\alpha}(0,\xi) = N^{\alpha}(0,\xi), \; N^{\beta}(x,0) = \frac{v_2^\star}{v_1^\star}N_2^{\beta\alpha}(x,0) \tag{50}$$
$$N^{\alpha}(x,0) = \frac{v_2^\star}{v_1^\star}N_2^{\alpha\alpha}(x,0), \; N^{\beta}(0,\xi) = \delta\frac{r_2}{r_1}N_1^{\beta\alpha}(0,\xi) + (1-\delta)r_2 N^{\alpha}(0,\xi), \tag{51}$$
$$N^{\alpha}(L,\xi) = \exp\left(-\frac{L}{\tau_2 v_2^\star}\right)\frac{1}{r_2}N^{\beta}(L,\xi), \; N_2^{\alpha\alpha}(L,\xi) = \exp\left(-\frac{L}{\tau_2 v_2^\star}\right)\frac{1}{r_2}N_2^{\beta\alpha}(L,\xi). \tag{52}$$

Equations (45)–(52) admit a unique solution, as proved in [17]. It has been shown in [17] that such an observer guarantees the convergence of the estimated states to the real states, that is the error states converge to zero. The error system (that will be useful to design our event-triggered control law) rewrites as

$$\partial_t \tilde{w}_1(t,x) + v_1^\star \partial_x \tilde{w}_1(t,x) = \mu_1(x)\tilde{w}_1(t,L), \tag{53}$$
$$\partial_t \tilde{v}_1(t,x) - (\gamma_1 p_1^\star - v_1^\star)\partial_x \tilde{v}_1(t,x) = c_1(x)\tilde{w}_1(t,x) + \nu_1(x)\tilde{w}_1(t,L), \tag{54}$$
$$\partial_t \tilde{w}_2(t,x) - v_2^\star \partial_x \tilde{w}_2(t,x) = \mu_2(x)\tilde{w}_1(t,L), \tag{55}$$
$$\partial_t \tilde{v}_2(t,x) + (\gamma_2 p_2^\star - v_2^\star)\partial_x \tilde{v}_2(t,x) = c_2(x)\tilde{w}_2(t,x) + \nu_2(x)\tilde{w}_1(t,L), \tag{56}$$
$$\tilde{w}_1(t,0) = \tilde{w}_2(t,0), \tag{57}$$
$$\tilde{v}_1(t,L) = r_1 \exp\left(\frac{-L}{\tau_1 v_1^\star}\right)\tilde{w}_1(t,L), \tag{58}$$
$$\tilde{w}_2(t,L) = \exp\left(\frac{-L}{\tau_2 v_2^\star}\right)\frac{1}{r_2}\tilde{v}_2(t,L), \tag{59}$$
$$\tilde{v}_2(t,0) = \delta\frac{r_2}{r_1}\tilde{v}_1(t,0) + (1-\delta)r_2\tilde{w}_2(0) \tag{60}$$

To show that this error system is exponentially stable, we shall consider the following backstepping transformation

$$\begin{pmatrix} \tilde{w}_1(t,x) \\ \tilde{v}_1(t,x) \\ \tilde{w}_2(t,x) \\ \tilde{v}_2(t,x) \end{pmatrix} := \left(\mathcal{N}\begin{pmatrix} \tilde{\alpha}_1(t,x) \\ \tilde{\beta}_1(t,x) \\ \tilde{\alpha}_2(t,x) \\ \tilde{\beta}_2(t,x) \end{pmatrix}\right)(x) \tag{61}$$
$$= \begin{pmatrix} \tilde{\alpha}_1(t,x) \\ \tilde{\beta}_1(t,x) \\ \tilde{\alpha}_2(t,x) \\ \tilde{\beta}_2(t,x) \end{pmatrix} - \int_0^L \begin{pmatrix} N_1^{\alpha\alpha}(x,\xi)\mathbb{1}_{[x,L]}(\xi) & 0 & 0 & 0 \\ N_1^{\beta\alpha}(x,\xi)\mathbb{1}_{[x,L]}(\xi) & 0 & 0 & 0 \\ N^{\alpha}(x,\xi) & 0 & N_2^{\alpha\alpha}(x,\xi)\mathbb{1}_{[0,x]}(\xi) & 0 \\ N^{\beta}(x,\xi) & 0 & N_2^{\beta\alpha}(x,\xi)\mathbb{1}_{[0,x]}(\xi) & 0 \end{pmatrix} \begin{pmatrix} \tilde{\alpha}_1(t,\xi) \\ \tilde{\beta}_1(t,\xi) \\ \tilde{\alpha}_2(t,\xi) \\ \tilde{\beta}_2(t,\xi) \end{pmatrix} d\xi.$$

where the different kernels are defined by (45)–(52). The transformation (61) is invertible. This can be seen, noticing first that the part acting on the states $\tilde{\alpha}_1$ and $\tilde{\beta}_1$ corresponds to a Volterra transformation (which is always invertible [16]). Then, the part acting on the states $\tilde{\alpha}_2$ and $\tilde{\beta}_2$ corresponds to Volterra transformation to which is added an affine term that depends on $\tilde{\alpha}_1$ and $\tilde{\beta}_1$. This transformation, maps the error system (53)–(60) to the system

$$\partial_t \tilde{\alpha}_i(t,x) + v_i^\star \partial_x \tilde{\alpha}_i(t,x) = 0, \tag{62}$$
$$\partial_t \tilde{\beta}_i(t,x) - (\gamma_i p_i^\star - v_i^\star)\partial_x \tilde{\beta}_i(t,x) = 0, \tag{63}$$
$$\tilde{\alpha}_1(t,0) = \tilde{\alpha}_2(t,0), \tag{64}$$
$$\tilde{\beta}_1(t,L) = r_1 \exp\left(\frac{-L}{\tau_1 v_1^\star}\right)\tilde{\alpha}_1(t,L), \tag{65}$$
$$\tilde{\alpha}_2(t,L) = \exp\left(\frac{-L}{\tau_2 v_2^\star}\right)\frac{1}{r_2}\tilde{\beta}_2(t,L), \tag{66}$$
$$\tilde{\beta}_2(t,0) = \delta\frac{r_2}{r_1}\tilde{\beta}_1(t,0) + (1-\delta)r_2\tilde{\alpha}_2(0), \tag{67}$$

This target system is exponentially stable due to Assumption 1. The design of our event-triggered procedure requires, the inverse transformation of (61). More precisely, we denote $\mathcal{R}$ the corresponding inverse transformation. It satisfies

$$\begin{pmatrix} \tilde{\alpha}_1(t,x) \\ \tilde{\beta}_1(t,x) \\ \tilde{\alpha}_2(t,x) \\ \tilde{\beta}_2(t,x) \end{pmatrix} := \left(\mathcal{R} \begin{pmatrix} \tilde{w}_1(t,\cdot) \\ \tilde{v}_1(t,\cdot) \\ \tilde{w}_2(t,\cdot) \\ \tilde{v}_2(t,\cdot) \end{pmatrix} \right)(x) \tag{68}$$
$$= \begin{pmatrix} \tilde{w}_1(t,x) \\ \tilde{v}_1(t,x) \\ \tilde{w}_2(t,x) \\ \tilde{v}_2(t,x) \end{pmatrix} - \int_0^L \begin{pmatrix} R_1^{ww}(x,\xi)\mathbb{1}_{[x,L]}(\xi) & 0 & 0 & 0 \\ R_1^{vw}(x,\xi)\mathbb{1}_{[x,L]}(\xi) & 0 & 0 & 0 \\ R^w(x,\xi) & 0 & R_2^{ww}(x,\xi)\mathbb{1}_{[0,x]}(\xi) & 0 \\ R^v(x,\xi) & 0 & R_2^{wv}(x,\xi)\mathbb{1}_{[0,x]}(\xi) & 0 \end{pmatrix} \begin{pmatrix} \tilde{w}_1(t,\xi) \\ \tilde{v}_1(t,\xi) \\ \tilde{w}_2(t,\xi) \\ \tilde{v}_2(t,\xi) \end{pmatrix} d\xi$$

Again, the different kernels are bounded functions defined on $\bar{\mathcal{T}}_1$ (kernels $R_1^{\cdot\cdot}$), $\bar{\mathcal{T}}_2$ (kernels $R_2^{\cdot\cdot}$) or $\bar{\mathcal{T}}_3$ (kernels $R^{\cdot}$). They satisfy the following set of equations on their respective domain of definition

$$\partial_x R_i^{ww}(x,\xi) + \partial_\xi R_i^{ww}(x,\xi) = 0, \tag{69}$$
$$(\gamma_i p_i^\star - v_i^\star)\partial_x R_i^{vw}(x,\xi) - v_i^\star \partial_\xi R_i^{wv}(x,\xi) = 0, \tag{70}$$
$$v_2^\star \partial_x R^w(x,\xi) - v_1^\star \partial_\xi R^w(x,\xi) = 0, \tag{71}$$
$$(\gamma_2 p_2^\star - v_2^\star)\partial_x R^v(x,\xi) + v_2^\star \partial_\xi R^v(x,\xi) = 0, \tag{72}$$

with the boundary conditions

$$R_i^{vw}(x,x) = \tfrac{c_i(x)}{\gamma_i p_i^\star}, \; R_1^{ww}(0,\xi) = R^w(0,\xi), \tag{73}$$
$$R^w(x,0) = \tfrac{v_2^\star}{v_1^\star} R_2^{ww}(x,0), \; R^v(x,0) = \tfrac{v_2^\star}{v_1^\star} R_2^{vw}(x,0), \tag{74}$$
$$R^v(0,\xi) = \delta \tfrac{r_2}{r_1} R_1^{vw}(0,\xi) + (1-\delta) r_2 R^w(0,\xi), \tag{75}$$
$$R^w(L,\xi) = \exp(-\tfrac{L}{\tau_2 v_2^\star}) \tfrac{1}{r_2} R^v(L,\xi), \tag{76}$$
$$R_2^{ww}(L,\xi) = \exp(-\tfrac{L}{\tau_2 v_2^\star}) \tfrac{1}{r_2} R_2^{vw}(L,\xi), \tag{77}$$

3.2 Output Feedback Control Law (Nominal)

Using the proposed observer, we can now design an output-feedback control law. More precisely, the following control law has been proposed in [17]

$$\begin{aligned} U_{\mathrm{nom}}(t) = \; & \tfrac{\rho_1^\star}{1-r_1} \Big(\int_0^L K_1^{vw}(L,\xi)\hat{w}_1(\xi,t)d\xi + K_1^{vv}(L,\xi)\hat{v}_1(\xi,t)d\xi \\ & + \int_0^L K^w(L,\xi)\hat{w}_2(\xi,t)d\xi \int_0^L K^v(L,\xi)\hat{v}_2(\xi,t)d\xi \Big), \end{aligned} \tag{78}$$

where the kernels K_1^{vw} and K_1^{vv} that are bounded functions defined on $\bar{\mathcal{T}}_2$, the kernels K_2^{vw} and K_2^{vv} are bounded functions defined on $\bar{\mathcal{T}}_1$, and the kernels K^w and K^v that are bounded functions defined on $\bar{\mathcal{T}}_3$. On their corresponding domain of definition, they verify

$$(\gamma_1 p_1^\star - v_1^\star)\partial_x K_1^{vw}(x,\xi) - v_1^\star \partial_\xi K_1^{vw}(x,\xi) = c_1(\xi) K_1^{vv}(x,\xi), \tag{79}$$
$$(\gamma_2 p_2^\star - v_2^\star)\partial_x K_2^{vw}(x,\xi) - v_2^\star \partial_\xi K_2^{vw}(x,\xi) = -c_2(\xi) K_2^{vv}(x,\xi) \tag{80}$$
$$\partial_x K_i^{vv}(x,\xi) + \partial_\xi K_i^{vv}(x,\xi) = 0, \tag{81}$$
$$\left(\gamma_1 p_1^\star - v_1^\star\right)\partial_x K^v(x,\xi) - \left(\gamma_2 p_2^\star - v_2^\star\right)\partial_\xi K^v(x,\xi) = 0, \tag{82}$$
$$\left(\gamma_1 p_1^\star - v_1^\star\right)\partial_x K^w(x,\xi) + v_2^\star \partial_\xi K^w(x,\xi) = c_2(\xi) K^v(x,\xi), \tag{83}$$

with the boundary conditions

$$K_i^{vw}(x,x) = -\frac{c_i(x)}{\gamma_i p_i^\star},\ K_1^{vv}(x,0) = \frac{v_2^\star}{v_1^\star}\delta K^v(x,0), \tag{84}$$
$$K^v(0,\xi) = \frac{r_1}{\delta r_2} K_2^{vv}(0,\xi),\ K^v(x,L) = -\exp\left(\frac{-L}{\tau_2 v_2^\star}\right) K^w(x,L), \tag{85}$$
$$K_2^{vv}(x,L) = -\exp\left(\frac{-L}{\tau_2 v_2^\star}\right) K_2^{vw}(x,L),\ K^w(0,\xi) = \frac{r_1}{\delta r_2} K_2^{vw}(0,\xi), \tag{86}$$
$$K^w(x,0) = -(1-\delta)K^v(x,0) + \frac{v_1^\star}{v_2^\star} K_1^{vw}(x,0). \tag{87}$$

It has been shown in [17] that the kernels equations (79)–(87) admit a unique solution. Since the kernels are bounded functions, our control operator is a linear bounded operator, and the control law U_{nom} is continuous. It is also strictly proper as it is only composed of integral terms. The exponential stability of the closed-loop system has been shown in [17]. Following the ideas of [2], we can prove that it is robust with respect to delays in the actuation and uncertainties on the parameters. For practical implementation of the ramp metering control input, we need to modulate the changing frequency of the on-ramp traffic light. This is one additional motivation for using an event-triggered procedure which is a way to implement the continuous-time controllers into digital forms by updating the input values only when needed. To show that the closed-loop system (24)–(31) with the control law (78) is exponentially stable in the sense of the L^2 norm, we can use the backstepping transformation

$$\begin{aligned} \begin{pmatrix} \hat{\alpha}_1(t,x) \\ \hat{\beta}_1(t,x) \\ \hat{\alpha}_2(t,x) \\ \hat{\beta}_2(t,x) \end{pmatrix} &:= \left(\mathcal{K}\begin{pmatrix} \hat{w}_1(t,\cdot) \\ \hat{v}_1(t,\cdot) \\ \hat{w}_2(t,\cdot) \\ \hat{v}_2(t,\cdot) \end{pmatrix}\right)(x) \\ &= \begin{pmatrix} \hat{w}_1(t,\cdot) \\ \hat{v}_1(t,\cdot) \\ \hat{w}_2(t,\cdot) \\ \hat{v}_2(t,\cdot) \end{pmatrix} - \int_0^L \begin{pmatrix} 0 & K_1^{vw}(x,\xi)\mathbb{1}_{[0,x]}(\xi) & 0 & 0 \\ 0 & K_1^{vv}(x,\xi)\mathbb{1}_{[0,x]}(\xi) & 0 & 0 \\ 0 & K^w(x,\xi) & 0 & K_2^{vw}(x,\xi)\mathbb{1}_{[x,L]}(\xi) \\ 0 & K^v(x,\xi) & 0) & K_2^{vv}(x,\xi)\mathbb{1}_{[x,L]}(\xi) \end{pmatrix}^T \begin{pmatrix} \hat{w}_1(t,\xi) \\ \hat{v}_1(t,\xi) \\ \hat{w}_2(t,\xi) \\ \hat{v}_2(t,\xi) \end{pmatrix} d\xi \end{aligned} \tag{88}$$

We can show that transformation (88) is invertible following the arguments we have used to show the invertibility of the transformation (61). The inverse transformation (useful to design our event-triggered controller) is given by

$$
\begin{aligned}
\begin{pmatrix} \hat{w}_1(t,x) \\ \hat{v}_1(t,x) \\ \hat{w}_2(t,x) \\ \hat{v}_2(t,x) \end{pmatrix} &:= \left(\mathcal{L} \begin{pmatrix} \hat{\alpha}_1(t,\cdot) \\ \hat{\beta}_1(t,\cdot) \\ \hat{\alpha}_2(t,\cdot) \\ \hat{\beta}_2(t,\cdot) \end{pmatrix} \right)(x) \qquad (89) \\
&= \begin{pmatrix} \hat{\alpha}_1(t,\cdot) \\ \hat{\beta}_1(t,\cdot) \\ \hat{\alpha}_2(t,\cdot) \\ \hat{\beta}_2(t,\cdot) \end{pmatrix} + \int_0^L \begin{pmatrix} 0 & L_1^{\beta\alpha}(x,\xi)\mathbb{1}_{[0,x]}(\xi) & 0 & 0 \\ 0 & L_1^{\beta\beta}(x,\xi)\mathbb{1}_{[0,x]}(\xi) & 0 & 0 \\ 0 & L^{\alpha}(x,\xi) & 0 & L_2^{\beta\alpha}(x,\xi)\mathbb{1}_{[x,L]}(\xi) \\ 0 & L^{\beta}(x,\xi) & 0 & L_2^{\beta\beta}(x,\xi)\mathbb{1}_{[x,L]}(\xi) \end{pmatrix}^T \begin{pmatrix} \hat{\alpha}_1(t,\xi) \\ \hat{\beta}_1(t,\xi) \\ \hat{\alpha}_2(t,\xi) \\ \hat{\beta}_2(t,\xi) \end{pmatrix} d\xi,
\end{aligned}
$$

where the different kernels are bounded functions defined on $\bar{\mathcal{T}}_1$ (kernels $L_2^{\cdot\cdot}$), $\bar{\mathcal{T}}_2$ (kernels $L_1^{\cdot\cdot}$) or $\bar{\mathcal{T}}_3$ (kernels $L^{\cdot}$). They satisfy the following set of equations on their respective domain of definition

$$(\gamma_i p_i^\star - v_i^\star)\partial_x L_i^{\beta\alpha}(x,\xi) - v_i^\star \partial_\xi L_i^{\beta\alpha}(x,\xi) = 0 \qquad (90)$$

$$\partial_x L_i^{\beta\beta}(x,\xi) + \partial_\xi L_i^{\beta\beta}(x,\xi) = 0 \qquad (91)$$

$$(\gamma_1 p_1^\star - v_1^\star)\,\partial_x L^{\beta}(x,\xi) - (\gamma_2 p_2^\star - v_2^\star)\,\partial_\xi L^{\beta}(x,\xi) = 0 \qquad (92)$$

$$(\gamma_1 p_1^\star - v_1^\star)\,\partial_x L^{\alpha}(x,\xi) + v_2^\star \partial_\xi L^{\alpha}(x,\xi) = 0, \qquad (93)$$

with the boundary conditions

$$L_i^{\beta\alpha}(x,x) = -\frac{c_i(x)}{\gamma_i p_i^\star},\ L_1^{\beta\beta}(x,0) = \frac{v_2^\star}{v_1^\star}\delta L^{\beta}(x,0), \qquad (94)$$

$$L^{\beta}(0,\xi) = \frac{r_1}{\delta r_2} L_2^{\beta\beta}(0,\xi),\ L^{\alpha}(0,\xi) = \frac{r_1}{\delta r_2} L_2^{\beta\alpha}(0,\xi) \qquad (95)$$

$$L_2^{\beta\beta}(x,L) = -\exp\left(\frac{-L}{\tau_2 v_2^\star}\right) L_2^{\beta\alpha}(x,L), \qquad (96)$$

$$L^{\beta}(x,L) = -\exp\left(\frac{-L}{\tau_2 v_2^\star}\right) L^{\alpha}(x,L), \qquad (97)$$

$$L^{\alpha}(x,0) = -(1-\delta)L^{\beta}(x,0) + \frac{v_1^\star}{v_2^\star} L_1^{\beta\alpha}(x,0). \qquad (98)$$

3.3 Emulation of the Output Control Law

As aforementioned, we now want to hold the nominal control $U_{\mathrm{nom}}(t)$ constant between two consecutive triggering times. Thus, the emulated version is given as $U_{\mathrm{nom}}(t_k) = U_{\mathrm{nom}}(t) + d(t)$, for all $t \in [t_k, t_{k+1})$. Notice that the output feedback affects the boundary condition (40) and (29) when introducing what we call "deviation of actuation" $d(t)$. Since we need to assess the impact of the deviation $d(t)$

to the closed-loop system under the event-triggered implementation, we use transformations (88) and (98) so that we can work on a suitable target system exhibiting the deviation $d(t)$ at the boundary, and from which, we can perform an easier Lyapunov stability analysis. Let us first apply the backstepping transformation (88) on the observer system (35)–(42) and compute the corresponding target system. After straightforward computations, we obtain

$$\partial_t\hat{\alpha}_1(t,x) + v_1^\star\partial_x\hat{\alpha}_1(t,x) = p_{\mu_1}(x)\tilde{\alpha}_1(t,L) \tag{99}$$

$$\partial_t\hat{\beta}_1(t,x) - (\gamma_1 p_1^\star - v_1^\star)\partial_x\hat{\beta}_1(t,x) = p_{\nu_1}(x)\tilde{\alpha}_1(t,L), \tag{100}$$

$$\partial_t\hat{\alpha}_2(t,x) - v_2^\star\partial_x\hat{\alpha}_2(t,x) = p_{\mu_2}(x)\tilde{\alpha}_1(t,L) \tag{101}$$

$$\partial_t\hat{\beta}_2(t,x) + (\gamma_2 p_2^\star - v_2^\star)\partial_x\hat{\beta}_2(t,x) = p_{\nu_2}(x)\tilde{\alpha}_1(t,L), \tag{102}$$

$$\hat{\alpha}_1(t,0) = \hat{\alpha}_2(t,0), \tag{103}$$

$$\hat{\beta}_1(t,L) = r_1\exp\left(\frac{-L}{\tau_1 v_1^\star}\right)\hat{\alpha}_1(t,L) + \tfrac{1-r_1}{\rho_1^\star}d, \tag{104}$$

$$\hat{\alpha}_2(t,L)\exp\left(\frac{-L}{\tau_2 v_2^\star}\right)\frac{1}{r_2}\hat{\beta}_2(t,L), \tag{105}$$

$$\hat{\beta}_2(t,0) = \delta\frac{r_2}{r_1}\hat{\beta}_1(t,0) + (1-\delta)r_2\hat{\alpha}_2(t,0), \tag{106}$$

where

$$p_{\mu_1}(x) = -\mu_1(x), \tag{107}$$

$$p_{\nu_1}(x) = -\nu_1(x) + \int_0^x K_1^{vw}(x,\xi)\mu_1(\xi) + K_1^{vv}(x,\xi)\nu_1(\xi)d\xi,$$
$$+\int_0^L \bar{K}^w(x,\xi)\mu_2(\xi) + K^v(x,\xi)\nu_2(\xi)d\xi, \tag{108}$$

$$p_{\mu_2}(x) = -\mu_2(x), \tag{109}$$

$$p_{\nu_2}(x) = -\nu_2(x) + \int_x^L K_2^{vw}(x,\xi)\mu_2(\xi) + K_2^{vv}(x,\xi)\nu_2(\xi)d\xi, \tag{110}$$

Note that the functions p_{ν_1} and p_{ν_2} are well-defined since they are solutions of Volterra equations [16]. Using the inverse transformation (89), we can now rewrite the nominal control law U_{nom} defined by (78) as a function of the states $\hat{\alpha}_i$ and $\hat{\beta}_i$

$$U_{\text{nom}}(t) = \tfrac{\rho_1^\star}{1-r_1}\left(\int_0^L \begin{pmatrix} L_1^{\beta\alpha}(L,\xi) & L_1^{\beta\beta}(L,\xi) & L^\alpha(L,\xi) & L^\beta(L,\xi)\end{pmatrix}\begin{pmatrix}\hat{\alpha}_1(t,\xi)\\ \hat{\beta}_1(t,\xi)\\ \hat{\alpha}_2(t,\xi)\\ \hat{\beta}_2(t,\xi)\end{pmatrix}d\xi\right) \tag{111}$$

an the corresponding emulated version

$$U_{\text{nom}}(t_k) = \frac{\rho_1^\star}{1-r_1}\left(\int_0^L \left(L_1^{\beta\alpha}(L,\xi)\; L_1^{\beta\beta}(L,\xi)\; L^{\alpha}(L,\xi)\; L^{\beta}(L,\xi) \right) \begin{pmatrix} \hat{\alpha}_1(t_k,\xi) \\ \hat{\beta}_1(t_k,\xi) \\ \hat{\alpha}_2(t_k,\xi) \\ \hat{\beta}_2(t_k,\xi) \end{pmatrix} d\xi \right) \tag{112}$$

for all $t \in [t_k, t_{k+1})$. We recall that $U_{\text{nom}}(t_k) = U_{\text{nom}}(t) + d(t)$ where d is given by

$$d(t) = \frac{\rho_1^\star}{1-r_1}\left(\int_0^L \left(L_1^{\beta\alpha}(L,\xi)\; L_1^{\beta\beta}(L,\xi)\; L^{\alpha}(L,\xi)\; L^{\beta}(L,\xi) \right) \begin{pmatrix} \hat{\alpha}_1(t_k,\xi)-\hat{\alpha}_1(t,\xi) \\ \hat{\beta}_1(t_k,\xi)-\hat{\beta}_1(t,\xi) \\ \hat{\alpha}_2(t_k,\xi)-\hat{\alpha}_2(t,\xi) \\ \hat{\beta}_2(t_k,\xi)-\hat{\beta}_2(t,\xi) \end{pmatrix} d\xi \right) \tag{113}$$

The function d (which will be fully characterized along with $(t_k)_{k\in\mathbb{N}}$ in the next section) can be viewed as an actuation deviation between the continuous controller and the event-triggered one. Notice that the nominal control, as well as its emulated version are expressed in terms of the kernels of transformation (89) and the states of the new target system (99)–(110). One of the main advantages of such an expression is that it can be easier to work with the target system (particularly when considering the input-to-state stability ISS properties of the system with respect to the deviation d) as well as an easier study of the *growth-in-time* of the deviation of actuation $d(t)$ (which is crucial to prove the avoidance of the so-called Zeno phenomenon). This is of specific interest when emulating the control law and finding conditions that guarantee the closed-loop stability under any event-triggered strategy. This methodology has been used in e.g. [6–8].

4 Observer-Based Event-Triggered Boundary Control Strategy

In this section we study the observer-based event-triggered boundary control strategy proposed in this chapter. It encloses an event-trigger mechanism containing a suitable triggering condition (which determines the time instant at which the controller needs to be updated) and the output backstepping feedback controller (112). Before we proceed with the definition of the observer-based event-triggered control, we rewrite first the target systems (62)–(67) and (99)–(106) in the following compact forms:

$$\partial_t \tilde{y}(t,x) + \Lambda \partial_x \tilde{y}(t,x) = 0, \tag{114}$$

$$\partial_t \hat{y}(t,x) + \Lambda \partial_x \hat{y}(t,x) = F(x)\tilde{\alpha}_1(t,L), \quad t \in \mathbb{R}^+, x \in [0,1] \tag{115}$$

where $\Lambda = \operatorname{diag}\left(\Lambda^+, -\Lambda^-\right)$ with

$$\Lambda^+ = \operatorname{diag}\Big(v_1^\star, (\gamma_2 p_2^\star - v_2^\star)\Big), \quad \Lambda^- = \operatorname{diag}\Big((\gamma_1 p_1^\star - v_1^\star), v_2^\star\Big), \tag{116}$$

and $F(x) = \left(p_{\mu_1}(x), p_{\nu_2}(x), p_{\nu_1}(x), p_{\mu_2}(x)\right)^\top$. We use the notation $\tilde{y} := \begin{pmatrix}\tilde{y}^+\\ \tilde{y}^-\end{pmatrix}$ with $\tilde{y}^+ := \begin{pmatrix}\tilde{\alpha}_1\\ \tilde{\beta}_2\end{pmatrix}$, $\tilde{y}^- := \begin{pmatrix}\tilde{\beta}_1\\ \tilde{\alpha}_2\end{pmatrix}$ and $\hat{y} := \begin{pmatrix}\hat{y}^+\\ \hat{y}^-\end{pmatrix}$ with $\hat{y}^+ := \begin{pmatrix}\hat{\alpha}_1\\ \hat{\beta}_2\end{pmatrix}$, $\hat{y}^- := \begin{pmatrix}\hat{\beta}_1\\ \hat{\alpha}_2\end{pmatrix}$. In addition, the boundary conditions can be rewritten as

$$\begin{pmatrix}\tilde{y}^+(t,0)\\ \tilde{y}^-(t,L)\end{pmatrix} = G\begin{pmatrix}\tilde{y}^+(t,L)\\ \tilde{y}^-(t,0)\end{pmatrix}, \quad \begin{pmatrix}\hat{y}^+(t,0)\\ \hat{y}^-(t,L)\end{pmatrix} = G\begin{pmatrix}\hat{y}^+(t,L)\\ \hat{y}^-(t,0)\end{pmatrix} + Bd(t), \tag{117}$$

where $B = \left(0, 0, \frac{1-r_1}{\rho_1^\star}, 0\right)^\top$, $G := \begin{pmatrix}0_{2,2} & G^+\\ G^- & 0_{2,2}\end{pmatrix}$ with $G^+ = \begin{pmatrix}0 & 1\\ \delta\frac{r_2}{r_1} & (1-\delta)r_2\end{pmatrix}$ and

$$G^- = \begin{pmatrix} r_1\exp\left(\frac{-L}{\tau_1 v_1^\star}\right) & 0\\ 0 & \frac{1}{r_2}\exp\left(\frac{-L}{\tau_2 v_2^\star}\right)\end{pmatrix}.$$

In addition, using the solutions to (94)–(98) along with (107)–(110), we introduce the following variables:

$$\kappa_{\hat{\alpha}_1} := \frac{\rho_1^\star}{1-r_1}\left(v_1^\star L_1^{\beta\alpha}(L,L) - (\gamma_1 p_1^\star - v_1^\star) r_1 \exp\left(\frac{-L}{\tau_1 v_1^\star}\right) L_1^{\beta\beta}(L,L)\right), \tag{118}$$

$$\kappa_{\hat{\beta}_1} := \frac{\rho_1^\star}{1-r_1}\left((\gamma_1 p_1^\star - v_1^\star) L_1^{\beta\beta}(L,0) - (\gamma_2 p_2^\star - v_2^\star)\delta\frac{r_2}{r_1} L^\beta(L,0)\right), \tag{119}$$

$$\begin{aligned}\kappa_{\hat{\alpha}_2} := &\frac{\rho_1^\star}{1-r_1}(-v_1^\star L_1^{\beta\alpha}(L,0) + v_2^\star L^\alpha(L,0)\\ &- (\gamma_2 p_2^\star - v_2^\star)(1-\delta) r_2 L^\beta(L,0)),\end{aligned} \tag{120}$$

$$\kappa_{\hat{\beta}_2} := \frac{\rho_1^\star}{1-r_1}\left(-v_2^\star \frac{1}{r_2}\exp\left(\frac{-L}{\tau_2 v_2^\star}\right) L^\alpha(L,L) + (\gamma_2 p_2^\star - v_2^\star) L^\beta(L,L)\right), \tag{121}$$

$$\kappa_{\tilde{\alpha}_1} := \frac{\rho_1^\star}{1-r_1}\left(\int_0^L \begin{pmatrix} L_1^{\beta\alpha}(L,\xi) & L_1^{\beta\beta}(L,\xi) & L^\alpha(L,\xi) & L^\beta(L,\xi)\end{pmatrix} \cdot \begin{pmatrix} p_{\mu_1}(\xi)\\ p_{\nu_1}(\xi)\\ p_{\mu_2}(\xi)\\ p_{\nu_2}(\xi)\end{pmatrix} d\xi\right), \tag{122}$$

$$\begin{aligned}\varepsilon_0 := 2\left(\frac{\rho_1^\star}{1-r_1}\right)^2 \max\Bigg\{ &\int_0^L \left(v_1^\star \partial_\xi L_1^{\beta\alpha}(L,\xi)\right)^2, \int_0^L \left((\gamma_1 p_1^\star - v_1^\star)\partial_\xi L_1^{\beta\beta}(L,\xi)\right)^2,\\ &\int_0^L \left(v_2^\star \partial_\xi L^\alpha(L,\xi)\right)^2 d\xi, \int_0^L \left((\gamma_2 p_2^\star - v_2^\star)\partial_\xi L^\beta(L,\xi)\right)^2 d\xi\Bigg\},\end{aligned} \tag{123}$$

$$\varepsilon_1 := 4\left((\gamma_1 p_1^\star - v_1^\star) L_1^{\beta\beta}(L,L)\right)^2, \tag{124}$$

$$\varepsilon_2 := 4\kappa_{\tilde{\alpha}_1}^2 \tag{125}$$

$$\mathcal{D}_0 := \mathrm{diag}\left(8\kappa_{\hat{\alpha}_1}^2, 8\kappa_{\hat{\beta}_2}^2, 8\kappa_{\hat{\beta}_1}^2, 8\kappa_{\hat{\alpha}_2}^2\right). \tag{126}$$

4.1 Definition of Observer-Based Event-Triggered Boundary Controller

The event-triggering condition is based on the evolution of the square of the actuation deviation (113) and of a dynamic variable satisfying a suitable ODE. It relies, in turn, on a Lyapunov function candidate for the target systems (114)–(115) that we define as follows (see [4, Sect. 5], [13, 15]):

$$V(\tilde{y}, \hat{y}) := V_1(\tilde{y}) + CV_2(\hat{y}) = \int_0^L \tilde{y}^\top(x)\mathcal{Q}(x)\tilde{y}(x)dx + C\int_0^L \hat{y}^\top(x)\mathcal{Q}(x)\hat{y}(x)dx, \tag{127}$$

where $\mathcal{Q}(x) = \mathrm{diag}[\mathcal{Q}^+(x), \mathcal{Q}^-(x)] = \mathrm{diag}[e^{-\mu x}Q^+, e^{\mu x}Q^-]$ with $\mu > 0$, $C > 0$, and diagonal positive definite matrices $Q^- \in \mathbb{R}^{2\times 2}$ and $Q^+ \in \mathbb{R}^{2\times 2}$ and such that $\mathcal{Q}(x)\Lambda = \Lambda\mathcal{Q}(x)$. Moreover, we use the fact that there exist $\underline{\varrho}$, $\overline{\varrho}$ (depending on the eigenvalues of Q^+, Q^- and on μ) such that $\underline{\varrho}\|\hat{y}(t,\cdot)\|_{L^2}^2 \leq V_2(\hat{y}(t,\cdot)) \leq \overline{\varrho}\|\hat{y}(t,\cdot)\|_{L^2}^2$.

Definition 1 Let $\theta_0, \theta_1 > 0$, $\eta > 0$, $\nu > 0$, $\sigma \in (0,1)$. Let ε_2 and $\mathcal{D}_0$ be given by (125) and (126), respectively. Let $t \mapsto V(\tilde{y}(t,\cdot), \hat{y}(t,\cdot))$ be given by (127) with diagonal positive definite matrices $Q^- \in \mathbb{R}^{2\times 2}$ and $Q^+ \in \mathbb{R}^{2\times 2}$ and $C > 0$. The observer-based event-triggered boundary control is defined by considering the following components:

(I) (The event-trigger) The times of the events $t_k \geq 0$ with $t_0 = 0$ form a finite or countable set of times which is determined by the following rules for some $k \geq 0$:

(a) if $\{t \in \mathbb{R}^+ | t > t_k \wedge \theta_1 C\left(\frac{1-r_1}{\rho_1^\star}\right)^2 d^2(t) \geq \frac{\nu}{2}\sigma V(t) - \frac{1}{\theta_0}m(t)\} = \emptyset$ then the set of the times of the events is $\{t_0, ..., t_k\}$.

(b) if $\{t \in \mathbb{R}^+ | t > t_k \wedge \theta_1 C\left(\frac{1-r_1}{\rho_1^\star}\right)^2 d^2(t) \geq \frac{\nu}{2}\sigma V(t) - \frac{1}{\theta_0}m(t)\} \neq \emptyset$, then the next event time is given by:

$$t_{k+1} = \inf\{t \in \mathbb{R}^+ | t > t_k \wedge \theta_1 C\left(\tfrac{1-r_1}{\rho_1^\star}\right)^2 d^2(t) \geq \frac{\nu}{2}\sigma V(t) - \frac{1}{\theta_0}m(t)\}. \tag{128}$$

where the actuation deviation $d(t)$ is given by (113) for all $t \in [t_k, t_{k+1})$, and m satisfies the ordinary differential equation,

$$\begin{aligned}\dot{m}(t) = -\eta m(t) + \Bigg(\theta_1 C\left(\tfrac{1-r_1}{\rho_1^\star}\right)^2 d^2(t) - \frac{\nu}{2}\sigma V(t) \qquad\qquad (129)\\ -2\theta_0\theta_1 C\left(\tfrac{1-r_1}{\rho_1^\star}\right)^2\left(\begin{pmatrix}\hat{y}^+(t,L)\\ \hat{y}^-(t,0)\end{pmatrix}^\top \mathcal{D}_0\begin{pmatrix}\hat{y}^+(t,L)\\ \hat{y}^-(t,0)\end{pmatrix} + \varepsilon_2\tilde{\alpha}^2(t,L)\right)\Bigg),\end{aligned}$$

for all $t \in (t_k, t_{k+1})$ with $\eta \geq \frac{\nu}{2}(1-\sigma)$, $m(0) = m^0 < 0$, and $m(t_k^-) = m(t_k) = m(t_k^+)$.

(II) (the control action) The output boundary feedback law is defined by

$$U_{\text{nom}}(t_k) = \tfrac{\rho_1^\star}{1-r_1}\left(\int_0^L \begin{pmatrix} L_1^{\beta\alpha}(L,\xi) & L_1^{\beta\beta}(L,\xi) & L^{\alpha}(L,\xi) & L^{\beta}(L,\xi) \end{pmatrix} \begin{pmatrix} \hat{\alpha}_1(t_k,\xi) \\ \hat{\beta}_1(t_k,\xi) \\ \hat{\alpha}_2(t_k,\xi) \\ \hat{\beta}_2(t_k,\xi) \end{pmatrix} d\xi \right), \tag{130}$$

for all $t \in [t_k, t_{k+1})$.

Remark 1 Although the function $V(t)$ and the function $m(t)$ depends on $\tilde{\alpha}_i(t,\cdot)$ and $\tilde{\beta}_i(t,\cdot)$ (which are a priori unknown), this is not a problem as these functions can be expressed as delayed functions of the measurement $yL(t)$ and of the observer state. Indeed, we have $\tilde{\alpha}(t,L) = \tilde{w}(t,L) = yL(t) - \hat{w}(t,L)$, which means that the function $\tilde{\alpha}(t,1)$ can be computed from the measurement. From (62)–(67), we immediately have for all $x \in [0,1]$

$$\tilde{\beta}_1(t,x) = r_1 \exp\left(\frac{-L}{\tau_1 v_1^\star}\right)\tilde{\alpha}_1(t - \frac{(L-x)}{\gamma_1 p_1^\star - v_1^\star}, L), \tag{131}$$

which means that we can also compute the function $\beta_1(t,x)$ from the measurement. Consider now the function $\alpha_2(t,0)$. We have (using the method of characteristics)

$$\begin{aligned} \tilde{\alpha}_2(t,0) = {} & \exp\left(\frac{-L}{\tau_2 v_2^\star}\right)(1-\delta)\tilde{\alpha}_2(t - \frac{1}{v_2^\star} - \frac{1}{\gamma_2 p_2^\star - v_2^\star}, 0) \\ & + \frac{\delta}{r_1}\exp\left(\frac{-L}{\tau_2 v_2^\star}\right)\tilde{\beta}_1(t - \frac{1}{v_2^\star} - \frac{1}{\gamma_1 p_1^\star - v_1^\star}, 0). \end{aligned} \tag{132}$$

Applying the method of characteristics on the term $\alpha_2(t,0)$ that appear on the right side of the above equation, and iterating N_0 times the procedure, we obtain

$$\begin{aligned} \tilde{\alpha}_2(t,0) = {} & (\exp\left(\frac{-L}{\tau_2 v_2^\star}\right)(1-\delta))^N \tilde{\alpha}_2(t - N_0(\frac{1}{v_2^\star} + \frac{1}{\gamma_2 p_2^\star - v_2^\star}), 0) \\ & + F(\tilde{\beta}_1(t,0)), \end{aligned} \tag{133}$$

where the function F only depends on delayed values of $F(\tilde{\beta}_1(t,0))$. Choosing N such that $N(\frac{1}{v_2^\star} + \frac{\Lambda}{\gamma_2 p_2^\star - v_2^\star}) - \frac{1}{v_1^\star} > 0$ and using Eq. (64), we obtain

$$\begin{aligned} \tilde{\alpha}_2(t,0) = {} & (\exp\left(\frac{-L}{\tau_2 v_2^\star}\right)(1-\delta))^N \tilde{\alpha}_1(t - N_0(\frac{1}{v_2^\star} + \frac{1}{\gamma_2 p_2^\star - v_2^\star}) + \frac{1}{v_1^\star}, L) \\ & + F(\tilde{\beta}_1(t,0)). \end{aligned} \tag{134}$$

Thus, we can compute the function $\tilde{\alpha}_2(t,0)$ using the available measurement. Using the method of the characteristics, it becomes straightforward to express $\tilde{\alpha}_i(t,x)$

and $\tilde{\beta}_i(t,x)$ as delayed functions of the available measurements. Consequently, the proposed event-triggered strategy is implementable simply using the available measurement (and the observer state).

We directly have the following lemma.

Lemma 1 *Under the definition of the observer-based event triggered boundary control* (130) *with the dynamic trigger condition* (128), *it holds that* $\theta_1 C\left(\frac{1-r_1}{\rho_1^\star}\right)^2 d^2(t) - \frac{\nu}{2}\sigma V(t) + m(t) < 0$ *and* $m(t) < 0$ *for* $t \in [0, T)$ *where* $T = \lim_{k\to\infty}(t_k)$.

Proof The proof follows the same lines of [7, Lemma 1]. □

The following result is useful to analyze the growth-in-time of the actuation deviation. A suitable characterization is given in the following lemma which is instrumental to derive the existence of a minimal dwell-time.

Lemma 2 *For* $d(t)$ *given by* (113), *it holds for all* $t \in (t_k, t_{k+1})$,

$$(\dot{d}(t))^2 \leq \varepsilon_0 \frac{1}{\underline{\varrho} C} V(t) + \varepsilon_1 d^2(t) + \varepsilon_2 \tilde{\alpha}^2(t,L) + \begin{pmatrix} \hat{y}^+(t,L) \\ \hat{y}^-(t,0) \end{pmatrix}^\top \mathcal{D}_0 \begin{pmatrix} \hat{y}^+(t,L) \\ \hat{y}^-(t,0) \end{pmatrix}, \tag{135}$$

for some $\underline{\varrho} > 0$, *and with* ε_0, ε_1, ε_2 *and* $\mathcal{D}_0$ *given by* (123), (124), (125) *and* (126), *respectively.*

Proof The proof follows the same lines of [7, Lemma 2]. □

5 Main Results

In this section we present our main results: the avoidance of the Zeno phenomenon and the exponential convergence in L^2-norm of the closed-loop system.

5.1 Avoidance of the Zeno Phenomenon

We first prove the avoidance of the Zeno phenomenon.

Theorem 1 *Under the event-triggered boundary control* (130)–(128) *in Definition 1, with parameters satisfying*

$$\theta_0\theta_1 < \underline{\varrho}\left(\frac{\rho_1^\star}{1-r_1}\right)^2 \frac{\nu\sigma}{4\varepsilon_0}, \tag{136}$$

there exists a minimal dwell-time $\tau^\star > 0$ *between two triggering times, i.e. there exists a constant* $\tau^\star > 0$ *(independent of the initial conditions) such that* $t_{k+1} - t_k \geq \tau^\star$, *for all* $k \geq 0$. *Moreover,* $\tau^\star$ *can be given by*

$$\tau^\star = \int_0^1 \frac{1}{a_0 + a_1 s + a_2 s^2} ds, \tag{137}$$

with $a_0 = 1 + \varepsilon_1 + \frac{1}{2\theta_0} + \eta$, $a_1 = 1 + \varepsilon_1 + \frac{1}{2\theta_0} + \eta$ *and* $a_2 = \frac{1}{2\theta_0}$.

Proof It follows the methodology employed in [6, 7] and makes uses of estimate (135) in Lemma 2. □

Remark 2 Since there is a minimal dwell-time (which is uniform and does not depend on initial conditions), no Zeno solution can appear. This has a very important consequence as it allows to guarantee the existence and uniqueness of the closed-loop solution. The solution, can be constructed by the step method. We omit the details of well-posedness in this chapter, but we refer to [6, 13] for further details on the notion of the considered solutions.

5.2 Lyapunov-Based Analysis

We perform a Lyapunov-based analysis on the target systems written in compact form i.e. (114)–(115) and we take into account the proposed event-triggered control strategy (128)–(130). In what follows, we define the following variables

$$\bar{\mathcal{D}}_0 = 2\theta_0\theta_1 \left(\frac{1-r_1}{\rho_1^\star}\right)^2 \mathcal{D}_0, \tag{138}$$

with $\mathcal{D}_0$ given by (126);

$$\mathcal{D}_1 = \begin{pmatrix} \mathcal{Q}^+(0)\Lambda^+ & 0_{2,2} \\ 0_{2,2} & \mathcal{Q}^-(L)\Lambda^- \end{pmatrix}, \tag{139}$$

$$\mathcal{D}_2 = \begin{pmatrix} \mathcal{Q}^+(L)\Lambda^+ & 0_{2,2} \\ 0_{2,2} & \mathcal{Q}^-(0)\Lambda^- \end{pmatrix}, \tag{140}$$

for $\mathcal{Q}(x)$ as in (127), and

$$\mathcal{D}_3 = \begin{pmatrix} \frac{\theta_2}{2} + 2\theta_0\theta_1 C \left(\frac{1-r_1}{\rho_1^\star}\right)^2 \varepsilon_2 & 0_{1,3} \\ 0_{3,1} & 0_{3,3} \end{pmatrix}, \tag{141}$$

with some $\theta_2 > 0$. The notation $0_{i,j}$ stands for the matrix with i rows and j columns whose all components are zero.

Theorem 2 *Let* $\theta_0 > 0$, $\sigma \in (0, 1)$, ε_0, ε_2 *be given by* (123), (125), *respectively and* $\mathcal{D}_0$ *be given by* (138). *Let* $\mathcal{D}_1$, $\mathcal{D}_2$ *and* $\mathcal{D}_3$ *be given by* (139),(140) *and* (141), *respectively. Let* (136) *hold. If there exist* $\theta_1, \theta_2 > 0$, $\mu > 0$, $\nu > 0$ *(thus there exists*

$\eta \geq \frac{\nu}{2}(1-\sigma)$) *and diagonal positive definite matrices* $Q^- \in \mathbb{R}^{2\times 2}$ *and* $Q^+ \in \mathbb{R}^{2\times 2}$ *such that for* $\mathcal{Q}(x)$ *given in* (127) *the following conditions hold,*

$$G^\top \mathcal{D}_1 G - \mathcal{D}_2 + \mathcal{D}_3 < 0, \tag{142}$$

$$\begin{pmatrix} \bar{\mathcal{D}}_0 + G^\top \mathcal{D}_1 G - \mathcal{D}_2 & G^\top \mathcal{D}_1 \\ (G^\top \mathcal{D}_1)^\top & \mathcal{D}_1 - \theta_1 I_{4,4} \end{pmatrix} < 0, \tag{143}$$

then, the closed-loop system (24)–(31) *with event-triggered control* (130), (128) *is exponentially convergent in the* L^2*-norm.*

Proof It follows by considering the following Lyapunov function candidate for the target systems (114)–(115) along with (129), defined for all $\tilde{y} \in L^2((0,L);\mathbb{R}^4)$, $\hat{y} \in L^2((0,L);\mathbb{R}^4)$ and $m \in \mathbb{R}^-$ by

$$W(\tilde{y},\hat{y},m) := V(\tilde{y},\hat{y}) - m \tag{144}$$

and establishing that the time derivative of (144) along the solutions is upper estimated (after using the sufficient conditions (142)–(143) and Lemma 1) as follows:

$$\dot{W}(t) \leq -\tfrac{\nu}{2}(1-\sigma)W(t). \tag{145}$$

Using the comparison principle and the bounded invertibility of the backstepping transformations we obtain the exponential convergence in the L^2-norm. □

Remark 3 In Theorem 2, we have established the exponential convergence of the closed-loop system to the equilibrium point. We could have obtained exponential stability if we set $m^0 = 0$. However, if $m^0 = 0$, then $m(t) \leq 0$. This specific issue may affect the conclusion on the existence of a minimal-dwell as stated in Theorem 1. Hence, we opted to choose m^0 strictly negative.

6 Numerical Simulations

The length of each freeway segment is chosen to be $L = 1$ km so the total length of the two connected segments are 2 km. The simulation time is $T = 16$ min. The maximum speed limit is $v_m = 40$ m/s $= 144$ km/h. We consider 6 lanes for the downstream freeway segment 1. Assuming the average vehicle length is 5 m plus the minimum safety distance of 50% vehicle length, the maximum density of the road is obtained as $\rho_{m,1} = 6/7.5$ vehicles/m $= 800$ vehicles/km. The upstream segment has less functional lanes thus its maximum density is $\rho_{m,2} = 700$ vehicles/km. We take $\gamma_i = 0.5$. The steady states $(\rho_1^\star, v_1^\star)$ and $(\rho_2^\star, v_2^\star)$ are chosen respectively as (600 vehicles/km, 19.4 km/h) and (488.6 vehicles/km, 23.8 km/h), both of which are in the congested regime. The constant flow rate is $q^\star = \rho_1^\star v_1^\star = \rho_2^\star v_2^\star =$ 11640 vehicles/h, same for the two segments. If we consider the segment 1 with 6

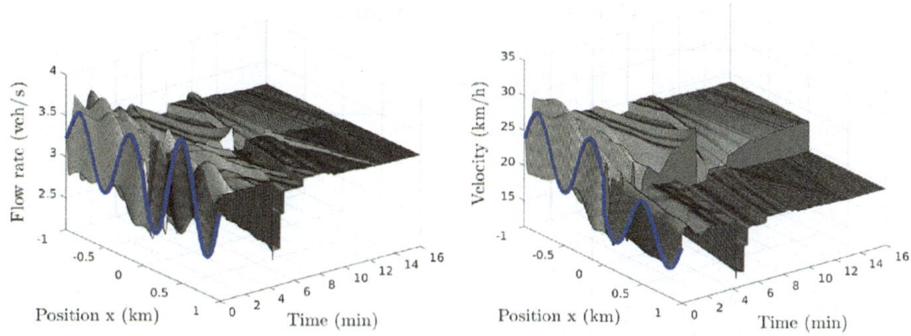

Fig. 3 Numerical solution of the flow rate and velocity with the ramp meetering event-triggered output control $U_{\text{nom}}(t_k)$ which is updated according to the observer-based event-triggered output control (128)

lanes, then the averaged flow rate of each lane is 1940 vehicles/h/lane. The equilibrium steady state of the downstream road has higher density and lower velocity, thus is more congested than the upstream road. The relaxation time is $\tau_1 = 90$ s and $\tau_2 = 60$ s. We use sinusoid initial conditions for flow rate and velocity field which represent the initial stop-and-go oscillations on the connected freeway. We perform the simulation on a time horizon of 16 min.

Event-Triggered Implementation and Closed-Loop Simulation

The parameters involved in the observer-based event-triggering boundary controller are $r_1 = -0.44$, $r_2 = -0.64$, $\varepsilon_0 = 6.4 \times 10^{-3}$, $\varepsilon_1 = 4.08 \times 10^{-3}$, $\varepsilon_2 = 1.73 \times 10^3$, $\kappa_{\hat{\alpha}_1} = 0.41$, $\kappa_{\hat{\beta}_1} = 3.45$, $\kappa_{\hat{\alpha}_2} = -1.45$, $\kappa_{\hat{\beta}_2} = -0.94$, $\kappa_{\tilde{\alpha}_1} = 22.19$. Then, we select $\theta_0 = 500$, $\sigma = 0.09$. Moreover, we perform a line search on μ while finding θ_1, θ_2, Q^+ and Q^- must verify conditions (136), (142) and (143) (using e.g. semi-definite programming combined with line search algorithm). We obtain $\mu = 1.01$, $\theta_1 = 0.061$, $\theta_2 = 1.3 \times 10^{-3}$, $Q^+ = \begin{pmatrix} 0.835 & 0 \\ 0 & 0.476 \end{pmatrix}$, $Q^- = \begin{pmatrix} 1.58 & 0 \\ 0 & 1.80 \end{pmatrix}$. We obtain also $\nu = 5.4 \times 10^{-3}$, $\eta = 0.0493$, $C = 1.76 \times 10^{-4}$. Hence, Theorem 2 applies. Moreover, we compute the minimal dwell-time between two triggering times according to (137), that is $\tau^\star = 0.65$. We stabilize the system on events under the event-triggered boundary control (130), (128). Figure 3 shows the numerical solution of flow rate and velocity with the ramp meetering event-triggered output control $U_{\text{nom}}(t_k)$ which is updated according to the observer-based event-triggered output control (128). Figure 4 shows the time-evolution of the control signal (recall that designed controller U_{nom} is the flow rate perturbation around a nominal flow rate), where we can observe that the updating is aperiodically, only when needed.

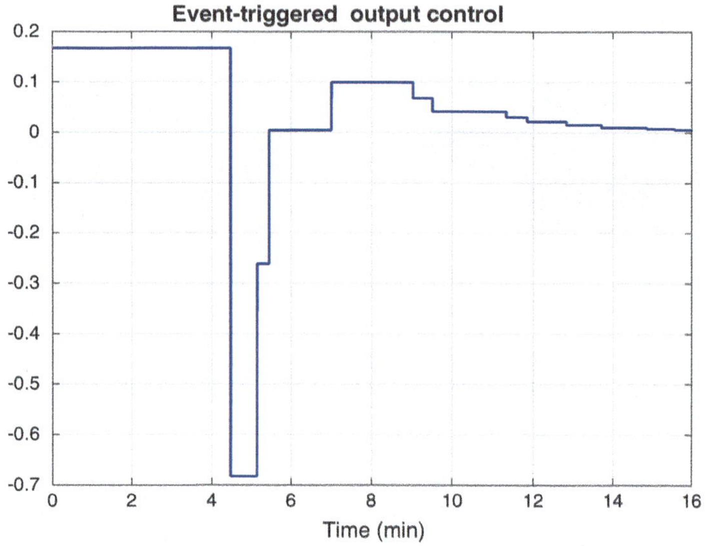

Fig. 4 Time-evolution of the event-triggered output control. The updating is aperiodically, according with (128)

7 Conclusion

In this chapter, we developed an event-triggered boundary output feedback control for simultaneous stabilization of traffic flow on connected roads. We built on the linearized ARZ model and designed a ramp metering strategy as a boundary control, through the backstepping method. The updating of the control signal is done according to a suitable dynamic triggering condition. We proved that under this strategy, there exists a uniform minimal dwell-time (independent of initial conditions), thus avoiding the Zeno phenomenon and we guaranteed the exponential convergence of the closed-loop system under the proposed event-triggered boundary control. Future work includes the design of periodic event-triggered control strategy to monitor the triggering condition periodically, hence, saving computational resources.

References

1. Auriol, J., Di Meglio, F.: An explicit mapping from linear first order hyperbolic PDEs to difference systems. Syst. Control Lett. **123**, 144–150 (2019)
2. Auriol, J., Di Meglio, F.: Robust output feedback stabilization for two heterodirectional linear coupled hyperbolic PDEs. Automatica **115**, 108896 (2020)
3. Aw, A., Rascle, M.: Resurrection of "second order" models of traffic flow. SIAM J. Appl. Math. **60**(3), 916–938 (2000)

4. Bastin, G., Coron, J.-M.: Stability and Boundary Stabilization of 1-D Hyperbolic Systems. Birkhäuser Basel (2016)
5. Coron, J.-M., Hu, L., Olive, G.: Stabilization and controllability of first-order integro-differential hyperbolic equations. J. Funct. Anal. **271**, 554–3587 (2016)
6. Espitia, N.: Observer-based event-triggered boundary control of a linear 2 × 2 hyperbolic systems. Syst. Control Lett. **138** (2020)
7. Espitia, N., Girard, A., Marchand, N., Prieur, C.: Event-based boundary control of a linear 2 × 2 hyperbolic system via backstepping approach. IEEE Trans. Autom. Control **63**(8), 2686–2693 (2018)
8. Espitia, N., Yu, H., Krstic, M.: Event-triggered varying speed limit control for stop-and-go traffic. In: Proceedings of the IFAC World Congress, Berlin (2020)
9. Fan, S., Seibold, B.: Data-fitted first-order traffic models and their second-order generalizations: comparison by trajectory and sensor data. Transp. Res. Rec. **2391**(1), 32–43 (2013)
10. Garavello, M., Piccoli, B.: Traffic flow on a road network using the aw-rascle model. Commun. Partial Differ. Equ. **31**(2), 243–275 (2006)
11. Herty, M., Rascle, M.: Coupling conditions for a class of second-order models for traffic flow. SIAM J. Math. Anal. **38**(2), 595–616 (2006)
12. Karafyllis, I., Papageorgiou, M.: Feedback control of scalar conservation laws with application to density control in freeways by means of variable speed limits. Automatica **105**, 228–236 (2019)
13. Prieur, C., Girard, A., Witrant, E.: Stability of switched linear hyperbolic systems by Lyapunov techniques. IEEE Trans. Autom. Control **59**(8), 2196–2202 (2014)
14. Smoller, J.: Shock Waves and Reaction-diffusion Equations, vol. 258. Springer Science & Business Media (2012)
15. Vazquez, R.: Coron, J.-M., Krstic, M., Bastin, G.: Collocated output-feedback stabilization of a 2 × 2 quasilinear hyperbolic system using backstepping. In: American Control Conference (ACC), 2012, pp. 2202–2207, Fairmont Queen Elizabeth, Montreal, Canada (2012)
16. Yoshida, K.: Lectures on Differential and Integral Equations, vol. 10. Interscience Publishers (1960)
17. Yu, H., Auriol, J., Krstic, M.: Simultaneous downstream and upstream output-feedback stabilization of cascaded freeway traffic. Automatica **125**, 109379 (2021)
18. Yu, H., Krstic, M.: Varying speed limit control of Aw-Rascle-Zhang traffic model. In: Proceedings of the 21st International conference on Intelligent Transportation Systems (ITSC), pp. 1846–1851, Maui, HI, USA (2018)
19. Yu, H., Krstic, M.: Traffic congestion control for Aw-Rascle-Zhang model. Automatica **100**, 38–51 (2019)
20. Zhang, H.M.: A non-equilibrium traffic model devoid of gas-like behavior. Transp. Res. Part B: Methodol. **36**(3), 275–290 (2002)
21. Zhang, L., Prieur, C., Qiao, J.: Pi boundary control of linear hyperbolic balance laws with stabilization of arz traffic flow models. Syst. Control Lett. **123**, 85–91 (2019)

Computational Modeling for One-Dimensional Distributed Parameter Systems

Daniela Danciu

Abstract The current chapter considers the class of distributed parameter systems (DPS) arising from natural sciences and control engineering applications, which are modeled by hyperbolic or parabolic partial differential equations (PDEs) with non-standard boundary conditions (nsBCs). The nsBCs are defined in opposition to the standard ones – Dirichlet, Neumann and Robin, which do not contain time derivatives for the boundary values of the state variables. Due to the complexity of their mathematical models, nowadays a feasible and practical alternative to the analytical solution and system behavior evaluation are the numerical and computational solutions. Consequently, the accuracy of the computational procedures used for deriving the approximate solution is of noticeable importance. We present a systematic computational procedure to solve the class of systems referred above. This procedure, based on a "convergent Method of Lines", ensures the convergence of the approximate numerical solution and also the preservation of the basic properties of the "true" solution as well as its Lyapunov stability. Thus, the approximate computational model allows numerical quantitative and qualitative analysis relevant to a specific problem. The computational efficiency of this procedure is ensured by its implementation by means of (massively) parallel structured devices belonging to the Artificial Intelligence field – the cell-based recurrent neural networks. An illustrative control engineering application is provided in the last technical section of this chapter, which ends with concluding remarks.

This project has received funding from the European Union's Horizon 2020 research and innovation programme under the Marie Sklodowska-Curie grant agreement No 873178.

D. Danciu (✉)
University of Craiova, 13, A.I. Cuza str., Craiova, Romania
e-mail: daniela.danciu@edu.ucv.ro

J. Auriol et al. (eds.), *Advances in Distributed Parameter Systems*, Advances in Delays and Dynamics 14, https://doi.org/10.1007/978-3-030-94766-8_12

1 Introduction

Since the 1970s, the field of distributed parameter systems (DSP) has known a fast advance in the development of modeling and control theory as well as in building practical computational tools [2, 30]. Within this context, it is worthwhile to point out here a fundamental idea emphasized by J.A. Burns in [2]:

> All computational tools are based on approximations of some type, and *the analysis of these numerical methods requires a mathematical framework to address issues such as accuracy and convergence*.

Also, another key idea, regarding the development of new methods and tools for solving real world control problems for systems with distributed parameters, emerges from the reflections of J.A. Burns about scientific success [2]:

> When one discusses "accomplishments" in a specific area such as DP control, it is worthwhile to keep in mind that there are several ways to measure success. A line of research that produces a fundamental contribution to the field itself is certainly considered an accomplishment. In addition, *research can be highly successful if it leads to significant practical tools for solving important real world control problems*. There is a third measure of success that is often overlooked by researchers working in the field. In some cases it happens that *research in one field generates tools and ideas that, when applied to another (very different) field, yield key breakthroughs*. [...] *The importance of staying close to an application area can not be overstated*. [...] *A good application can* lead to exciting new mathematics and *pave the way for major breakthroughs in computational algorithms*.

The work presented throughout this chapter can give an answer to some of the issues pointed out by these two quotations about the field of Distributed Parameter Control Systems (DPCS). More precisely, as will become apparent in the sequel, we refer to the accuracy and convergence of the approximation as well as the use of tools, ideas and frameworks from different scientific areas, combined in order to yield new approaches and effective tools for model reduction and problem solving (by means of computational modeling) of a wide class of DPS.

Class of systems addressed. For computational modeling and model reduction of partial differential equations (PDEs), several approaches are reported within literature, most of them suitable for parabolic or elliptic PDEs (pPDEs, ePDEs). It is known that the hyperbolic PDEs (hPDEs) are more difficult to handle due to the phenomenon of discontinuity propagation, which arises in the case of a "mismatch" between the initial and the boundary conditions. This aspect further leads to problems regarding the convergence of the approximate solution to the solution of the original problem of hPDEs. The usual academic examples encountered within literature for hPDEs consider problems with known solutions, as well as examples of initial value problems (IVP), or boundary value problems (BVP) with standard BCs, or initial boundary problems (IBVP) with standard BCs (see, for instance, [29, 33] and references herein). Nevertheless, the distributed parameter control systems for propagation and transport phenomena encountered in real world processes are more complex and usually imply a mutual control, *via* nsBCs, between PDEs and various sub-systems having complicated dynamics.

This chapter addresses the class of DPS described by mixed initial boundary value problems in one dimension for either hPDEs or pPDEs with non-standard boundary conditions. These types of systems, with complex space-time dynamical behavior, occur in different domains from natural sciences (physics, chemistry, biology etc.) as well as from control engineering applications and various industry processes (chemistry, aerospace, hydrodynamics, biotechnology, communication, mechanics etc.).

Numerical modeling. From the analytical point of view, the infinite-dimensional mathematical models of the real world applications for DPS are difficult to handle, if not even are they intractable. Consequently, several tools have been developed for obtaining reduced-order systems which approximate the original ones with distributed parameters. In this way, the approximate solutions obtained using the *Computational Mathematics* methods can be analyzed, interpreted and visualized by making use of different technologies (analog, digital or software) for the approximate problem/solution implementation.

The Method of Lines (MoL) is a well-known computational approach (more a general guide) used for approximating PDEs by a possibly high-order system of ordinary differential equations (ODEs), for the solution of which there are already developed various high-performance software packages. It is important to point out here that not every approach of MoL is able to guarantee for a correct discretization ensuring proper convergence of the approach. Moreover, only a correct application of the MoL will ensure the preservation of the Lyapunov stability property of the solution [17]. Concerning this aspect, we followed an approach of MoL which takes all these issues into consideration, as will be described in the next section.

Computational modeling. Since the first decade of the 2000s, the field of *Neural Networks* (NN) has evolved to include a new branch of research concerned with solving of both weak-formalized and formalized problems (see, for example, [21, 22, 25]) arising from applied mathematics, physics and engineering applications. This new field of research, named *Neural Mathematics*, can be also viewed as a new branch of the *Computational Mathematics* field. Within this context, we consider here the issue of computational modeling of spatio-temporal phenomena by means of Recurrent Neural Networks (RNNs) – devices of elementary computational elements (the artificial neurons), massively interconnected in a network of possibly huge number of parallel layers displaying feedback connections at the neurons level as well as at the level of the inter-layers connections. The key idea of *Neural Mathematics* followed here is to utilize such NN in order to make use of the natural parallelism of the problem itself and to cast it in the form of a certain RNN of equivalent dynamics and topology [22, 23]. More precisely, the solution is obtained *via* problem implementation by means of a model-driven neural network whose parameters (weights, biases and inputs) are given by the problem itself, not by a "learning" process based on experimental data.

Purpose and approach. The procedure, previously introduced in [17] and presented in this chapter, aims at computational modeling of problems arising from physics, engineering and industry applications of the aforementioned class of DPS and is mainly based on two seminal results:

(i) a rigorously proven result by Halanay and V. Răsvan and its "Applications to the foundations of the Method of Lines" (Part II. in [27]),
(ii) the Cellular Neural Networks (CNNs) (or cell-based neural networks) paradigm introduced by L.O. Chua and L. Yang [4, 5] – extremely versatile devices belonging to the *Artificial Intelligence* field and whose dynamics was rigorously proven [23] to be richer than that of PDEs.

The approach used for applying and combining the two "tools" – a "convergent Method of Lines" resulted from (i) and the CNNs paradigm of (ii) – gives rise to a well-structured and effective algorithm for computationally solving mixed initial boundary value problems for hyperbolic as well as parabolic PDEs with possibly non-standard BCs. Moreover, this approach ensures the convergence of the approximate solution to the solution of the original differential problem (pointwise or average, as will be emphasized in Sect. 2), as well as the preservation of the basic properties of this original solution (existence, uniqueness and continuous dependance of the parameters) together with its Lyapunov stability property. The implementation by means of CNNs allows reduction of rounding errors, capacity storage as well as of computational effort and time. This procedure has been introduced for hPDEs in [17] and then extended in several papers [12–14, 18, 19] for various subclasses of DPS, including pPDEs and different nsBCs.

The remaining of this chapter is organized as follows. Section 2 introduces the methodological approach and the mathematical framework at the foundation of CMoL. Section 3 synthesizes the main stages and steps of the computational procedure. Section 4 presents an illustrative application of the procedure, along with discussions on both the simulated experiments and the graphical representations. The chapter ends with a concluding section.

2 Methodological Approach and Mathematical Framework

The methodological approach of the procedure is suggested by the results in [27] regarding, on one hand, the approximation by ODEs of the class of time-delay systems which particularly arise from mixed initial boundary value problems for hPDEs, and on the other hand, the CMoL approach which makes use of these findings.

As already mentioned, in the case of control engineering problems for DPS encountered in different industry areas, the control takes place on the boundaries of the space variable domain and moreover, these boundary conditions are non-standard. We define non-standard BCs in opposition to standard ones. The standard boundary conditions are Dirichlet, Neumann and Robin and they *do not contain time derivatives for the boundary values of the state variables* – see e.g. [26]. In the case of a general one-dimensional second-order linear hyperbolic problem of the form

$$v_t + c_1 w_s = 0,\ \ w_t + c_2 v_s = 0,\ \ t \geq 0,\ \ 0 \leq s \leq L, c_i > 0,\ \ i = 1, 2 \qquad (1)$$

with the initial conditions

$$v(s, 0) = v_0(s),\ \ w(s, 0) = w_0(s),\ \ 0 \leq s \leq L \qquad (2)$$

and the boundary conditions of Dirichlet type

$$v(0, t) + \beta_1 w(0, t) = \varphi_1(t),\ \ w(L, t) + \beta_2 v(L, t) = \varphi_2(t),\ \ \beta_1, \beta_2 \in \mathbb{R} \qquad (3)$$

one can enumerate the following nsBCs which can replace the BCs (3) [17, 27]:

- *BCs controlled by a system of ODEs which, in turn, is controlled by the BCs*

$$\begin{aligned} &v(0, t) + \beta_1 w(0, t) = c_1^T x(t) - \beta_3 \varphi(c_0^T x(t)) + \varphi_1(t),\ \ \beta_1, \beta_3 \in \mathbb{R} \\ &w(L, t) - \beta_2 v(L, t) = c_2^T x(t) - \beta_4 \varphi(c_0^T x(t)) + \varphi_2(t),\ \ \beta_2, \beta_4 \in \mathbb{R} \\ &\dot{x}(t) = Ax(t) + b_{11} v(0, t) + b_{12} w(0, t) + b_{21} v(L, t) + b_{22} w(L, t) \\ &\qquad -b_0 \varphi(c_0' x(t)) + f(t),\ \ x(0) = x_0 \end{aligned} \qquad (4)$$

- *derivative BCs, as considered, for instance, in* [6]

$$\begin{aligned} &\sum_0^{L_1} a_{j1} \frac{\mathrm{d}^j}{\mathrm{d}t^j} v(0, t) + \sum_0^{K_2} a_{j2} \frac{\mathrm{d}^j}{\mathrm{d}t^j} w(0, t) = \varphi_1(t) \\ &\sum_0^{K_1} b_{j1} \frac{\mathrm{d}^j}{\mathrm{d}t^j} v(L, t) + \sum_0^{L_2} b_{j2} \frac{\mathrm{d}^j}{\mathrm{d}t^j} w(L, t) = \varphi_2(t) \end{aligned} \qquad (5)$$

- *BCs described by certain nonlinear Voltera operators* [1].

2.1 Methodological Approach

The overall methodological approach is that suggested by the MoL "guide" as already described in Sect. 1. First, a discretization with respect to the space variable is performed – using, for example, methods such as finite differences, finite elements etc. This operation associates a large-scale system of ODEs – whose dimension depends on the number N of the discretization points – to the original infinite-dimensional problem. The approximate solution *via* the ODEs system can be then computed by means of different technical tools, devices or approaches. As will be detailed in the sequel, the specificity of our procedure is given by the particular approaches we use at its different stages/steps.

The methodology idea at the foundation of CMoL, as resulted from Part II in [27], takes into consideration that any high-order hPDE or system of hPDEs, written in the symmetric Friedrichs form, may be converted *via* a nonsingular linear transformation T into the normal form of the Riemann invariants. More precisely, according to [24, 28], the $2D$ hyperbolic system

$$A(x, y, t)\frac{\partial u}{\partial t} + B(x, y, t)\frac{\partial u}{\partial x} + C(x, y, t)\frac{\partial u}{\partial y} = f(x, y, t) \tag{6}$$

is said to be *symmetric t-hyperbolic* (in the sense of K. O. Friedrichs) if matrices A, B, C are symmetric and if A is positive definite. In the case of $1D$ systems, a similarity transformation by a nonsingular matrix T can given to the system the canonical form

$$I\frac{\partial v}{\partial t} + K\frac{\partial v}{\partial x} + Mv = g(x, t) \tag{7}$$

with I the identity matrix (hence, positive definite) and K - diagonal. The components of the new state vector v are the Riemann invariants of the system.

In the case of modeling lossless or distortionless propagation phenomena, the hPDEs written in the Riemann variables are completely decoupled, their coupling being made through the boundary conditions only. Further, integration of the Riemann invariants along the characteristics allows association of certain equations with deviated arguments to the original problem for hPDEs: "difference equations for the BCs (3), coupled differential and difference equations for the BCs (4), delay-differential equations or neutral functional differential equations for the BCs (5), etc." [27]. In this way, a one-to-one correspondence between the solutions of the equations with deviated arguments and the solutions of the original boundary value problem for hPDEs is established. Consequently, the basic theory and stability properties obtained for one mathematical object are projected back on the other one. On the other hand, the equations with deviated argument are approximated by ODEs, as it is rigorously proven in Lemma of [27] which also gives the convergence rate – see Lemma 1 in the next subsection.

2.2 Mathematical Framework. Foundation of the CMoL

Following Part I in [27], we recall in the sequel the main results which give the mathematical framework at the foundation of the CMoL. The development of CMoL in [27] strongly relies on the aforementioned (Sect. 2.1) one-to-one correspondence between a non-standard BVP for hPDE and certain systems of functional differential equations having the form of coupled delay differential and difference equations with the rather general form

$$\begin{aligned}\dot{x}(t) &= f(t, x(t), y(t-h)) \\ y(t) &= g(t, x(t)) + A_0 y(t-h).\end{aligned} \tag{8}$$

The approximation approach – known as "delay approximation by ordinary differential equations" is the reflex of the formal approximation formula in the complex domain

$$e^{-hs} \approx \frac{1}{(1+\frac{h}{N}s)^N} \tag{9}$$

which is given a sense in the time domain (here $s \in \mathbb{C}$, $h > 0$ is the time delay and N is a positive integer sufficiently large). Engineers call the above formula "approximation of a pure time delay by a chain of first order elements". The basic result is Lemma 1 of [27].

Lemma 1 ([27]) *Consider the system of linear differential equations*

$$\begin{aligned} \dot{y}_1(t) &= \frac{N}{h}\,[x(t) - y_1] \\ \dot{y}_j(t) &= \frac{N}{h}\left[y_{j-1} - y_j\right],\ j = \overline{2, N} \end{aligned} \tag{10}$$

with the initial conditions

$$y_j(t_0) = \frac{N}{h}\int_{t_0 - j\frac{h}{N}}^{t_0-(j-1)\frac{h}{N}} x(s)\mathrm{d}s,\ j = \overline{1, N} \tag{11}$$

where $x \in L^1(t_0 - h, t_0 + T; \mathbb{R}^n)$ and t_0, $T > 0$ are arbitrary. Then, the following results are true:

$$\int_{t_0}^{t_0+T} \left| y_j(s) - x\left(s - j\frac{h}{N}\right)\right| \mathrm{d}s \le \gamma_1\left(\frac{h}{N}\right),\ j = \overline{1, N} \tag{12}$$

$$\sum_1^N \int_{-j\frac{h}{N}}^{-(j-1)\frac{h}{N}} \left|y_j(t) - x(t+s)\right| \mathrm{d}s \le \gamma_2\left(\frac{h}{N}\right),\ t_0 \le t \le t_0 + T. \tag{13}$$

If $x \in C(t_0 - h, t_0 + T; \mathbb{R}^n)$, then

$$\left| y_j(t) - x\left(t - j\frac{h}{N}\right)\right| \le \gamma_3\left(\frac{h}{N}\right),\ j = \overline{1, N} \tag{14}$$

for $t_0 \le t \le t_0 + T$, and this is true also if the initial conditions (11) are replaced by

$$y_j(t_0) = x\left(t_0 - j\frac{h}{N}\right),\ j = \overline{1, N}. \tag{15}$$

All these estimates are not dependent on t_0. Here $\gamma(r)$ are Kamke-Massera $\mathcal{K}$-functions, i.e. monotonically increasing and subject to $\lim_{r\to 0} \gamma_i(r) = 0$.

The properties of the $\gamma(r)$ functions are obtained along the proof of the Lemma in [27].

Lemma 1 is next applied to linear forced difference equations, which represent a subsystem of the system of coupled delay differential and difference equations. The solution of the linear difference equations can be constructed by steps on intervals $(kh, (k+1)h)$, where $k > 0$ is a positive integer; this solution has the same smoothness as the initial condition and/or the forcing term and displays finite discontinuities at $t = kh$ for $k > 0$. The solution may be also constructed backwards provided A_0 is non-singular. Considering the usual notation $x_t(s) = x(t+s)$, $-h \le s \le 0$, the next result of [27] is formulated as follows.

Theorem 1 ([27]) *Consider the difference equation*

$$x(t) = Ax(t-h) + g(t), \tag{16}$$

with $g \in L^1(t_0, t_0+T; \mathbb{R}^n)$ *integrable on* (t_0, t_0+T) *and the initial condition* $x_{t_0} = \varphi$, $\varphi \in L^1(-h, 0; \mathbb{R}^n)$. *Suppose that the eigenvalues of A are inside the unit circle.*

If $y_j(t)$ *are solutions of*

$$\begin{aligned} \dot{y}_1(t) &= \frac{N}{h}\left[Ay_N + g(t) - y_1\right] \\ \dot{y}_j(t) &= \frac{N}{h}\left[y_{j-1} - y_j\right],\ \ j = \overline{2, N} \end{aligned} \tag{17}$$

with the initial conditions

$$y_j(t_0) = \frac{N}{h}\int_{-j\frac{h}{N}}^{-(j-1)\frac{h}{N}} \varphi(s)\mathrm{d}s,\ \ j = \overline{1, N}, \tag{18}$$

then:

$$\int_{t_0}^{t_0+T} \left|y_j(s) - x\left(s - j\frac{h}{N}\right)\right| \mathrm{d}s \le \gamma_4\left(\frac{h}{N}\right),\ \ j = \overline{1, N} \tag{19}$$

$$\sum_1^N \int_{-j\frac{h}{N}}^{-(j-1)\frac{h}{N}} \left|y_j(t) - x(t+s)\right| \mathrm{d}s \le \gamma_5\left(\frac{h}{N}\right),\ \ t_0 \le t \le t_0 + T. \tag{20}$$

If $\varphi \in C(-h, 0; \mathbb{R}^n)$, $g(t)$ *is continuous on* $[t_0, t_0+T]$ *and if the following condition of "matching" is fulfilled*

$$\varphi(0) = A\varphi(-h) + g(t_0), \tag{21}$$

then

$$\left|y_j(t) - x\left(t - j\frac{h}{N}\right)\right| \le \gamma_6\left(\frac{h}{N}\right),\ \ j = \overline{1, N}, t_0 \le t \le t_0 + T \tag{22}$$

and this is true also if the initial conditions (18) are replaced by the following ones:

$$y_j(t_0) = \varphi\left(-j\frac{N}{h}\right), \quad j = \overline{1, N}. \tag{23}$$

The functions γ_i are of the same type as those of the Lemma 1 and the estimates are independent of t_0.

The next step is to apply the procedure to the system of coupled delay differential and difference equations. Its solution is constructed by steps on the same intervals $(kh, (k+1)h)$ as follows: using the initial condition for the differential equation $x(t_0)$ and for the difference equation on $(t_0 - h, t_0)$, $x(t)$ is obtained on $(t_0, t_0 + h)$, resulting e.g. absolutely continuous. Next, $y(t)$ is obtained from the difference equation using the initial condition on $(t_0 - h, t_0)$ for $y(t)$ and the resulted $x(t)$ on $(t_0, t_0 + h)$, having the same smoothness as its initial condition. The procedure is next iterated on $(t_0 + kh, t_0 + (k+1)h)$. Obviously, $y(t)$ will result with finite discontinuities at $t = t_0 + kh$, regardless the smoothness of the initial condition.

Theorem 2 ([27]) *Consider the coupled system*

$$\begin{aligned} \dot{x}(t) &= f(t, x(t), y(t-h)), \\ y(t) &= g(t, x(t)) + A_0 y(t-h) \end{aligned} \tag{24}$$

with the initial conditions $x(t_0) = x_0$, $y_{t_0} = \varphi$, where $\varphi \in L^1(-h, 0; \mathbb{R}^m)$. Assume that $f(t, x, y)$ is globally Lipschitz with respect to x and y uniformly for $t_0 \le t \le t_0 + T$ and, $g(t, x)$ uniformly globally Lipschitz with respect to x for $t_0 \le t \le t_0 + T$, that is

$$|f(t, x_2, y_2) - f(t, x_1, y_1)| \le L_1|x_2 - x_1| + L_2|y_2 - y_1|$$

$$|g(t, x_2) - g(t, x_1)| \le L_3|x_2 - x_1|$$

for any $x \in \mathbb{R}^n$, $y \in \mathbb{R}^m$, where L_k are the Lipschitz constants. Assume also that the eigenvalues of A_0 are inside the unit circle.

If $y_j(t)$ are solutions of

$$\begin{aligned} \dot{y}_0(t) &= f(t, y_0, y_N), \\ \dot{y}_1(t) &= \frac{N}{h}\left[g(t, y_0) + A_0 y_N - y_1\right], \\ \dot{y}_j(t) &= \frac{N}{h}\left[y_{j-1} - y_j\right], \quad j = \overline{2, N} \end{aligned} \tag{25}$$

with the initial conditions

$$y_0(t_0) = x_0, \quad y_j(t_0) = \frac{N}{h}\int_{-j\frac{h}{N}}^{-(j-1)\frac{h}{N}} \varphi(s)\mathrm{d}s, \quad j = \overline{1, N}, \tag{26}$$

then:

$$|y_0(t) - x(t)| \leq \gamma_7\left(\frac{h}{N}\right), \quad t_0 \leq t \leq t_0 + T \tag{27}$$

$$\int_{t_0}^{t_0+T}\left|y_j(s) - y\left(s - j\frac{h}{N}\right)\right| ds \leq \gamma_8\left(\frac{h}{N}\right), \quad j = \overline{1, N} \tag{28}$$

$$\sum_1^N \int_{-j\frac{h}{N}}^{-(j-1)\frac{h}{N}} |y_j(t) - y(t+s)|\, ds \leq \gamma_9\left(\frac{h}{N}\right), \quad t_0 \leq t \leq t_0 + T. \tag{29}$$

If $\varphi \in C(-h, 0; \mathbb{R}^m)$ and if the following condition of "matching" is fulfilled

$$\varphi(0) = g(t_0, x_0) + A_0\varphi(-h), \tag{30}$$

then

$$\left|y_j(t) - y\left(t - j\frac{h}{N}\right)\right| \leq \gamma_{10}\left(\frac{h}{N}\right), \quad j = \overline{1, N}, t_0 \leq t \leq t_0 + T \tag{31}$$

and this is true also if the initial conditions (26) are replaced by

$$y_0(t_0) = x_0, \quad y_j(t_0) = \varphi\left(-j\frac{N}{h}\right), \quad j = \overline{1, N}. \tag{32}$$

The functions γ_i have the properties from the previous cases and are independent of t_0.

Theorem 3 ([27]) *Consider the neutral functional differential equations (FDE)*

$$\dot{y}(t) - A_0\dot{y}(t-h) = f(t, y(t), y(t-h)), \tag{33}$$

with the initial condition $y_{t_0} = \varphi$, where $\varphi : [-h, 0] \to \mathbb{R}^n$ is absolutely continuous. Suppose that the eigenvalues of A_0 are inside the unit circle and, denoting

$$x(t) = y(t) - A_0 y(t-h), \tag{34}$$

we have

$$|f(t, x_2, y_2) - f(t, x_1, y_1)| \leq L_1|x_2 - x_1| + L_2|y_2 - y_1|$$

for any $x, y \in \mathbb{R}^n$, where L_k are the Lipschitz constants.

Then system

$$\begin{aligned} \dot{y}_0 &= f(t, y_0, y_N), \\ \dot{y}_1 &= \frac{N}{h}\left[y_0 + A_0 y_N - y_1\right], \\ \dot{y}_j &= \frac{N}{h}\left[y_{j-1} - y_j\right], \quad j = \overline{2, N} \end{aligned} \tag{35}$$

with the initial conditions

$$\begin{aligned} y_0(t_0) &= x(t_0) = \varphi(0) - A_0\varphi(-h) \\ y_j(t_0) &= \varphi\left(-j\frac{N}{h}\right), \quad j = \overline{1, N}. \end{aligned} \tag{36}$$

approximates (33) in the sense that

$$|y_0(t) + A_0 y_N(t) - y(t)| \le \gamma_{11}\left(\frac{h}{N}\right), \quad t_0 \le t \le t_0 + T. \tag{37}$$

The equation of neutral type considered in Theorem 3 can be reduced to the system considered in Theorem 2. Indeed, denoting $z(t) := y(t) - A_0 y(t-h)$ the following system is obtained

$$\begin{aligned} \dot{z}(t) &= f(t, z(t) + A_0 y(t-h), y(t-h)) \\ y(t) &= z(t) + A_0 y(t-h). \end{aligned} \tag{38}$$

Since the initial condition for $y(t)$ is absolutely continuous, this degree of smoothness is propagated *via* the construction by steps. Therefore, $y(t)$ can be differentiated, but its derivative will display finite discontinuities at $t_0 + kh$.

Summarizing the above results, some remarks are to be made concerned the right application of the CMoL.

- In order to obtain convergent approximations through the use of Riemann invariants, the space discretization has to be performed following the Courant-Isaacson-Rees rule introduced in [7]: use forward schemes for discretizing backward waves and backward schemes for the forward waves. According to [27], "the Courant-Isaacson Rees rule for discretization corresponds, in the framework for the CMoL described above, to the approximation of time-delays by ODEs".
- The role of the symmetric Friedrichs form in this approach is to ensure, by diagonalization, the obtaining of the normal hyperbolic form having as variables the Riemann invariants.
- If the "matching" condition between the initial and the boundary conditions is fulfilled, then the approximation is pointwise. Otherwise, the "unmatching" is propagated (propagation of singularities) and thus only average approximation can be obtained [27]. More precisely, it has been mentioned that between the solutions of the IBVP for hPDE and the solutions of the associated system of functional differential equations (coupled delay differential and difference equations) there is a one-to-one correspondence. The mismatch of the initial and boundary conditions of the IBVP for hPDE is transferred to the system of functional differential equations as a singularity – a discontinuity at the initial point t_0 for the difference subsystem. This discontinuity is propagated to each point $t_0 + kh$. Consequently, the approximation by ODEs is *average* – see [27]. The solution of the IBVP will be a *classical discontinuous one* in the sense given by [1].

- The assumption on the eigenvalues of A and A_0 is technical, used by the authors of the cited reference to prove the theorems given here for the sake of completeness. It corresponds however, when projected back on the hPDE *via* the neutral functional differential equations (see Refs. [1, 6]), to the so-called *dissipative boundary conditions* as defined in [24].
- Preservation of the stability through approximation was not followed either in [27] or in subsequent papers. For the approximation of the delays in delay-differential equations, there are known the results on stability preservation in [32]. For neutral equations which are connected to the method of lines for hPDE, preservation of stability through the *present approximation by ordinary differential equations* was not studied in [27]. It is however suggested to follow the approach in [32] which is nothing more but the approach of E. A. Barbashin of the "two epsilons".

The aforementioned results are derived for hPDEs models of the DPS displaying lossless and/or distortionless phenomena: only in these cases the Riemann invariants can be decoupled by a corresponding linear nonsingular transformation and such functional differential equations containing pure delay terms $z(t-\tau)$ can be attached.

In more general cases, the functional differential equations associated to the mixed IBVP are rather complicated [1]. In these cases, the convergence of the MoL can be proven provided one turns to a fundamental text where the basics for PDEs are considered starting from a computational point of view [24]. The equations, as well as the BCs are discretized with respect to both space and time variables and then there is proven the convergence of the solutions of the fully discretized difference system to a solution of the boundary value problem.

On the other hand, as already mentioned, the formal application of the MoL yields a system of ODEs. Thus, by adapting the approach of [24], Chap. II, Sects. 15–18, it is not difficult to prove that the fully discretized difference system solutions converge also to the solutions of the system of ODEs resulting from the MoL. Combining the two results will give convergence of the MoL to the solution of the mixed problem for the hyperbolic PDEs without the lossless/distortionless propagation assumption.

To conclude this section, we remark that however, this kind of construction gives an additional insight on the MoL: in comparison to a complete discretization, within the Method of Lines the time discretization is finer than space discretization and, this is particularly important when the analysis aims at dynamics and transients [17].

3 A Hybrid Computational Procedure

The computational procedure we present relies on the two powerful tools mentioned in **Introduction**, and consists in three main stages:

1. *analytical stage* which involves the application of the CMoL to the IBVP for the hPDEs giving rise to three systems of ODEs: one for the forward wave, one for the backward wave and one for the boundary conditions;

2. *preprocessing stage* which aims at making use of the parallelism of the new problem described by a high-dimensional system of ODEs, as well as of the sparsity of the Jacobian matrix, in order to identify the so-called "cloning templates" specific to the Cellular Neural Networks paradigm;
3. *implementation stage* where, based on the findings of the previous stage, the purpose is to cast the large-scale system of ODEs in the form of a CNN.

3.1 Cellular Neural Networks - The Paradigm

Before introducing the procedure, let us to highlight those basic features of a CNN that make it a powerful and versatile device for high-speed VLSI chips, digital, as well as software implementations for modeling various dynamical systems [3, 18]: (a) only local interconnections among cells, (b) different types of inputs for a cell, (c) representation of the interactions within the network by regularly distributed "cloning templates" (small number of arrays of parameters, which are identical for almost all the CNN's cells), (d) different topologies for layers interconnections.

The cell, the artificial neuron of a CNN, is a nonlinear elementary dynamical system with several inputs and one output. Within a CNN, the cells can be interconnected in different regular manners, for instance as 1D, 2D and 3D arrays. The ith cell dynamics in a 1D array of M cells is described by the canonical equation [3, 23] (see also [18])

$$\dot{x}_i(t) = \sum_{j\in N_r(i)} T_{ij}^A y_j + \sum_{j\in N_r(i)} T_{ij}^B u_j + \sum_{j\in N_r(i)} T_{ij}^C x_j + I_i, \quad y_i(t) = f(x_i) \tag{39}$$

where the notations are as follows: x_i – state variable of the ith cell, y_i – output variable of the ith cell, u_j – control variable from the neighboring cells, $N_r(i)$ – r-neighborhood of interaction of the ith cell, T_{ij}^A – element of the output-feedback cloning template $\mathbf{T}^A$, T_{ij}^B – element of the control cloning template $\mathbf{T}^B$, T_{ij}^C – element of the state-feedback cloning template $\mathbf{T}^C$, I_i – bias or external stimulus; it is assumed that the templates $\mathbf{T}^A$, $\mathbf{T}^B$, $\mathbf{T}^C$ are linear and space-invariant. The nonlinearity $f(\cdot)$ is the unit bipolar ramp function [5] $f : \mathbb{R} \to \mathbb{R}$

$$f(x_i) = \frac{1}{2}(|x_i + 1| - |x_i - 1|) \tag{40}$$

a continuous, bounded, nondecreasing, piecewise-linear and globally Lipschitzian function, with the Lipschitz constant $L = 1$; it also verifies

$$0 < \frac{f(x_i)}{x_i} \le L, \quad f(0) = 0. \tag{41}$$

The recurrent interconnections of CNNs, at both the cell and the r-neighborhood levels, lead to a very rich space-time dynamical behavior. Thus, CNNs are able to model DPS described by PDEs, as it is studied e.g. in [5, 13, 18, 23, 29, 33, 34]. Also, there are reported several studies concerning the stability property of CNNs, viewed as nonlinear dynamical systems with or without time-delays, as well as for various cloning templates – see e.g. [4, 11, 20] to give a few examples.

3.2 Computational Procedure for hPDEs - Stages and Steps

In order to set up the general framework, let us consider a linear hyperbolic problem defined by the homogeneous one-dimensional wave equation

$$y_{tt} - c^2 y_{ss} = 0, \quad s \in [0, L], t \in [0, T] \tag{42}$$

with the initial conditions

$$y(s, 0) = y_0(s), \quad y_t(s, 0) = y_1(s), \quad s \in [0, L], \tag{43}$$

and, for the sake of the simplicity, with the Dirichlet boundary conditions

$$\begin{aligned} y_t(0, t) + \alpha_1 y_s(0, t) &= \varphi_1(t) \\ y_s(L, t) - \alpha_2 y_t(L, t) &= \varphi_2(t) \end{aligned} \tag{44}$$

where $c > 0$ is the so-called wave speed. The Cauchy problem (42)–(43) is well-posed for $0 \le t \le T$, $y_0 \in C^2(\mathbb{R})$, $y_1 \in C^1(\mathbb{R})$ – see e.g. [31]. As already mentioned, the real-world applications for DPCS imply non-standard BCs; thus, the BCs (44) can be replaced by one of those BCs enumerated at the begining of Sect. 2. For improving the numerical conditioning, one can write the model with rated space variable by introducing $\sigma = s/L$, $\sigma \in [0, 1]$. The main stages and steps of the procedure are as follows [17, 18, 27].

S1. Analytical stage

S1.1. Introduce the symmetric Friedrichs form with the new distributed variables $v(\sigma, t) = y_t(\sigma, t)$ and $w(\sigma, t) = y_\sigma(\sigma, t)$; taking into account that $v_\sigma(\sigma, t) = y_{t\sigma}(\sigma, t) = y_{\sigma t}(\sigma, t) = w_t(\sigma, t)$, the system in the Friedrichs form reads as

$$V_t(\sigma, t) + AV_\sigma(\sigma, t) = 0\ , \quad V(\sigma, t) = [\, v(\sigma, t)\ w(\sigma, t)\,]^T . \tag{45}$$

S1.2. Write the system in the normal hyperbolic form of the Riemann invariants by introducing the new distributed variables $R(\sigma, t) = [\, r^+(\sigma, t)\ r^-(\sigma, t)\,]^T$, i.e. find a nonsingular linear transformation $T(\sigma)$ such that $V(\sigma, t) = T(\sigma)R(\sigma, t)$ and,

$$R_t(\sigma, t) + A_d R_\sigma(\sigma, t) = 0 \tag{46}$$

with $A_d = \text{diag}[\lambda^+, \lambda^-]$, where obviously $\lambda^\pm$ are the eigenvalues of A. We can identify now the forward and the backward waves for system (46): the forward wave corresponds to the variable $r^+(\sigma, t)$ which is multiplied by the positive eigenvalue denoted by λ^+ and the backward wave corresponds to the variable $r^-(\sigma, t)$ which is multiplied by the negative eigenvalue denoted by λ^-.

S1.3. Application of the *convergent Method of Lines* according to the results of [27].

S1.3.1. Discretize the domain [0, 1] of variation for the space variable σ into a grid with $N + 1$ equally spaced points; denote $h = 1/N$ and $\sigma_i = ih$, $i = \overline{0, N}$.

S1.3.2. Apply the Courant-Isaacson-Rees rule for approximating the derivatives with respect to the space variable σ:

– for $r^+(\sigma, t)$, use the backward Euler scheme

$$\frac{\partial r^+}{\partial \sigma}(\sigma_i, t) \approx \frac{1}{h}(r^+(\sigma_i, t) - r^+(\sigma_i - h, t)) \ , \ i = \overline{1, N} \tag{47}$$

– for $r^-(\sigma, t)$, use the forward Euler scheme

$$\frac{\partial r^-}{\partial \sigma}(\sigma_i, t) \approx \frac{1}{h}(r^-(\sigma_i + h, t) - r^-(\sigma_i, t)) \ , \ i = \overline{0, N-1} \tag{48}$$

S1.4. Write the approximating system of ODEs for the system written in the normal hyperbolic form (46).

S1.4.1. Considering the numerical schemes (47)–(48) and introducing the average approximation functions $\xi_i^+(t) \approx r^+(\sigma_i, t)$ for $i = \overline{1, N}$ and $\xi_i^-(t) \approx r^-(\sigma_i, t)$ for $i = \overline{0, N-1}$, we formally write the two approximate systems of ODEs (the normal Cauchy form) with respect to the new set of lumped variables $\{\xi_i^+(t), \xi_i^-(t)\}$. Remark that at this point, we have no approximations for $\xi_0^+(t)$ and $\xi_N^-(t)$ – they will be obtained from the BCs.

S1.4.2. Write the BCs of the normal form with respect to the new functions $\xi_i^\pm$ and derive the equations for $\xi_0^+(t)$ and $\xi_N^-(t)$. Replace the new expressions for $\xi_0^+(t)$ and $\xi_N^-(t)$ in the two ODEs systems as it is needed. In the case of BCs controlled by a system of ODEs, we obtain here the third system of ODEs for the approximate system. Also, according to the features of the BCs, it may be possible to additionally introduce variables in order to eventually obtain the whole approximate ODEs system for the original IBVP for hPDEs.

S1.4.3. Define the "matching" equations for the boundary and initial conditions in order to obtain a pointwise approximation, according to Theorem 1.

S1.4.4. Calculate the approximate solutions for the propagation phenomena, i.e. the approximating functions $\eta_i^v(t) \approx v(\sigma_i, t)$, $\eta_i^w(t) \approx w(\sigma_i, t)$, $i = \overline{0, N}$ by using the inverse transformation $T^{-1}(\sigma)$ and taking into account the "matching" requirements of ICs with BCs.

S2. Preprocessing stage
S2.1. Using a flow diagram for the whole approximate system of ODEs (consisting in (possibly) three distinct systems of ODEs) find the optimum arrangement of the equations in order to reveal the regularities which lead to the so-called *cloning templates* proper to the CNNs devices.
S2.2. Considering the whole approximate system of ODEs with the optimum arrangement, identify the parameters for building the CNN used for implementation: define the output-feedback, control and state-feedback cloning templates for both the inner and the boundary cells, as well as the external inputs.

S3. Implementation stage and approximate solution computing
S3.1. Usually, this stage starts with a software emulation of the cell-based neural network. There are several high-performance software packages for numerical integration of ODEs systems. Taking into account that the approximate system of ODEs for PDEs problems has a sparse Jacobian matrix, as well as the optimum arrangement performed in order to build the cloning templates, good choices are those solvers that take advantage of these features in order to reduce the computational time and effort, as well as the rounding errors. Other efficient implementations of CNNs are the locally connected analog and digital array-computing structures in VLSI technology [29].
S3.2. Computation of the approximate solution, as well as evaluation of the system behavior and control performances by means of 2D and 3D graphical representations.

Remark 1 It was explained the one-to-one correspondence between the solutions of the IBVP for hPDE and the solutions of an associated system of functional differential equations. A convergent approximation by ODEs has been obtained in [27]; the convergence rate was given by the Kamke-Massera $\mathcal{K}$-functions $\gamma_i(h/N)$, with N a large positive integer. The CMoL appeared as the counterpart of this discretization for the functional differential equations when projected back on the IBVP for hPDE. Therefore, the convergence of the delay approximation by ODEs is also projected back on the CMoL, being given by the same $\gamma_i(h/N)$, with N a large positive integer.

3.3 *Computational Modeling for Parabolic PDEs*

3.3.1 The Approach Related to CMoL

Consider the class of DPS modeled by a generic one-dimensional parabolic transport equation modeling the reaction-advection-diffusion phenomena.

$$\frac{\partial u}{\partial t} + v\frac{\partial u}{\partial x} - d\frac{\partial^2 u}{\partial x^2} = s, \quad t \in [0, T], \quad x \in [0, L] \tag{49}$$

with possibly nsBCs and some initial conditions $u(0, x) = u_0(x)$, $x \in [0, L]$.

The proposed procedure for hPDEs is adapted to the case of pPDEs at the step ***S1.3.2.*** concerning the correct application of the Courant-Isaacson-Rees rule for approximating the derivatives with respect to the space variable [13]:

S1.3.1p

– Advection (convection) term. In the case of transport processes the information moves forward in space and time, and there is no backward influence. This fact suggests, according to the Courant-Isaacson-Rees rule, the use of the first-order implicit backward Euler scheme for the approximation of the first derivative with respect to x (a scheme unconditionally stable which ensures the convergence in most cases):

$$\frac{\partial u}{\partial x}(x_i, t) \approx \frac{1}{h}(u(x_i, t) - u(x_i - h, t)) \; , \; i = \overline{1, N}. \tag{50}$$

– Dissipative term. In order to ensure a quite simple structure of the resulted approximated system for further implementation by means of CNNs, we shall use the usual central difference approximation for the second derivative with respect to x; for $i = \overline{1, N-1}$,

$$\frac{\partial^2 u}{\partial x^2}(x_i, t) \approx \frac{1}{h^2}(u(x_i - h, t) - 2u(x_i, t) + u(x_i + h, t)). \tag{51}$$

– Derivatives within BCs. Obviously, the derivative for $x = 0$ will be approximated by using the forward Euler scheme (52), while the derivative for $x = L$ will be approximated by using the backward Euler scheme (50),

$$\frac{\partial u}{\partial x}(x_i, t) \approx \frac{1}{h}(u(x_i + h, t) - u(x_i, t)) \; , \; i = \overline{0, N-1}. \tag{52}$$

4 Motivating Applications to Control Processes

Before discussing an illustrative application, let us briefly present first some real-world DPCS together with their mathematical description from the point of view of our approach.

- *the controlled torsional stick-slip oscillations of a drilling plant* [16] – the mathematical model corresponds to the wave equation with boundary control: the control is applied only at one boundary (the surface level) but the other boundary (the bit level of the drill string) is subjected to a nonlinear torque;
- *the controlled flexible arm of an ocean vessel riser* [10] – a model displaying disturbance signals, both distributed and lumped (at the boundaries), control signals (at the two boundaries as well) and a distributed damping;
- *the vibrational percussion system with boundary dissipation at one extremity and a normal compliance contact condition at the other extremity* [8];

- *a unified model from the* Contact Mechanics *belonging to the class of longitudinal vibrations of rods* [14] – a mixed IBVP for hPDEs with distributed axial force, nonlinear nsBCs of type (4) and two time-scales: the fast dynamics represented by the transients within the rod compared with the slow dynamics at its boundaries;
- *the overhead crane with nonhomogeneous cable* [17] – the system is described by IBVP for hPDEs, with space-varying parameters and controlled derivative nonlinear BCs; in this case, the equations of the normal hyperbolic form of the Riemann invariants are coupled;
- *a nonlinear bioreactor distributed parameter model with the specific growth rate modeled by the Hadane kinetic function* [13] – a parabolic transport system of the type reaction-advection-diffusion with non-homogeneous Robin BCs at $x = 0$ and Neumann BCs at $x = L$;
- *the distributed parameter control system for cogeneration* [9, 15, 18] – a system of two one-dimensional hyperbolic conservation laws with the BCs controlled by an ODEs system with bilinear terms;
- *the water hammer hydraulic phenomenon for a hydraulic plant with two conduits* [12] – a case with two boundary value problems (BVPs) for the hPDEs with nsBCs which include the differential equation of the surge tank water level, which is connected in an "internal feedback" to the BCs of the two BVPs.

Considering various scenarios for such control processes as those mentioned above, the visualization of simulations results, obtained for the approximate system by means of the computational procedure, can shead light on both the transient and asymptotic behaviors, giving also insight regarding the effectiveness of different control strategies.

4.1 Computational Modeling for a System of Two-Scale Hyperbolic PDEs

In order to illustrate the application of the procedure, we consider a representative example – the computational modeling of an unified model from *Contact Mechanics* belonging to the class of longitudinal vibrations of rods. This mathematical model can describe, for instance, phenomena which arise in percussion systems, as well as in human and robotic gait in the autonomous navigation.

The mathematical model. The mathematical model of the set-up is as follows [14]

$$\rho A u_{tt}(t,x) - AE u_{xx}(t,x) = f(t,x), \quad x \in (0,1), \quad t \in (0,T) \tag{53}$$

with the boundary conditions

$$
\begin{aligned}
&m_0 u_{tt}(t,0) + c_0 u_t(t,0) + k_0 u(t,0) - AE u_x(t,0) = g_0(t), \quad t \in (0,T) \\
&m_1 u_{tt}(t,1) + AE u_x(t,1) + p(c_1 u_t(t,1) + k_1(u(t,1) - d)) = 0, \quad t \in (0,T)
\end{aligned} \tag{54}
$$

and the initial conditions

$$
u(0,x) = u_0(x), \quad u_t(0,x) = v_0(x), \quad x \in (0,1), \tag{55}
$$

where the notations are the following: $m_0 > 0$ accounts for the inertia at the forced end $x = 0$ of the rod, $m_1 > 0$ is the (possibly non-negligible) payload mass at $x = 1$, $\rho > 0$ is the mass density, $A > 0$ is the rod cross-section area, E is the Young modulus; $u(t,x)$ is the longitudinal vibration amplitude, $f(t,x)$ is the distributed axial force, $u(t,1) - d$ accounts for a possible contact problem with an initial gap d between the end $x = 1$ and a deformable obstacle, as it is studied in [8]; $p : \mathbb{R} \to \mathbb{R}$, having the measure unit of stress, is defined by

$$
p(r) = \begin{cases} 0 & \text{if } r < 0 \\ r & \text{if } r \geq 0 \end{cases} \tag{56}
$$

this function being a possible relaxation of the so-called Signorini inequality condition, which is usual in *Contact Mechanics*.

If we consider the time constant of the dynamics within the rod, $T_a = \sqrt{\rho/E}$, we can remark that in metal rods $T_a << 1$. Thus, the set-up under consideration is a two-time-scales system, where the fast dynamics is represented by the transients within the rod and the slow dynamics is given by the transients at the rod boundaries. Consequently, for avoiding ill-conditioning of the problem, we define the stretched time scale $\tau := t/T_a$ and the new variable $\mu(\tau,x) := u(\tau T_a, x)$; thus, the system becomes [14]

$$
\mu_{\tau\tau}(\tau,x) - \mu_{xx}(\tau,x) = \frac{1}{AE} f(\tau T_a, x), \quad \tau \in (0, T/T_a) \tag{57}
$$

with the BCs

$$
\begin{aligned}
&\frac{m_0}{k_0 T_a^2}\mu_{\tau\tau}(\tau,0) + \frac{c_0}{k_0 T_a}\mu_\tau(\tau,0) + \mu(\tau,0) - \frac{AE}{k_0}\mu_x(\tau,0) = \frac{1}{k_0} g_0(\tau T_a) \\
&\frac{m_1}{k_1 T_a^2}\mu_{\tau\tau}(\tau,1) + \frac{1}{k_1} p\left(k_1\left(\frac{c_1}{k_1}\mu_\tau(\tau,1) + u(\tau,1) - d\right)\right) + \frac{AE}{k_1}\mu_x(\tau,1) = 0
\end{aligned} \tag{58}
$$

and the ICs

$$
\mu(0,x) = u_0(x), \quad \mu_\tau(0,x) = T_a v_0(x). \tag{59}
$$

The computational procedure applied. The ***analytical stage*** starts by deducing the symmetric Friedrichs form for system (57)–(59). Thus, by introducing the new distributed variables $v(\tau,x) := \mu_\tau(\tau,x)$, $w(\tau,x) := \mu_x(\tau,x)$, as well as the lumped variables for the ODEs system acting on BCs, $z_0(\tau) := \mu(\tau, 0)$, $z_1(\tau) := \mu(\tau, 1) - d$, the system written with respect to the new variables is as follows [14]

$$\begin{aligned} v_\tau(\tau, x) - w_x(\tau, x) &= \frac{1}{AE} f(\tau T_a, x) \\ w_\tau(\tau, x) - v_x(\tau, x) &= 0 \end{aligned} \tag{60}$$

with the boundary conditions

$$\begin{aligned} &\frac{m_0}{k_0 T_a^2} v_\tau(\tau, 0) + \frac{c_0}{k_0 T_a} v(\tau, 0) + z_0(\tau) - \frac{AE}{k_0} w(\tau, 0) = \frac{1}{k_0} g_0(\tau T_a) \\ &\frac{m_1}{k_1 T_a^2} v_\tau(\tau, 1) + \frac{1}{k_1} p\left(k_1 \left(\frac{c_1}{k_1} v(\tau, 1) + z_1(\tau)\right)\right) + \frac{AE}{k_1} w(\tau, 1) = 0 \\ &\dot{z}_0(\tau) = v(\tau, 0) \\ &\dot{z}_1(\tau) = v(\tau, 1) \end{aligned} \tag{61}$$

and the initial conditions

$$\begin{aligned} &w(0, x) = u_0'(x), \quad v(0, x) = T_a v_0(x) \\ &z_0(0) = \mu(0, 0) = u_0(0), \quad z_1(0) = \mu(0, 1) - d = u_0(1) - d. \end{aligned} \tag{62}$$

Concerning the equilibria of the system (60)–(62), after some straightforward manipulations we obtain

$$p(k_1 \bar{z}_1) + \bar{z}_0 = \bar{g}_0 - \int_0^1 f(\lambda) \mathrm{d}\lambda, \tag{63}$$

equation which suggests that the system has a set of equilibria. More precisely, if $p(r)$ is defined by (56), then for each right hand side of (63) there is defined a set of equilibria in plane $(\bar{z}_0, \bar{z}_1)$ [14].

There exists a nonsingular linear transformation T such that the Riemann invariants can be written as

$$r^{\pm}(\tau, x) = v(\tau, x) \mp w(\tau, x) \tag{64}$$

which leads to the system's normal hyperbolic form with decoupled equations [14]

$$\begin{aligned} r_\tau^+(\tau, x) + r_x^+(\tau, x) &= \frac{1}{AE} f(\tau T_a, x) \\ r_\tau^-(\tau, x) - r_x^-(\tau, x) &= \frac{1}{AE} f(\tau T_a, x) \end{aligned} \tag{65}$$

having the BCs

$$\frac{m_0}{k_0 T_a^2}\ddot{z}_0(\tau) + \frac{c_0}{k_0 T_a}\dot{z}_0(\tau) + z_0(\tau) - \frac{AE}{2k_0}(-r^+(\tau,0) + r^-(\tau,0)) = \frac{1}{k_0}g_0(\tau T_a)$$
$$\frac{m_1}{k_1 T_a^2}\ddot{z}_1(\tau) + \frac{AE}{2k_1}(-r^+(\tau,1) + r^-(\tau,1)) + \frac{1}{k_1}p\left(k_1\left(\frac{c_1}{k_1}\dot{z}_1(\tau) + z_1(\tau)\right)\right) = 0$$
$$\dot{z}_0(\tau) = \frac{1}{2}(r^+(\tau,0) + r^-(\tau,0))$$
$$\dot{z}_1(\tau) = \frac{1}{2}(r^+(\tau,1) + r^-(\tau,1)) \tag{66}$$

and the ICs

$$r^{\pm}(0,x) = v(0,x) \mp w(0,x) = T_a v_0(x) \mp u_0'(x). \tag{67}$$

The application of the CMoL presented in Sect. 3 – steps S1.3 and S1.4 – to the IBVP (64)–(67) requires firstly the discretization of the space domain [0, 1] in N equidistant intervals, with the discretization step $h = 1/N$ and then, the application of the Courant-Isaacson-Rees rule for approximation of the derivatives with respect to the space independent variable x. Next, by denoting $x_i := ih$ and introducing the functions $\xi_i^{\pm}(\tau) \approx r^{\pm}(\tau, x_i)$ for $i = \overline{0, N}$, one finally obtains the following approximate system of ODEs for the initial-boundary value problem (57)–(59) [14]

$$\begin{cases} \dot{\xi}_0^+(\tau) = -N\xi_1^- + (N + A_1)\xi_0^- - A_2\xi_0^+ - A_4 z_0 - \alpha f(\tau T_a, 0) + \beta_0 g_0(\tau T_a) \\ \dot{\xi}_i^+(\tau) = N\xi_{i-1}^+ - N\xi_i^+ + \alpha f(\tau T_a, x_i), \ \ i = \overline{1, N} \\ \dot{\xi}_i^-(\tau) = -N\xi_i^- + N\xi_{i+1}^- + \alpha f(\tau T_a, x_i), \ \ i = \overline{0, N-1} \\ \dot{\xi}_N^-(\tau) = -N\xi_{N-1}^+ + (N + A_3)\xi_N^+ - A_3\xi_N^- - \alpha f(\tau T_a, 1) \\ \qquad -\beta_1 p\left(k_1\left(\frac{c_1}{2k_1 T_a}(\xi_N^+ + \xi_N^-) + z_1\right)\right) \\ \dot{z}_0(\tau) = \frac{1}{2}\xi_0^+ + \frac{1}{2}\xi_0^- \\ \dot{z}_1(\tau) = \frac{1}{2}\xi_N^+ + \frac{1}{2}\xi_N^- \end{cases} \tag{68}$$

with the initial conditions

$$\begin{aligned} &\xi_i^{\pm}(0) = T_a v_0(x_i) \mp u_0'(x_i) \ \ i = \overline{1, N-1} \\ &\xi_0^-(0) = \xi_0^+(0) + \frac{2c_0}{AE}v_0(0) + \frac{2k_0}{AE}u_0(0) + \frac{2}{AE}g_0(0) \\ &\xi_N^-(0) = \xi_N^+(0) + \frac{2}{AE}p\left(k_1\left(\frac{c_1}{k_1}v_0(1) + u_0(1) - d\right)\right) \\ &z_0(0) = u_0(0), \ \ z_1(0) = u_0(1) - d \end{aligned} \tag{69}$$

where the notations in (68) are as follows:

$$A_1 = \frac{AET_a^2}{m_0} - \frac{c_0 T_a}{m_0},\ A_2 = \frac{AET_a^2}{m_0} + \frac{c_0 T_a}{m_0},\ A_3 = \frac{AET_a^2}{m_1},$$

$$A_4 = \frac{2k_0 T_a^2}{m_0},\ \alpha = \frac{1}{AE},\ \beta_0 = \frac{2T_a^2}{m_0},\ \beta_1 = \frac{2T_a^2}{m_1}.$$

Remark 2 The second and third equations in (69) reflect the "matching" of the ICs with BCs – condition which ensures pointwise approximations *via* the CMoL approach we have used in our procedure.

In the ***preprocessing stage*** we can see that the structure of the approximate ODEs system (68) suggests an one-dimensional CNN with $2N + 4$ cells for the computational implementation. Choosing the state vector $y \in \mathbb{R}^{2N+4}$ of the form

$$y^T = [\,\xi_0^+\ \xi_1^+\ \ldots\ \xi_N^+\ \xi_N^-\ \ldots\ \xi_1^-\ \xi_0^-\ z_0\ z_1\,] \tag{70}$$

we obtain two inner state-feedback cloning templates of the form

$$\begin{aligned} T_1^C &= [N \quad -N \quad 0]\,,\ i = \overline{2, N+1} \bigcup \overline{N+3, 2N+2} \\ T_2^C &= [-N \quad N + A_3 \quad -A_3]\,,\ i = N+2 \end{aligned} \tag{71}$$

and their dynamics described by the equations

$$\dot{y}_i(\tau) = T_1^C \begin{bmatrix} y_{i-1} \\ y_i \\ y_{i+1} \end{bmatrix},\ \dot{y}_{N+2}(\tau) = T_2^C \begin{bmatrix} y_N \\ y_{N+1} \\ y_{N+2} \end{bmatrix} + I_1(\tau T_a, y_{N+2}) + I_2(y) \tag{72}$$

with $I_1 = -\alpha f(\tau T_a, y_{N+2})$, $I_2 = -\beta_1 p(c^T y)$, $c^T = [0 \quad \ldots \quad 0 \quad \frac{c_1}{2T_a} \quad \frac{c_1}{2T_a} \quad 0 \quad \ldots \quad 0 \quad k_1]$; also, by comparison of (68) with (39) we observe that $T^A = T^B = 0$.

At this point we can make some preliminary considerations regarding the computational efficiency obtained by this approach of the procedure which combines CMoL with the CNNs paradigm for computationally modeling the mixed IBVP (53)–(55) for hPDEs with nsBCs. Considering the above two structures for the inner cloning templates and, taking into account that the entire structure has additionally three different templates for the CNNs' boundary cells C_1, C_{2N+3} and C_{2N+4}, we can conclude that this CNN structure used for the approximate ODEs system of $2N + 4$ equations leads to only five different ordinary differential equations to be computed, regardless of the number of the space-discretization points, N.

Once the solutions of the approximate system are computed, i.e. the forward and the backward waves $r^{\pm}(\tau, x_i)$ at the discrete space points $i = \overline{0, N}$, the approximations for the variables of the Friedrichs form can be obtained *via* the converse transformation

$$\begin{aligned} v(\tau, x) &= \frac{1}{2}(r^+(\tau, x) + r^-(\tau, x)) \\ w(\tau, x) &= \frac{1}{2}(-r^+(\tau, x) + r^-(\tau, x)). \end{aligned} \tag{73}$$

In the ***implementation stage***, a software implementation using a solver for sparse Jacobian matrices will drastically reduce the number of function evaluations. For the numerical integration we have used the MATLAB solver ode15s, recommended for stiff differential equations and sparse Jacobian matrices. For simulations we considered the following values for parameters: $E = 200 \cdot 10^9$ N/m^2, $\rho = 7.86 \cdot 10^3$ kg/m^3, $L = 1$ m, $A = 0.25 \cdot 10^{-4}$ m^2; $m_0 = 0.0655$ kg, $k_0 = 4.1667 \cdot 10^5$, $c_0 = 198.2423$; $m_1 = 0.0655$ kg, $k_1 = 1.8519 \cdot 10^5$, $c_1 = 110.1346$, $d = 0$ m. Several scenarios can be evaluated on this unified model. An illustrative example from [14] considers the initial conditions suggested by S. Timoshenko in [35] (Chap. VI): the rod "compressed by forces applied at its ends, is suddenly released of this compression at $t = 0$". This assumption gives the following initial conditions

$$u_0(x) = \begin{cases} \frac{e}{2} - ex, \ x \in [0, 1], & v_0(x) = 0 \\ 0, & \text{otherwise} \end{cases} \tag{74}$$

with the unit compression $e = 10^{-5}$. We consider the external force $g_0(t)$, which acts at the end $x = 0$ of the rod, of the form

$$g_0(\tau T_a) = \begin{cases} 1, \ \tau \in [0, 100), \\ 4, \ \tau \in [100, 300]. \end{cases} \tag{75}$$

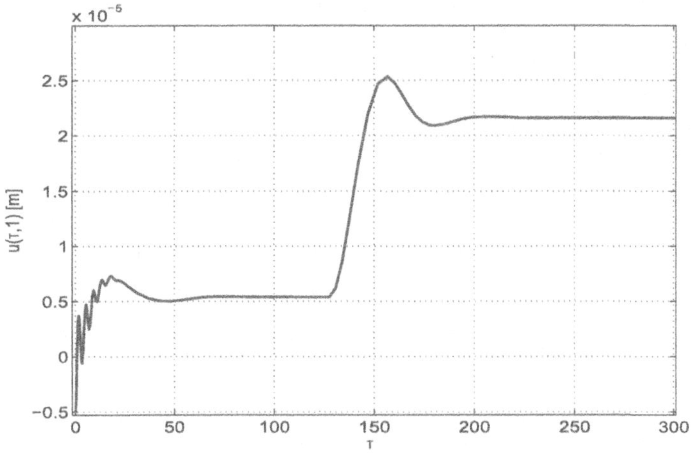

Fig. 1 The τ-evolution of the displacement u at $x = 1$ for the initial conditions (74), $p(k_1 z_1(\tau)) \neq 0$ and $g_0(\tau T_a)$ described by (75) [14]

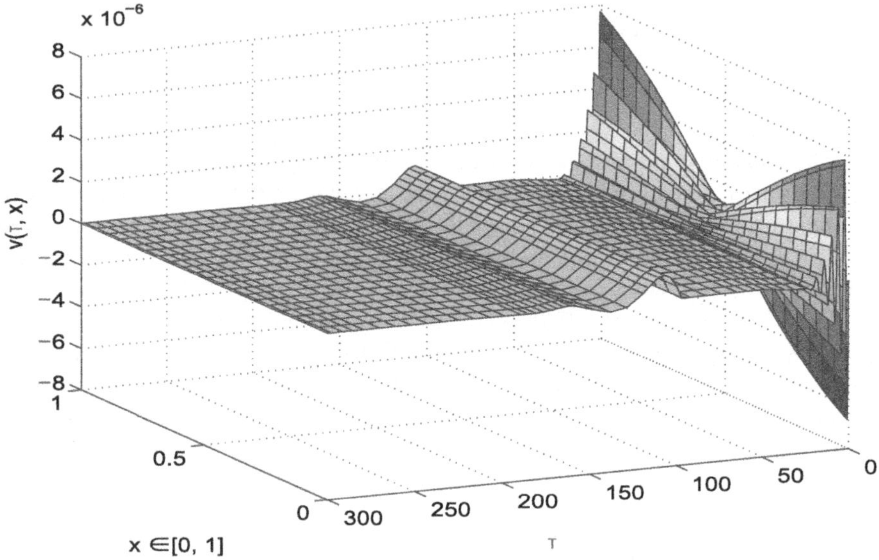

Fig. 2 3D representation of the velocity $v(\tau, x)$ for the initial conditions (74), $p(k_1 z_1(\tau)) \neq 0$ and $g_0(\tau T_a)$ described by (75) [14]

and the case $p(k_1 z_1(\tau)) \neq 0$. We can see in Figs. 1 and 2 the fast dynamics due to the initial conditions $u_0(x)$, which lasts about $25T_c$ and vanishes at $t = 0.00495s$, and the slow dynamics due to the external forces $g_0(t)$ at $x = 0$ and $p(t)$ at $x = 1$. Also, it can be seen the increasing velocity beginning from the middle of the rod as an effect of the Timoshenko initial conditions.

5 Conclusion

Distributed Parameters Systems for propagation or transport phenomena are ubiquitous and intensively studied as mathematical objects but also as challenging processes to be understood, modeled, predicted, controlled. In the case of Distributed Parameter Control Systems arising from engineering applications and modeled by hPDEs or pPDEs, the BCs are usually non-standard: in both open loop (control signal on boundary) and closed loop (the control synthesized on boundary) these mathematical models have dynamical BCs, i.e. described by differential, integral or integro-differential equations.

A good numerical-computational procedure has to fulfil the following necessary requirements: (a) Mathematical point of view: convergence of the approximation to the "true" (analytical) solution and preservation of the solution basic properties as well as its Lyapunov stability. (b) Computational point of view: efficiency in reducing the systematic/rounding errors as well as reducing the computational effort and, thus, the computational time. (c) Artificial Intelligence point of view: parallelization of the numerical algorithms and procedures used for solving mathematical problems.

Aiming to answer to the aforementioned issues, this chapter introduces a well-structured procedure for computationally solving mixed initial boundary values problems for both hyperbolic or parabolic PDEs with possibly non-standard boundary conditions. The proposed procedure combines a "convergent Method of Lines" approach with the paradigm of Cellular Neural Networks. Here we have recalled and discussed the foundation of the CMoL and the framework for the procedure application. Also, there are presented the stages and steps of the procedure. Among several applications listed throughout the chapter, we have chosen an illustrative application from *Contact Mechanics*.

Worth to point out at the end that namely the well-structured parallelism and interconnection of simple computational units in large networks offer a perspective of computationally tackling more complex models for various physical systems, e.g. systems of interconnected rods/strings in mechanics but not only – genuine challenges for research.

References

1. Abolinia, V.E., Myshkis, A.D.: Mixed problem for an almost linear hyperbolic system in the plane. Matem. Sbornik **50**(92), 423–442 (1960)
2. Burns, J.A.: Nonlinear distributed parameter control systems with non-normal linearizations: applications and approximations. In: Smith, R.C., Demetriou, M.A. (eds.) Reseach Directions in Distributed Parameters Systems, pp. 17–54. SIAM, Heidelberg (2003)
3. Chua, L., Roska, T.: The CNN paradigm. IEEE Trans. Circuits Syst. I **40**(3), 147–156 (1993)
4. Chua, L.O., Yang, L.: Cellular neural networks: theory. IEEE Trans. Circuits Syst. **35**(10), 1257–1272 (1988)
5. Chua, L.O., Yang, L.: Cellular neural networks: applications. IEEE Trans. Circuits Syst. **35**(10), 1273–1290 (1988)
6. Cooke, K.: A linear mixed problem with derivative boundary conditions. In: Seminar on Differential Equations and Dynamical Systems, III. Lecture series, vol. 51, pp. 11–17. University of Maryland (1970)
7. Courant, R., Isaacson, E., Rees, M.: On the solutions of nonlinear hyperbolic differential equations by finite differences. Comm. Pure Appl. Math. **5**(3), 243–255 (1952)
8. Danciu, D., Matei, A.C., Micu, S., Rovenţa, I.: Nonlinear feedback control and artificial intelligence computational methods applied to a dissipative dynamic contact problem. In: 2014 11th International Conference on Informatics in Control, Automation and Robotics (ICINCO), pp. 528–539 (2014). https://doi.org/10.5220/0005055005280539
9. Danciu, D., Popescu, D., Bobaşu, E.: Proceedings of the 2015 16th International Carpathian Control Conference (ICCC), pp. 100–105 (2015). https://doi.org/10.1109/CarpathianCC.2015.7145054
10. Danciu, D., Răsvan, V.: Delays and propagation: Control Liapunov functionals and computational issues. In: Seuret, A., Bonnet, C., Mounier, H., Özbay, H. (eds.) Low Complexity Controllers for Time-delay Systems. Advances in Delays and Dynamics, pp. 141–154. Springer (2014)
11. Danciu, D., Răsvan, V.: Stability results for cellular neural networks with time delays. In: Cabestany, J., Prieto, A., Sandoval, F. (eds.) Computational Intelligence and Bioinspired Systems. Ser. Lect. Notes in Computer Science, vol. 3512, pp. 366–373. Springer (2005)

12. Danciu, D., Stîngă, F.: Low-dimension robust computational algorithm for a linear distributed parameter water hammer model. In: 2019 23rd International Conference on System Theory, Control and Computing (ICSTCC), pp. 586–591 (2019). https://doi.org/10.1109/ICSTCC.2019.8885430
13. Danciu, D.: Computational and analog modeling of parabolic transport equations using Cellular Neural Networks. In: 2016 20th International Conference on System Theory, Control and Computing (ICSTCC), pp. 681–686 (2016). https://doi.org/10.1109/ICSTCC.2016.7790745
14. Danciu, D.: Distributed parameter system from contact mechanics: modelling and computational issues based on cellular neural networks paradigm. In: 2014 18th International Conference on System Theory, Control and Computing (ICSTCC), pp. 347–352 (2014). https://doi.org/10.1109/ICSTCC.2014.6982440
15. Danciu, D.: Feedback control of a nonlinear propagation system: numerical modeling and implementation. In: 2017 21st International Conference on System Theory, Control and Computing (ICSTCC), pp. 424–429 (2017). https://doi.org/10.1109/ICSTCC.2017.8107071
16. Danciu, D.: Numerics for hyperbolic partial differential equations (PDE) via cellular neural networks (CNN). In: 2nd International Conference on Systems and Computer Science, pp. 183–188 (2013). https://doi.org/10.1109/IcConSCS.2013.6632044
17. Danciu, D.: A CNN-based approach for a class of non-standard hyperbolic partial differential equations modeling distributed parameters (nonlinear) control systems. Neurocomputing **164**, 56–70 (2015). https://doi.org/10.1016/j.neucom.2014.12.092
18. Danciu, D., Popescu, D., Bobaşu, E.: Neural networks-based computational modeling of bilinear control systems for conservation laws: application to the control of cogeneration. IEEE Trans. Ind. Appl. **54**(6), 6498–6507 (2018). https://doi.org/10.1109/TIA.2018.2855171
19. Danciu, D., Popescu, D., Răsvan, V.: Control of a time delay system arising from linearized conservation laws. IEEE Access **7**, 48524–48542 (2019). https://doi.org/10.1109/ACCESS.2019.2910018
20. Di Marco, M., Forti, M., Grazzini, M., Pancioni, L.: Convergence of a class of cooperative standard cellular neural network arrays. IEEE Trans. Circuits Syst. I **59**(4), 772–783 (2012)
21. Filici, C.: On a neural approximator to ODEs. IEEE Trans. Neural Netw. **19**(3), 539–543 (2008)
22. Galushkin, A.I.: Neural Network Theory. Springer, Berlin Heidelberg (2010)
23. Gilli, M., Roska, T., Chua, L., Civalleri, P.: CNN dynamics represents a broader class than PDEs. I. J. Bifurc. Chaos **12**(10), 2051–2068 (2002)
24. Godounov, S.K.: Équations de la Physique Mathématique. Éditions Mir, Moscou (1973)
25. Guo, D., Zhang, Y.: Novel recurrent network for time-varying problems solving. IEEE Comp. Intellig. Mag. **7**(4), 61–65 (2012)
26. Gustafson, K.E.: Introduction to Partial Differential Equations and Hilbert Space Methods. Dover Publications, Mineola, New York (1999)
27. Halanay, A., Răsvan, V.: Approximations of delays by ordinary differential equations. In: Recent Advances in Differential Equations. Academic Press, pp. 155–197 (1981)
28. John, F.: Hyperbolic and parabolic equations. In: John, F., Schechter (eds.) Partial Differential Equations. Lectures in Applied Mathematics, vol. 3A, pp. 1–129. American Mathem. Society, Providence (1964)
29. Kozek, T., Chua, L.O., Roska, T., Wolf, D., Tetzlaff, R., Puffer, F., Lotz, K.: Simulating nonlinear waves and partial differential equations via CNN-Part II: typical examples. IEEE Trans. Circuits Syst. **42**(10), 816–820 (1995)
30. Lions, J.L.: Controle Optimal des Systèmes Gouvernés par des Équations aux Dérivées Partielles. Dunod, Paris (1969) (English translation: Springer, New York, 1971)
31. Pinchover, Y., Rubinstein, J.: An introduction to Partial Differential Equations. Cambridge (2012)
32. Repin, Y.M.: On the approximate replacement of systems with lag by ordinary dynamical systems. J. Appl. Math. Mech. **29**(2), 254–264 (1965)
33. Roska, T., Chua, L.O., Wolf, D., Kozek, T., Tetzlaff, R., Puffer, F.: Simulating nonlinear waves and partial differential equations via CNN-Part I: basic techniques. IEEE Trans. Circuits Syst. **42**(10), 807–815 (1995)

34. Szolgay, P., Voros, G., Eross, G.: On the applications of the cellular neural network paradigm in mechanical vibrating systems. IEEE Trans. Circuits Syst. **I**(40), 222–227 (1993)
35. Timoshenko, S.: Vibration Problems in Engineering. D. Van Nostrand Company (1937)

The manufacturer's authorised representative in the EU is Springer Nature Customer Service Centre GmbH, Europaplatz 3, 69115 Heidelberg, Germany. If you have any concerns regarding our products, please contact ProductSafety@springernature.com

Printed and bound by CPI Group (UK) Ltd, Croydon, CR0 4YY
30/09/2026
02225324-0009